MUSEO
Salvatore Ferragamo

EQUILIBRIUM

edited by Stefania Ricci and Sergio Risaliti

SKIRA

Graphic project
Studio Contri Toscano

Editorial coordination
Emma Cavazzini

Copy editor
Emanuela di Lallo

Translations
Sylvia Adrian Notini (Bernardi,
Natalini, Ricci, Ricci and Risaliti,
Vaccarino, interviews to Wanda
Miletti Ferragamo, James
Ferragamo, Eleonora Abbagnato
and Reinhold Messner, List
of works exhibited).
Sergio Knipe (Demetrio, Enria,
Guadagnini, Risaliti) and Paul
Metcalfe (Arensi, Pievani, Sisi)
on behalf of *Scriptum*, Rome

First published in Italy in 2014 by
Skira editore S.p.A.
Palazzo Casati Stampa
Via Torino 61
20123 Milan, Italy
www.skira.net

Printed and bound in Italy
First edition

ISBN: 978-88-572-2229-5

Distributed in USA, Canada,
Central & South America by
Rizzoli International Publications,
Inc., 300 Park Avenue South,
New York, NY 10010, USA.
Distributed elsewhere in the
world by Thames and Hudson
Ltd., 181A High Holborn,
London WC1V 7QX, United
Kingdom.

EQUILIBRIUM

Florence
Museo Salvatore Ferragamo
Palazzo Spini Feroni
19 June 2014
12 April 2015

Under the Patronage of
Ministero dei Beni
e delle Attività Culturali
e del Turismo
Regione Toscana
Comune di Firenze

Catalogue edited by
Stefania Ricci
Sergio Risaliti

Graphic project
Studio Contri Toscano

Photographs
Arrigo Coppitz

Exhibition conceived
and curated by
Stefania Ricci
Sergio Risaliti
with the collaboration of
Emanuele Enria

Exhibition organized by
Museo Salvatore Ferragamo
with the collaboration of
Soprintendenza Speciale
per il Patrimonio
Storico, Artistico ed
Etnoantropologico
e per il Polo Museale
della città di Firenze
Fondazione Ferragamo

Coordinator of the
organizational secretariat
Paola Gusella

Catalogue coordinator
Domenico Giampà

Organizational secretariat
Chiara Bacci
Martina Cocchi
Federica Locatelli
Francesca Piani
Michela Tucci

Exhibition design
Silvia Cilembrini
Fabio Leoncini

Exhibition installation
Urban Production s.r.l.

The model *Australopithecines
on the move* was created
by Lorenzo Possenti

Technology
STS Communication s.r.l.

Filmed interviews
Francesco Fei

Video research and project
Daniele Tommaso

Shipping
Arterìa s.r.l.
Savitransport International

Insurance
Aon s.p.a. Fine & Jewellery
Division
CS Insurance Service
Axa Art

with the collaboration of  under the Patronage of

The exhibition curators
and editors of this catalogue
wish to thank:

Ministero dei Beni e delle Attività
Culturali e del Turismo
Regione Toscana
Comune di Firenze
Soprintendenza Speciale per
il Patrimonio Storico, Artistico,
Etnoantropologico e per il Polo
Museale della città di Firenze
Soprintendenza per i Beni
Archeologici della Toscana
Sovrintendenza Capitolina ai
Beni Culturali di Roma Capitale

Biblioteca Nazionale Centrale
di Firenze, Florence
Fondazione Canova onlus,
Museo e Gipsoteca Antonio
Canova, Possagno (Treviso)
Fondazione Ferragamo
Gabinetto Disegni e Stampe
degli Uffizi, Florence
Galleria d'Arte Moderna di
Palazzo Pitti, Florence
Musée Bourdelle, Paris
Musée des Beaux-Arts, Nantes
Musée d'Orsay, Paris
Musée Rodin, Paris
Musei Civici, Pavia
Museo Archeologico Nazionale,
Florence
Museo dei Fori Imperiali –
Trajan's Markets, Rome
Museo della Civiltà Romana,
Rome
Museo di Palazzo Pretorio, Prato
Museo di Scultura Antica
Giovanni Barracco, Rome
Museo di Storia Naturale e del
Territorio dell'Università di Pisa,
Calci (Pisa)
Museo Marino Marini, Florence
Museo Nacional Centro de Arte
Reina Sofía, Madrid

The State Hermitage Museum,
Saint Petersburg
Walker Art Center, Minneapolis

Archivi Alinari, Florence
Archivio Jerry Ferragamo,
Florence
Archivio Ugo Mulas, Milan
Balmond Studio, London
Beck & Eggeling International
Fine Art, Düsseldorf
Bill Viola Studio LLC, Long Beach
Cango – Cantieri Goldonetta,
Florence
Cathedral Church of Saint John
the Divine, New York
Collection Enrico Navarra, Paris
Collezione Monte dei Paschi
di Siena
Electronic Arts Intermix,
New York
Farmacia Porta Rossa, Florence
Galleria Christian Stein, Milan
Galleria Francesca Antonacci,
Rome
Galleria Lia Rumma, Milan/Naples
Galleria Tornabuoni Arte,
Florence
GAMA Galleria d'Arte Moderna,
Albenga
Marina Abramović Archives
and Murray Grigor
Messner Mountain Museum,
Bolzano
Scuola di danza Hamlyn,
Florence
Sikkema Jenkins & Co.,
New York
Stichting Lima, Amsterdam
Studio Giulio Paolini, Turin
The George and Helen Segal
Foundation, New York
The Milan Triennale
Yoshii Gallery, New York

The private collectors who
wish to remain anonymous

In particular, we wish to thank:

Giovanni Abatista, Marina
Abramović, Cristina Acidini,
Sara Adams, Clelia Albarello,
Alessandra Alberti, Silvia
Alessandri, Aurelio Amendola,
Paola Andreini, Sergey
Androsov, Giuseppe Anichini,
Silvana Annicchiarico, Francesca
Antonacci, Thérèse Barbanel,
Roberto Barbuti, Roberto
Barni, Massimo Barzagli, Surya
Bellandi, Duccio Benedetti,
Gianfranco Benedetti, Francesco
Bigazzi, Filippo Bigi, Matteo
Boddi, Cristina Bonetti,
Manuel Borja-Villel, Maria
Bottari, Laura Brazzini, Scott
Briscoe, Trisha Brown, Pino
Brugellis, Andrea Brugnoni,
Marco Brusamolin, Simone
Bruschi, Laura Buonocore,
Greg Burchard, Cinzia Bussotti,
Letizia Campana, Valentina
Cappa, Janis Carroll, Michele
Casamonti, Roberto Casamonti,
Michele Cecchini, Mario Ceroli,
Blandine Chavanne, Catherine
Chevillot, Jane Cho, Miriana
Ciacci, Maddalena Cima,
Giuseppina Carlotta Cianferoni,
Sofia Ciucchi, Rebecca Cleman,
Donata Clovis, Guy Cogeval,
Simonella Condemi, Cathy
Cordova, Andrea Coveri,
Alessandro Curotti, Jeanie
Deans, Bettina Della Casa,
Daniela De Lorenzo, Maria Elena
De Luca, Lucia De Siervo, Sara
De Tullio, Colette de Turville,
Audrey d'Hendecourt, Soledad
de Pablo Roberto, Casimiro
Di Crescenzo, Maddalena
Disch, Laura Doretto, Matthew
Droege, Barbara Dufty, Jennifer
Engle, Ruth Ennemoser, Carlo
Esposito, Jenni Evans, Marzia

Faietti, Dario Franceschini,
Sonia Francini, Marcela Frias De
Oliveira Alge, Gian Pietro Favaro,
Victoria Fernández-Layos
Moro, Erika Filipponi, Stefano
Fiordi, Luigi Fiore, Monica
Fiorini, Agathe Fischnaller,
Marco Fossi, Stefano Frasconi,
Simone Frosecchi, Roberto
Gaggioli, Giancarlo Galan, Elena
Ghidini, Daniela Giuliani, Sergio
Givone, Clémence Goldberger,
Harry Gordon, Daniele Gori,
Giuliano Gori, Antony Gormley,
Marco Graziano, Francesca
Grifoni, Sarah Grünberg, Mario
Guderzo, Chiara Guglielmi, Josef
Helfenstein, Rita Iacopino, Aldo
Innocenti, Cristina Intelisano,
Guido Iodice, Mario Iozzo,
Bobby Jablonski, Katarina
Jerinic, Amir G. Kabiri, Stefano
Karadjov, Corey Keller, Heather
Kowalski, Paul Lee, Paolo La
Morgia, Damiano Lapiccirella,
Perrine Latrive, Colin Lemoine,
Nick Lesley, Carlo Lisi, Caterina
Lucherini, Antonella Maggiorelli,
Aline Magnien, Maria Pia
Mannini, Jean-Philippe Manzano,
Francesca Marini, Mara Martini,
Caroline Mathieu, Alejandro
May, Marica Mercalli, Gianni
Mercurio, Massimiliano Migliorini,
Cherry Montejo, Laura Mori,
Luca Mugnaini, Sandro Nadalini
Ristori, Dario Nardella, Moti
Nativ, Bruce Nauman, Enrico
Navarra, Carlo Nesi, Rosella
Nesi, Antonella Nicola, Aurora
Nomellini, Eleonora Barbara
Nomellini, Kathy O'Donnell,
Chiara Onniboni, Giulio Paolini,
Demetrio Paparoni, Claudio Parisi
Presicce, Alberto Pecci, Susanna
Pelle, Kira Perov, Andrea Pessina,
Carla Pinzauti, Gianni Piolanti,
Mikhail Piotrovsky, Patrizia Pisani,

Giuseppe Poeta, Paola Potena,
Alessandra Pozzati, Nathalie
Prat, David Racz, Nilda Rivera,
Martine Robert, Alberto Rossetti,
Enrico Rossi, Albino Ruberti,
Sidney Russel, Alberto Salvadori,
Sergio Salerni, Stefano Salvatici,
Micaela Sambucco, Nicoletta
Santoro, Cristiana Savoia, Heather
Schweikhardt, Annalisa Scotton,
Maria Letizia Sebastiani, Virgilio
Sieni, Amélie Simier, Sebastiano
Soldi, Gabriella Sorelli, Christen
Sperry-Garcia, Elizabeth Steele
Basile, Daniela Tabò, Diane Tytgat,
Tommaso Tombelli, Veronica
Tonini, Laura Trambusti, Andrea
Tremolada, Federica Trerè, Guy
Tubiana, Lucrezia Ungaro, Milly
Valdevies, Giannantonio Vannetti,
Nicola Vascellari, Anna Vecchi,
Marco Vianello, Giuseppe Viesti,
Bill Viola, Massimo Vitta Zelman,
Francesco Vossilla, Barbara
Wagner, Lynn Wichern, Betty
Woodman, Kazuhito Yoshii,
Mauro Zaccariello, Susanna Zatti,
Filippo Zevi, Alessandro Zuri,
Theus Zwakhals

We also express our gratitude
to the authors of the essays
and of the films on show

A very special thank you
to the celebrities interviewed:
Eleonora Abbagnato, Cecil
Balmond, James Ferragamo,
Reinhold Messner, Wanda Miletti
Ferragamo, Philippe Petit, Will Self

We are grateful to Jerry Ferragamo
and Stefano Ferragamo for their
precious collaboration

and to our Sponsor

CONTENTS

Stefania Ricci, Sergio Risaliti
9 Equilibrium

Stefania Ricci
15 Feet don't lie

Sergio Risaliti
29 Standing, walking, moving in equilibrium
as a dancer or tight-rope walker

Telmo Pievani
71 Walking for six million years

Duccio Demetrio
83 Walking. A philosophy and lifestyle

Carlo Sisi
101 Neoclassical balance. Canova
and the rhythms of grace

Flavio Arensi
119 Volatile balance. Degas, Rodin,
Bourdelle and the dancing revolution

Elisa Guzzo Vaccarino
143 Divine equilibriums. Terpsichore
and Mercury

Sandro Bernardi
171 Chaos in order: Chaplin's walk

Emanuele Enria
191 Balances: learning to learn.
Feldenkrais' great lesson

Walter Guadagnini
217 A portrait of the photographer
as a tight-rope walker

Arabella Natalini
239 The present's fragile equilibrium

259 Transmitting the experience of equilibrium
edited by Emanuele Enria

263 Wanda Miletti Ferragamo,
James Ferragamo
A history of equilibrium across
the generations

269 Cecil Balmond
The dynamics of equilibrium

277 Reinhold Messner
Instinctive equilibrium

283 Will Self
Walking: perceiving space
and the sense of equilibrium

291 Eleonora Abbagnato
On tiptoe

295 Philippe Petit
Mind and body in equilibrium
on the wire

302 List of works exhibited
308 Bibliography

EQUILIBRIUM

STEFANIA RICCI, SERGIO RISALITI

Humanity has been "walking for six million years" is the title of the essay by Telmo Pievani on human evolution published here: it's a question of having good feet and keeping a difficult equilibrium. What was new about the first hominids, what made them different from all the other apes, and helped them develop the encephalon, was indeed their bipedalism. No longer crawling around on all fours but getting up on their feet with the centre of gravity far from the surface of the ground involved the difficult reorganization of the anatomy and a very high risk of falling. "I always have the impression or feeling" said Alberto Giacometti "of the frailty of living beings, as if at any moment it took a formidable amount of energy for them to remain standing, always threatened by collapse". In the act of walking the weight of the body shifts from one foot to the other in search of stability and balance, a dynamic sort of balance, as it is undermined immediately after in the next step taken.

But how is anyone supposed to maintain this state, carry the body's weight in movement, when one's feet are encased in shoes, that is, in forms that are not indispensable to a natural state, seeing that man was made to walk barefoot? This was the first concern of Salvatore Ferragamo, artist and shoemaker, as well as engineer, architect of the shoe, student of anatomy.

"When I began studying human anatomy", wrote Salvatore in his autobiography *Shoemaker of Dreams*, "I found my first clue to the problem in the distribution of the weight of the body over the joints of the foot. I discovered the interesting fact that the weight of our bodies when we are standing erect drops straight down on the arch of the foot".

Beige kid flat inspired by the ballet shoes with the famous "shell" sole patented by Ferragamo in 1957, 22 cm Florence, Museo Salvatore Ferragamo.

des choses saisissables tu
contractes avec la nature un
pacte de solidarité : c'est l'angle droit
Debout devant la mer vertical
te voilà sur tes jambes.

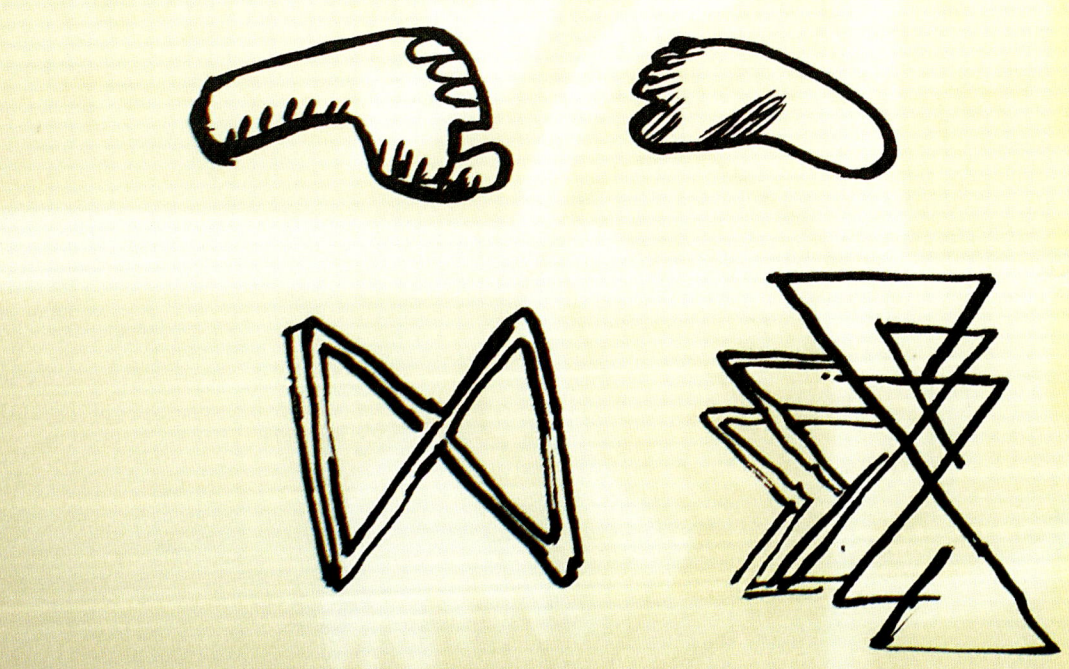

Ferragamo dedicated a lot of time to the study of the foot's mechanics, to its anatomy, and to the scientific laws that govern one's way of walking, the architecture of the skeleton and how the muscles function to be able to understand how the arch of the foot worked, to what extent and how much the golden section as well as the weight distribution between the centre of the arch and the ends (heel and toes) of the foot were important. That is, he saw *à-plomb*, a fundamental concept in dance, in architecture, as being of crucial importance. The task of the arch of the foot, said Ferragamo, goes well beyond carrying a stationary weight: it must carry our moving weight as we walk. Therefore Nature has provided the foot with joints and swivels that work together like the instruments in an orchestra, a complex whole of bowed stringed instruments in tension and compression "like no architect could ever build", says Cecil Balmond, an artist, architect and engineer of international renown interviewed for this exhibition.

It is thanks to an analysis of this theme, which was so meaningful to Ferragamo's story, and which was suggested to the curators by the contemporary dance expert Emanuele Enria, that the project called *Equilibrium* was born. Albeit far from being exhausted, we developed the topic by way of paradigms and suggestions, referring to the world of art, in all its manifestations, a world that has always – but especially during the modern age – been drawn to movement, walking, experimenting with the limits and measure of the human figure as it moves on its feet.

Edgar Degas studied the most complicated arabesques of dancers, just as Marino Marini showed his fascination with dance movements. Julio González was also enthralled by the theme of the dancer, expressed in his design of a metal figure in which the subject becomes a sculptural line, the expression of forces and ratios of balance in space. Rodin focused on the free gesture of the ballet dancer, as well as on "the walking man", a theme also cherished by Alberto Giacometti and by many contemporary artists, from Mario Ceroli to Roberto Barni, from Marina Abramović to Bruce Nauman, to Bill Viola. Pablo Picasso, Gino Severini, Fernand Léger and Charlie Chaplin were drawn to the precarious equilibrium of tight-rope walkers and acrobats, who take human possibility to its very limits before falling. Alexander Calder explored the world of the circus to craft his first *Mobiles*. The theme of equilibrium became of crucial importance to art between the nineteenth and twentieth centuries, as revealed in the works of the artists cited, as well as in those of Paul Klee, Wassily Kandinsky, Joan Miró, Fausto Melotti and Giulio Paolini, who focused on the most extreme poses of persons and objects. Photography is not exempted from this interest either, and this exhibition could indeed have been entirely based on this genre. Without mentioning, of course, the art of dance, which is wholly centred on keeping one's balance, and architecture, whose very essence lies in ratio and static.

Physical equilibrium influences the nervous system, it becomes "mental equilibrium", as Ferragamo put it. So walking is more than just an essential action. In his essay for this catalogue Duccio Demetrio writes that walking "is a philosophy, a style of life, a search into oneself and elsewhere". And Dante Alighieri was no doubt a good walker, having conceived the *Divine Comedy* "midway in the journey of our life". His journey through Hell, Purgatory and Paradise is made up of steps, falls, losses of consciousness. The history of humankind is determined by the many paths taken, in which the psychological, cultural, emotional, social, civic and religious attitudes of women and men are all acknowledged. Hence, this exhibition could not do without a section dedicated to the phenomenology of walking in the moving images of famous walks: from those of rulers, dictators and politicians to the slapstick comedy of Charlie Chaplin, whose Little Tramp persona celebrates its one hundredth anniversary this year, in 2014, to that of the first man on the Moon.

The debate on these themes is so alive in contemporary society – even going so far as to touch upon morality as well as the concepts of rectitude and justice – that as curators of the exhibition we found it necessary to include interviews with a number of international figures who have expressed their own opinions on equilibrium, the foot, the action of walking and how they relate to their respective jobs and life stories: Wanda Miletti Ferragamo, wife of Salvatore, entrepreneur and mother of six children, James Ferragamo, nephew of the founder and currently in charge of woman's shoes and leather at Ferragamo, Eleonora Abbagnato, étoile at the Paris Opéra, Reinhold Messner, world-class Italian mountain climber, Will Self, the British writer who has turned walking into a way of approaching contemporary society and space, and Philippe Petit, the renowned wire-walker whose feats have verged on the incredible, and who perfectly equates to the figure described by Théodore de Banville in *Odes funambulesques* (1856): "he who conquers the air experiences the dizziness of height, stares into the depths of the sky, soars above the amazed and awestruck heads of his counterparts who, enslaved to their weight, have remained anchored to the ground". We should once again mention the name of Cecil Balmond, who addressed our questions about equilibrium as it relates to architecture by specially realizing an art installation for this exhibition born from his reading of Ferragamo's autobiography and a knowledge of his studies on the anatomy of the foot.

The catalogue and the exhibition were a collaborative effort on the part of many experts and scholars on the subject, and of the many institutions, both public and private, of great international prestige that have embraced this project: The State Hermitage Museum in Saint Petersburg, Musée d'Orsay, Musée Bourdelle and Musée Rodin in Paris, Museo Nacional Centro de Arte Reina Sofía in Madrid, Musée des Beaux-Arts in Nantes, Walker Art Center in Minneapolis, George and Helen Segal Foundation in New York, Biblioteca Nazionale Centrale di Firenze, Gabinetto Disegni e Stampe degli Uffizi, Museo Archeologico, Galleria d'Arte Moderna di Palazzo Pitti and Museo Marino Marini in Florence, Museo Civico di Palazzo Pretorio in Prato, Raccolte Museali Fratelli Alinari in Florence, Museo Barracco, Museo della Civiltà Romana and Museo dei Fori Imperiali in Rome, Museo e Gipsoteca Antonio Canova in Possagno, Musei Civici in Pavia, as well as many private collections and galleries among which the Yoshii Gallery and Sikkema Gallery in New York, Collection Enrico Navarra in Paris, Beck & Eggeling International Fine Art in Düsseldorf, Galleria Francesca Antonacci in Rome, Galleria Christian Stein in Milan, Galleria d'Arte Moderna in Albenga, the collection of the Monte dei Paschi di Siena, Tornabuoni Arte in Florence. Lastly, some of the works and images came directly from the studios of the artists Giulio Paolini, Roberto Barni and Cecil Balmond.

We wish to thank the Ministry of Cultural Heritage and Activities and Tourism for its constant support, the Regional Government of Tuscany and the Municipality of Florence for their sponsorship, the Soprintendenza Speciale per il Patrimonio Storico, Artistico e Etno-antropologico e per il Polo Museale della Città di Firenze, the Soprintendenza per i Beni Archeologici della Toscana, the Soprintendenza Speciale per i Beni Archeologici di Roma.

FEET DON'T LIE

STEFANIA RICCI

All his life Salvatore Ferragamo proudly referred to himself as a shoemaker. Today he is remembered as an artist, a skilled craftsman, a refined entrepreneur, who succeeded in revolutionizing the very idea of making shoes. The most outstanding part of his whole universe is represented by the relationship between the foot and the shoe. This is where his figure takes on the fascinating dimension of a wizard, a healer, an architect, an engineer and a wise visionary.

In all the photographs in which we can see his hands as they touch feet, we realize that the unique quality of his shoes started from right there, from that ability to perceive and appreciate the beauty of the foot. "I love feet", Salvatore wrote in his memoirs. "They talk to me. As I take them in my hands I feel their strengths, weaknesses, their vitality or their failings. A good foot, its muscles firm, its arch strong, is a delight to touch, a masterpiece of divine workmanship".[1]

And: "What do I mean when I say that feet talk to me? Just that: they communicate the character of the person".[2]

Ferragamo's words take on new meaning if we think that not many people before him used a scientific method to analyse one of the most common and primary human actions: walking. As early as 1833, in his short treatise *Théorie de la démarche*, Honoré de Balzac wondered: "Why has man's walking always been worse off? Because man has always preferred to deal with the movement of stars".[3]

Why has so little attention been devoted to feet in general? Ferragamo explains it very clearly in his autobiography: "… ignorance. We simply do not know what to look for, and no one tells us. How often, when you have been feeling ill and you have

Viatica, 2012–13
Red patent leather pump,
22 x 11 cm
Contemporary realization of the
original model created by Salvatore
Ferragamo for Marilyn Monroe
Florence, Museo Salvatore Ferragamo.

consulted a doctor, has he taken your feet in his hands and examined the bone structure, probing to find out if they were good feet? ... There are three things we must do regularly for complete bodily heath: first we must breathe, then we must eat, finally we must walk. Walking stimulates the circulation of the blood and keeps us fit in body and mind".[4]

How can we explain the immediate success of an adolescent shoemaker who had just arrived in the United States from a small town in the Irpinia region of Italy, in search of fortune, work, an opportunity in life, if not by the fact that he knew how to reconcile the reasons of beauty with those of purpose, so that everyone could walk happily, wearing comfortable, elegant shoes?

This formula would be the quest of his entire life.

Ever since he began to work as a shoemaker, not much more than a boy, in the town of Bonito where he was born, Salvatore was obsessed by a mystery that no shoemaker seemed to be able to solve. However accurately and however rigorously a foot may have been measured,

was there anyone who could explain why shoes didn't always fit, or why they didn't fit the same way, and why so many feet were ruined by this? So Salvatore went to Naples and then on to the United States to broaden his horizons, but above all to give free rein to his thirst for knowledge. He had only recently opened shop in Santa Barbara, California with his brothers, where shoes were repaired and custom-made shoes were crafted, and already he was attracting his first demanding, exclusive customers: these were the actors and actresses of the nascent art of the cinema. There was no shortage of orders for his shoes, and already the word had spread about this young shoemaker of great skill. His brothers were satisfied, but Salvatore wasn't. As soon as he had learned English better, he enrolled in the night school courses on anatomy at the University of California, Los Angeles. Three or four days a week he would travel 60 kilometres to attend lessons on the human skeletal system, and he would devote every free moment from work to his studies. Each day in the shop

In his autobiography Ferragamo remarked: "I discovered the interesting fact that the weight of our bodies when we are standing erect drops straight down on the arch of the foot, as shown by a plumb-line" (*Shoemaker of Dreams. The Autobiography of Salvatore Ferragamo* [1957], Florence: Giunti, 1985, p. 67).

top
Patent 1,479,536 for "orthopaedic device" granted in the United States on 1 January 1924.

Patent 1,399,606 for "surgical instrument" granted in the United States on 6 December 1921.

bottom
System to reinforce the arch of the sole of a shoe, known as the waist. Patent 281241 granted on 7 January 1931.

Multiple "X" reinforcement. Patent 71816 granted on 21 March 1959.

he would observe his customers' feet, examine them and make a diagnosis. He soon realized that the flaws in a foot are not hereditary; if this were so, certain problems would also occur in the populations that walk barefoot. However, a child could inherit from one of its parents a type of joint and bone formation that, when compressed into bad shoes, produce defects. Ferragamo would later say: "Advancing age, also, does not itself produce deformities of the foot, though the development of diseases like arthritis and rheumatism will twist bones and joints. In fact, the foot that is correctly shod does not age. The face and the figure may show the tell-tale signs of advancing years; the feet remain youthful and beautiful. … Feet don't lie".[5]

So the problem lay in the system used to measure the foot or in the internal construction of shoes: "When I began studying human anatomy I found my first clue to the problem in the distribution of the weight of the body over the joints of the foot. I discovered the interesting fact that the weight of our bodies when we are standing erect drops straight down on the arch of the foot. A small area of between one and a half and two inches on each foot carries all our weight. As we walk the weight of our bodies is swung from one foot to the other".[6]

In traditional shoemaking, both artisanal and industrial, shoes were made keeping in mind the need for adequate support for the toe knuckle and the heel. The arch was left suspended, the way it is in nature. This system seemed to make sense. If we look at the print left by a bare foot in wet sand we can see the space under the arch: we can identify the marks left by the toes, the forefoot and the heel, uniquely connected – in a normal foot – by a small surface, the outside edge of the sole. The rest of the foot does not touch the ground.

"Yet the fact remains that many feet are injured by shoes. Does the answer lie, then, in the fact that when the foot is inside the shoe it is no longer allowed to perform its natural functions? Is it imprisoned like a bird in a cage, unable to work properly? If that is so does this imprisonment affect the arch? Again, if this is so, does this mean that the arch not only should but must be supported? Nature, the supreme architect from whom Man has borrowed and adapted so many of his

System used in the 1950s to measure lasts
Florence, Jerry Ferragamo Archive.

Ferragamo's shoes were famous for their lightness also thanks to the use of a stainless steel blade supporting the arch of the foot.

ideas, has created the human foot in that shape and not allowed it to develop without an arch because, as any architect will tell you ... an arch can carry more weight than a flat surface. This arch, however, has to do more than carry our moving weight as we walk. Therefore Nature has provided the foot with joints and swivels to allow us to walk in comfort. ... This simple mechanism moves and stretches as you walk barefoot; the joints and the toes perform their duties freely, falling back into their natural positions at the end of each step, ready for the next. You feel comfortable and free, as indeed you should. These are natural movements".[7]

Thanks to his research, Ferragamo created revolutionary forms that, by using a steel blade (shank) to support the plantar arch, allow the foot to move like an inverted pendulum. By exploiting the plumb line the same way architects and builders of cathedrals and triumphal arches had always done, Salvatore would calculate the median line that from the top of the body (the head and barycentre) drops down towards the horizontal plane (the ground) where the sole of the foot rests, the axis of equilibrium that falls at the centre of the plantar arch. The metatarsal joints and the foot no longer support any weight, so the feet guide the equilibrium of the body as it walks instead of opposing it. A space is created under the metatarsus to make room for the flexion of the forefoot, so that the joint can go back into place when the foot takes a step. Only by assuring the necessary support to the arch of the foot can the shoe maintain the equilibrium that the foot conserves when it is bare, both when it is still and when it is in motion. Equilibrium, Ferragamo believed, is not static but dynamic, not just physical, but mental as well. "Tiredness and physical weakness affect our nerves. We become irritable, bad-tempered, nervy – yes, even mentally unbalanced".[8]

Over the years Salvatore patented several versions of this original steel support. In the mid-1930s, upon returning to Italy, at a time when good quality steel was hard to come by, he invented the shoe of the century, the orthopaedic shoe with a cork sole or wedge heel to fill the area between the heel and the forefoot. It was absolute perfection. The foot found a stable support over the whole surface and was relaxed, without the risk of yielding to the however small flexion that always exists with metal. And when the wedge heel is made from natural cork, the walker's health is assured.

The famous wooden last custom-made by Salvatore Ferragamo for Marilyn Monroe.

pp. 20–21
Black satin sandal with layered cork wedge in gold and silver calfskin, 1938, 26 x 7 cm
Florence, Museo Salvatore Ferragamo.

Black satin sandal with shaped, chiselled and grooved wedge, covered in gold kid, 1940, 20 x 6 cm
Florence, Museo Salvatore Ferragamo.

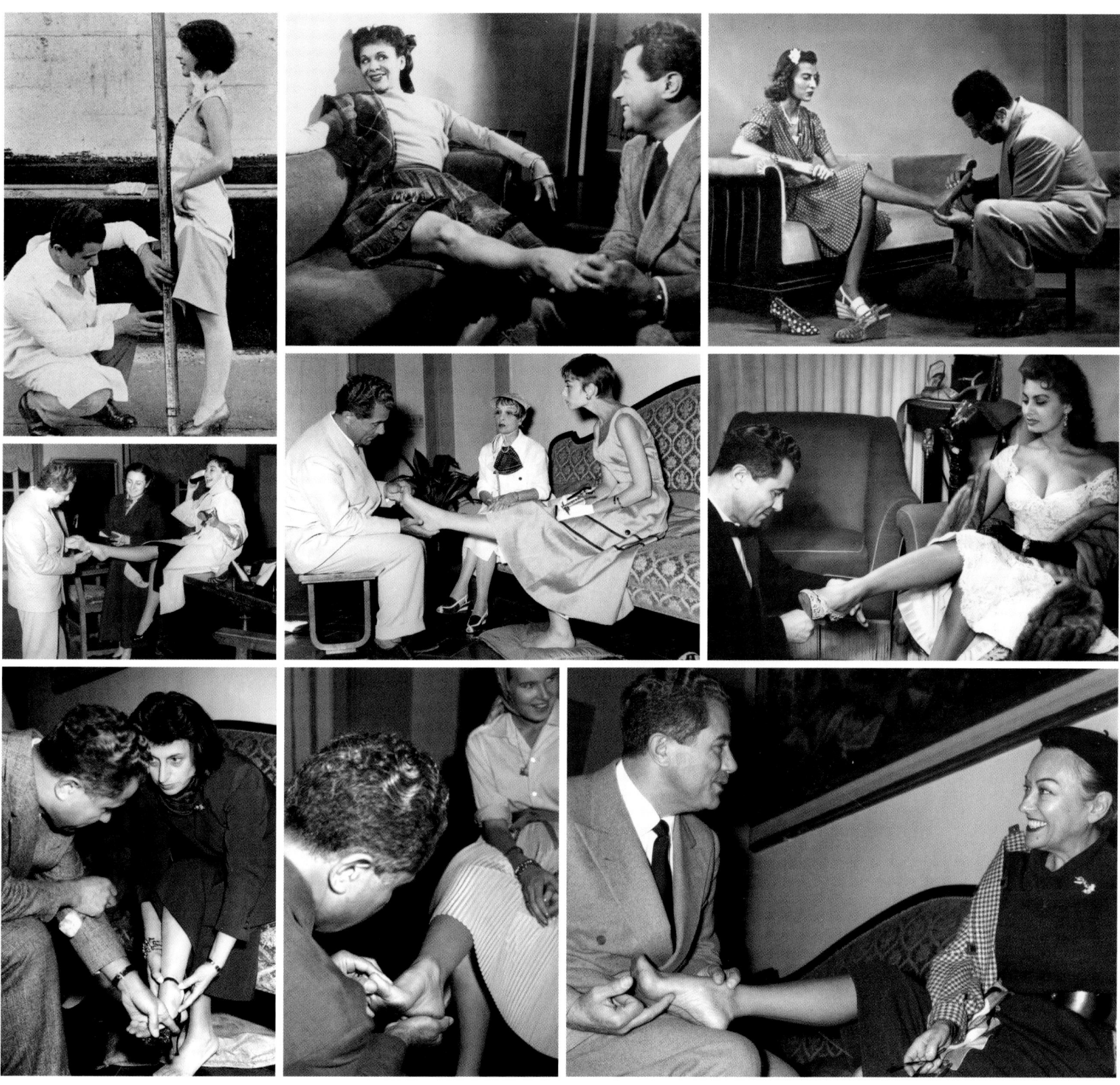

Salvatore Ferragamo touching the feet of some of his clients. From top left: a client, Katherine Dunham, a client, Paulette Goddard, Audrey Hepburn, Sophia Loren, Anna Magnani, Gabriella of Savoy and Gloria Swanson. Ferragamo described Swanson's feet as follows:

"When I took her feet into my hands I found them as youthful and beautiful as when I first shod them more than thirty years ago" (*Shoemaker of Dreams. The Autobiography of Salvatore Ferragamo* [1957], Florence: Giunti, 1985, p. 64).

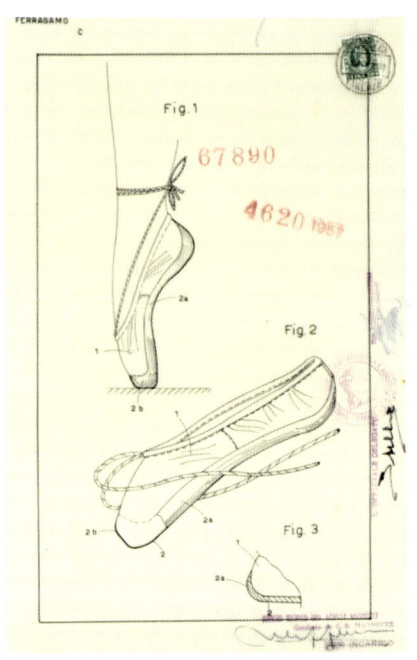

Ferragamo's research into the construction of shoes was also accompanied by experiments on the treatment of the leather required to make the perfect shoe. The young shoemaker enrolled in a correspondence course at the Faculty of Chemical Engineering at Berkeley, California. He studied for three or four years and then earned his degree.[9]

Such determination also led him to invent and patent a "surgical instrument" in 1921 – even before his shoes. After a serious car accident that caused the death of his brother, Salvatore was hospitalized with several leg fractures. For six months he was bedridden with the fear of having to walk with a limp for the rest of his life. In this situation of forced meditation he honed his project, which consisted of a cylinder several centimetres longer than the leg it contained. One end of the cylinder exerted pressure on the pelvic joints and the tension of the limb was achieved thanks to a device placed at the end of the foot, which could either be stretched or relaxed, but that, once it was fixed, maintained constant traction. This alleviated the pain and improved the effectiveness of the treatment.

The "Ferragamo device" was later produced and sold by an orthopaedic company in Chicago.[10]

Salvatore's studies on the way a shoe fits were crucial to the development of the Ferragamo method. During the time he spent in the United States, he also worked, for just fifteen days, for the Queen Quality Shoe Company, a shoe factory in Boston where he learned about systems used to measure and produce industrial shoe forms. Later, when he began to open his first shops, at first in Santa Barbara and later in Hollywood, he had the opportunity to handle hundreds of feet: long, short, thin, wide, very slender and arched like those of the Spaniards and the Mexicans, more tapered than the Americans', or feet with small heels like those of the South Africans. He realized that feet were all different and that they could vary not just in length but in width as well, depending on people's physical build and whether they are more

or less long-limbed. He also eventually realized that it's impossible to use just one measurement system, and developed different shapes, which for every foot length could be combined with six different sizes denoted by the letters of the alphabet, from the narrowest, AAAA, to the widest, D.

But the transition from theory to practice took time. It wasn't easy to make a series of wooden forms on which to build the same model in different fits. The methods used by Americans, who were much more sensitive than Europeans to function rather than aesthetics, were not suited to the philosophy of craftsmanship: indeed, these were methods that had been created to suit the industrial system. Conversely, in Italy the makers of forms seemed to understand even less about the young shoemaker's revolutionary work. Ferragamo's comment was: "Shoemakers throughout the world use everything I offer, except one. They use my patterns, my designs, my styles, my ideas. Well, I don't mind. There are plenty more where they came from and, in a way, it is the supreme compliment. Besides, the man who is busy copying the work of others has no time to be original for himself. So they steal everything ephemeral, everything that tomorrow or the next season or next year will be dead and gone, passed into the sphere of the old-fashioned until it is resurrected and modified in the course of the evolutionary cycle of fashion. Yet they refuse to look at the one unchanging feature of my shoes, the reason why my shoes fit and others do not. They not only refuse to look, they deny its value".[11]

Luckily, Ferragamo's production was addressed to special orders, and the lack of productive supply chains was compensated for by making custom-made shoes, by working on the individual foot, cor-

Patent 67890, granted on 22 April 1958. Dancer's pointe shoe with sole featuring a tapered border glued to the sides of the upper and forming a reinforcement at the toe.

Black satin dance shoe created by Ferragamo in 1957, 23 cm. The shoe exemplifies the patent Florence, Museo Salvatore Ferragamo.

recting the form directly, adding and removing plaster. "It was very complicated. To make a single pair of shoes was like making a structure for a thousand pairs of shoes", recalls Salvatore's nephew Jerry Ferragamo, who at the age of twenty-one had the chance to work with his uncle, assisting him in this important aspect of his production: the construction of the form, the anatomical study of the foot and the shoe fit.

Ferragamo immediately discovered the effects of his research on the feet of his famous clients. He saw the results on the feet of actors like John Barrymore, who was slightly flat-footed because of the bad shoes he'd worn as a child, on those of Benito Mussolini years later, whose feet were affected with calluses and hard skin, or those of Queen Helen of Italy, who had long aristocratic feet that were difficult to find shoes for, on those of Princess Maria José of Savoy or of the Duke of Savoy-Aosta; for the latter client he was able to perform a miracle by making shoes that were capable of treating an ugly sore that had been caused by a motorboat accident.[12] Lots of stories, lots of anecdotes that you won't find in fashion books but that certainly contributed to Salvatore Ferragamo's fame. Many inventions were added to the first ones – special soles, inner parts, heels of different shapes, some of which very high, reinforcements: 400 patents whose purpose was to combine aesthetics with comfort.

When Salvatore Ferragamo passed away it immediately became evident that it was going to be hard to carry on with the Ferragamo method. The artist, the image, the guide and the great designer was gone. Fiamma, his nineteen-year-old daughter who gave up school to go work for the company, took over in the style department. Jerry, his nephew, worked in production; not as yet experienced or independent enough, he was nonetheless driven by his passion, and he had had the good fortune of being trained directly by his uncle. But there was no time to lose: just a few months before he passed away, Salvatore managed to get an order for 15,000 pairs of shoes from Saks Fifth Avenue, a large American department store; it was a huge amount for those days, when daily production was barely 200 pairs. But it wasn't easy to make shoes manufactured in a series fit American feet, and in Italy there still wasn't a form factory capable of competing with American manufacturers. Jerry acquired several forms in Germany, others in the United States; but the productive systems of the Ferragamo company, of an artisanal kind, were hard to adapt to the concepts with which those forms had been built. So it was necessary to study new models, find shoe sizes that responded not to made-to-measure creations, but to series production, made equally well in all sizes. What was needed was a shoe that could fit any Cinderella perfectly, and to later use that Cinderella as a basis for the overall development.

The shoe was finally created, and it is the shoe that made it possible to carry over into the present the values and the studies of Salvatore Ferragamo, making this statement evermore real and up to date: "It is upon this discovery – not design, style, or handicraft, but in the foot comfort of the hundreds and thousands of people for whom I have made shoes – that I have founded my fortune".[13]

[1] *Shoemaker of Dreams. The Autobiography of Salvatore Ferragamo* [1957] (Florence: Giunti, 1985), p. 67.
[2] Ibid., p. 193.
[3] Honoré de Balzac, "Théorie de la démarche", in *Oeuvres diverses de Honoré de Balzac*, vol. 2 [1830–35] (Paris: Louis Conard, 1938), pp. 613–43. Balzac's "Theory of Walking" was originally published in 1833.
[4] *Shoemaker of Dreams* 1985, p. 157.
[5] Ibid., p. 64.
[6] Ibid., p. 67.
[7] Ibid., pp. 69–70.
[8] Ibid., p. 157.
[9] Ibid., p. 63.
[10] Patent 1,399,606 of 6 December 1921 and patent 1,479,536 of 1 January 1924, registered in the United States.
[11] *Shoemaker of Dreams* 1985, p. 67.
[12] Ibid., p. 161.
[13] Ibid., p. 65.

Fiamma, Salvatore and Wanda Ferragamo's eldest daughter, after the death of her father took over the helm of the company's leather department and became the image of Ferragamo shoes in the world. Here she is in her studio in a photograph by Locchi from 1966.

STANDING, WALKING, MOVING IN EQUILIBRIUM AS A DANCER OR TIGHT-ROPE WALKER

SERGIO RISALITI

LA GRANDE PARADE

"L'homme, premier des primates à adapter la station debout, est aussì le premier des saltimbanques. Osant se dresser sur le sol, hésiter, s'avancer puis, sans trébucher, trouver son chemin sur deux pieds, il fut, parmi les mammifères, le premier funambule à parcourir le fil invisible de son existence. Là où les autres des son espèce, collées au sol, rampaient, sautillaient, claudiquaient, il s'élança."[1] Jean Clair used this dizzying sequence of images and concepts to introduce *La Grande Parade: Portrait de l'artiste en clown*, the exhibition that was held in Paris in 2004 in which the first human walking was compared to the feat of a tight-rope performer, a key figure for many famous artists of the twentieth century. In the works of Degas, Picasso, Klee and Léger, acrobats and tight-rope walkers offer a mythology alternative to that of the classical tradition. Figures performing stunts and balancing feats, including a ballerina with her *attitudes* and *arabesques*, stand halfway between a monkey and a soaring angel, an orangutan and the Apollo Belvedere. True to his comparison, Clair launched what is a crucial theme for us: the endless roaming of our species across planet Earth. Looking back, we can envisage the distance covered since the beginning and count the steps taken by our species in order to change place and survive, escape and reach new destinations.

Who is simultaneously a biped, a triped, and a quadruped?
Diodorus Siculus

Paul Klee, *Tight-rope Walker*, 1923
Lithograph on paper, 52.7 x 37.5 cm
Düsseldorf, Beck & Eggeling
International Fine Art.

Fernand Léger, original lithograph
from *Cirque* (Paris: Tériade, 1950),
44 x 34 x 4.2 cm
Florence, Biblioteca Nazionale
Centrale di Firenze.

Roman art, *Right Foot in Gilded Bronze Belonging to a Nike*, from the Forum of Augustus, inaugurated in year 2 BC
Gilded bronze (cast) on bronze solid tenon joint and iron hinge,
76 x 40 x 49 cm
Sovrintendenza Capitolina ai Beni Culturali
Rome, Museo dei Fori Imperiali – Trajan's Markets.

Henri Matisse, *Study of a Foot*, c. 1909
Bronze, h 30 cm
Saint Petersburg, The State Hermitage Museum.

According to the leading scientist André Leroi-Gourhan, the history of mankind begins with man's feet. Over the course of this adventure, the arch of the foot has exercised an essential function: supporting the human body, whether in a static or dynamic position, motionless or on the move. In this respect, the foot may be regarded as the cornerstone of human history, even more so than the club or Prometheus's fire – long before the wheel and plough. Artistic expressions enable us to study the evolution of our species by focusing on the feet rather than stomach, sex organs, or hands. The anatomy of the lower limbs, as this developed and took shape from the age of the earliest hominins onwards, may be used to learn about the destiny of man and his specialization compared to other living creatures. Even more so than words or objects, what played a leading role were the steps taken across space and time. As scientists explain, it all started millions of years ago, when in the Rift Valley, between Syria and Mozambique, the human species made a crucial evolutionary leap. The footprints of our African ancestors are the necessary and founding premise to Paul Klee's *Tight-rope Walker*, the icon of a season of the early twentieth century – as creative as it was free – that entrusted the painter with a demanding task. After moving away from the world of perfect geometries, beautiful proportions, and tidy cities and countrysides, the artist could only walk on the verge of an abyss by tracing unpredictable lines according to discordant harmonies – like those explored in the musical sphere by Igor Stravinsky with *Le Sacre du printemps*, or by Arnold Schoenberg with *Pierrot lunaire*. Possibly also in order to transcend the heap of ruins that Klee's *Angelus Novus* had left behind: remains and ruins that only increase after each new step in the march of progress.

THAT'S ONE SMALL STEP FOR [A] MAN, ONE GIANT LEAP FOR MANKIND

Ernst Jünger wrote: "the foot embodies the plant element and the idea of rootedness, what is stable and permanent in man, his connection with the firm and solid element".[2] A person cannot act with his hands if he loses his footing: "*Fußen* means standing in a position of firm balance – what in English is called a *footing*. This position depends on the foot and its support; it can never be based on sheer action. Among all creatures and all forms of the natural world, the foot is what maintains a contact with the ground and ensures the stability of one's position".[3] When at rest, the foot embodies a base, the place where we stand. As soon as we move and walk, the idea of time comes into play – even if this occurs faraway from the Earth, on the surface of the Moon. According to Jünger, "walking is that solemn and joyous contact of the feet with the ground; the spirit of the Earth manifests itself as a social and civic spirit. As such, walking shows itself in the triumph of the victorious warrior who advances with his head held high, within the procession of the mimes in a tragedy (the primitive satyrs of a dithyramb), as the sacrificing priest moves towards the altar".[4] The feet may come in contact with forces from below, as when the hero roams the land of the dead. Or they can lead us upwards, as when we climb trees grown overnight, reaching the clouds. With our feet we risk our lives by walking the tight-rope and savour the light joy of dancing to exhaustion. According to Jünger, "dancing is the freest manifestation of chthonic life. It brings together everything that the energy of the Earth has to offer: the rhythm of sowing and reaping, the deep enjoyment of wine and sex. The dance expressed by different peoples seizes and seduces us; we can almost witness a playful re-enactment of our origins, of the strange becoming that erupts from deep within and catches the imagination of man".[5] We thus catch a glimpse of something sacred in dancing, something awesome that leads us from intoxication to the brink of ecstasy.

WALKING FOR THE SAKE OF WALKING

There is a primordial form of mobility that is functional to the survival of our species. But there is also an equally instinctual form of mobility related to energy, to sexuality, which runs through the body and pervades it, to the point of leading it to dissipation. By engaging in acrobatic feats and pirouettes, in wild dance moves, endless walks and aimless runs, the human being has deployed his superabundant energy in the aesthetic and spiritual field. In dancing, for instance, the dissipation of energy is organized as a mode of representation and expression that, like music, unfolds over time and, like sculpture, is located in space, with the aim of bringing pleasure to oneself and others. The vital movement that is conveyed or withheld by the dancer acts as a barrier against entropy, restraining pleasure as far as possible, to prevent the intensity from degenerating into pain or dissatisfaction – into unbearable feelings. Paul Valéry, the author of *Degas danse dessin*, had perfectly grasped the meaning of these poses and movements of dissipation. These are physical actions that only come to a halt through a cause external to their own being, figure, and species; and that instead of being subject "to principles of economy, almost appear to have *dissipation* itself as their object".[6]

And again: "The leaps and capers of a child, or dog, and *walking for the sake of walking*, or *swimming for the sake of swimming*, are activities that have the sole aim of altering our perception of energy, of engendering a given state for this perception. The acts within a category of this sort can and must multiply themselves, until a circumstance brought about by them – and utterly different from an external alteration – comes into play. This can be *any* circumstance in relation to them: *tiredness*, for instance, or *convention*. These movements, which have their aim in themselves, namely to engender a given state, spring from the need to be performed or from an occasion that triggers them; yet these impulses assign them no direction in space".[7]

These movements inevitably prove disorderly. "An animal tired of forced immobility will run off, shake off and flee from a feeling, not a thing; it will break out in rapid, excessive movements. A man in whom joy, anger or agitation, or sudden intellectual ebullience, releases an energy whose cause no specific action can absorb or exhaust, will walk off with a frantic pace; without realizing it, in the space he covers he is obeying this superabundant energy".[8]

Over time, mankind has learned to manage this superabundance by turning it into a controlled form of dissipation in the pursuit of aesthetic, spiritual, magical and apotropaic goals. Paul Valéry shifts the focus on dancing, arguing that its merit lies in the capacity to order and arrange these movements of dissipation, which upon finding release in space become bound to time. Before being adequately corrected and integrated through civilized art forms, before being accepted within the ranks of the academic fine arts, ever since the Middle Ages dancing had been symbolically associated with lust. When in pagan antiquity dancing and acrobatics ceased exercising a ritual and votive function, many of the movements and exercises related to the display of the body came to be regarded as indecent and unseemly. The frantic motions of the maenads and bacchants accompanying Dionysus, or hunting down Pentheus, survived Pan's death, resurfacing many centuries later. A kind of mental disorder and erotic dissipation also pervade the young Florentine women portrayed by Pollaiolo, Ghirlandaio and Botticelli in their religious paintings. In these modern nymphs, with their distinctive traits (serpentine flowing locks, dresses billowing in the wind, bare feet), Aby Warburg saw the rebirth of the erotic forces of the pagan world. These nymphs advance through dance steps and radiate indescribable feelings and engaging passions. Their postures betray alienation and hysteria, a form of desire caught between heavenly lightness and chthonic violence. This "Dionysiac" mode of existence is visible behind the sophisticated harmony of the Renaissance and the dignified gait of earthly "madonnas", amid

Pericle Fazzini,
The Dancer, 1936–37
Wood, 164 x 95 x 34.5 cm
Siena, Collection of Banca Monte
dei Paschi di Siena.

Raphael (school), *Dancer*,
fifteenth century
Pen on burnished white paper,
23 x 13.9 cm
Florence, Gabinetto Disegni e Stampe
degli Uffizi.

streets and palaces that are still recognizable to-day. When discussing a drawing by Botticelli in the Musée Condé in Chantilly, Warburg coined the expression "brise imaginaire". The same fey beauty is exuded by one of the female figures from Ghirlandaio's fresco in the Tornabuoni Chapel in the church of Santa Maria Novella, executed between 1485 and 1490. Enchanted, Warburg described her as a "pagan petrel". The scene in question is the *Birth of Saint John the Baptist*: in the room where the saint's mother lies after having just given birth, a young Florentine woman advances with a basket of flowers and fruit on her head; as if carried by a breeze, her dress billowing, the figure moves in manifest frenzy, dancing rather than walking. Warburg, a real diviner when it comes to exposing old enemies lurking behind iconographic layers, detects some sinister and disquieting presences here: in his *Mnemosyne Atlas*, nymphs keep close company with maenads, and are followed by "headhunters" – Botticelli's Judith and Salome.[9] Warburg sees them everywhere, within a world open to the pagan era; time, in his view, is reversible and permeable: his research

constantly shifts between antiquity and modernity. In Warburg's eyes, Judith is more than just a Biblical heroine: when she returns from the enemy camp, she rejoices and trembles with pleasure, still tasting the violence perpetrated, the blood spilled. By arousing sexual desire, she has defended her virginity and the freedom of her people. Judith is filled with excitement and on her way back she seems to be driven by the wind – a little mad, a little reckless.

Nymphs made a comeback in the late eighteenth century. Lady Hamilton, the muse behind many of Canova's creations, was a genuine modern bacchant who "constantly lived between stone and aether, flux and stability".[10] Then there is Maria Medina, Salvatore Viganò's wife, who is best remembered for her Grecian attire, with bare arms and breast, sandals, and hair gathered in tresses around the head. Finally, there is Isadora Duncan, the living image of the Gradiva, this "protean and disturbingly alien heroine".[11] In the eyes of Rodin, Bourdelle and Nomellini, she embodied "the hidden similarities through which all ages suddenly dance together and all possible embodiments mingle, like in a dream".[12] Duncan had chosen to establish a direct contact with the sources of rhythm by doing away with the rhetoric of the romantic ballerina. She had thus turned to the pagan world, to Greek inspiration that, like a modern nymph, she brought to the stage: barefoot, she would charm and bewilder her public of refined men of letters, musicians, poets and artists. Plinio Nomellini portrayed her as a maenad spirited away by the wind. In dozens of drawings we see her moving in the open air, drunk with the spirit of the Mediterranean, entranced by the light and the rhythm of the waves (see here on pp. 153–55). The setting for the encounter between the painter and the dancer was Versilia. Duncan had moved to Tuscany to find some peace of mind after the tragic loss of her son. Eleonora Duse welcomed the forlorn mother in her villa by the sea; she witnessed her dance on the shore – a happy moment of oblivion for a woman who longed to forget everything.

Domenico Ghirlandaio,
Birth of Saint John the Baptist,
1485–90 (detail)
Fresco
Florence, church of Santa Maria
Novella, Tornabuoni Chapel.

Greek art, *Relief with Dancing Maenads*, first century AD
Marble, 57 x 69 x 2.5 cm
Rome, Museo di Scultura Antica
Giovanni Barracco.

A CABLE STRETCHED
BETWEEN BEAST AND SUPERMAN

In the century of the Enlightenment and of early Romantic upheavals, Italian art witnessed the development of a different and possibly alternative mythology through a painter with a keen interest in satire, the theatre and pantomime. In his new world, centaurs, satyrs and Pulcinellas took the place of a fallen, or rather decayed, aristocracy. With the frescoes of Villa Zianigo (c. 1759–97), now in Ca' Rezzonico in Venice, and with his *Divertimento per li ragazzi*, 104 cards drawn and painted in watercolour after 1791, Giandomenico Tiepolo sought to replace the most noble tragedies with street theatre and comedy. Behind libertinage he glimpsed the state of nature; behind the images of gods and demigods, the popular masks from Naples and Venice. Pulcinella on a swing or tight-rope, dancing with dogs or performing acrobatic tricks, marked the dawn of a new age, no longer dominated by the values which Giandomenico's father Giambattista Tiepolo had still looked up to. The civilization, history and culture of humanism survived in precarious balance; everything was about to topple, as mankind grotesquely advanced, driven by wild instincts. Progress did not prevail on the original nature of man, which remained what it had always been. Like a centaur, the man of the world, the gentleman, found himself caught between instinct and reason, nature and culture. After all, man himself is not a perfectly symmetrical creature. As Jünger wrote, the human being is a centaur: "High and low, head and foot, are distinguished in an even more hierarchically defined way than the right hand and the left. We are accustomed to drawing a distinction between what pertains to the head and what pertains to the feet. Just as right and wrong branch out to form a symmetrical structure, so high and low stand in opposition according to the centaur model. The former distinction leads us to the sphere of good and evil, the latter provides the criteria of good and bad quality".[13]

Genetic otherness is of as much interest to artists, men of letters and poets as mental derangement – the presence of another self deep within. From Goya to Géricault, from Shelley and Poe to Hugo and Rimbaud, the "je" invariably also coincides with the "autre". With Charles Darwin's *On the Origins of the Species*, things took a drastic turn: the idea of the heavenly origin of the human body was rejected, since the latter, according to science, is not made in the image and likeness of a god, but is rather related to orangutans. An increased awareness of evolution reduced the gap between man and his fellow creatures – primates. Even the ethereal, enchanted and noble world of ballet started featuring human beings behaving in grotesque, disorderly and animal-like ways. Degas's female dancers, for instance, show a degree of primeval brutality: under certain angles, or in certain situations, they struggle to conceal their true origins. In Degas's later works – those made after 1875 – the anatomical details no longer correspond to reality but rather to the artist's figurative intentions, as he makes conscious mistakes in order to affirm his own personal interpretation. Degas seems to catch his ballerinas just as their phantasmatic bodies are opening up to the repressed – the female body expressing a vestige of animality that survives at the anatomical level or buried in the unconscious. In *Dancers at the Barre* – a work executed around 1900 (Washington, DC, The Phillips Collection) – two figures are seen warming up between the floor and the barre; yet they do so without the gracefulness one would expect, and this makes them resemble apes. Their anatomy is reduced

Giandomenico Tiepolo, *Pulcinella on a Swing*, 1793
Detached fresco, 200 x 170 cm
Venice, Ca' Rezzonico, Museo del Settecento Veneziano.

to a mere sketch, as the emphasis is on colour – orange and grey hues. The arrangement of the two figures side by side engenders something beastly, almost monstrous. The tutu belongs to both dancers; only two feet are resting on the ground, whereas in the upper part of the painting an ill-defined number of limbs – partly reduced to mere stumps – move about like the pincers of a two-headed crab. From Edgar Degas's journal we can excerpt some interesting thoughts on his work method and his approach to different subjects, including dancers and horses: "Perform simple operations, such as drawing a profile that does not move, as we ourselves move … studying a foreshortened figure or object, or anything else … Leave much out: of a dancer, trace the arms or legs, or lower back; trace the ballet slippers, the hands of the woman doing her hair, the cropped hair, the bare feet in the act of dancing".[14] Degas is ruthless: with a surgeon's eye, he selects his preys, probes those "things" and coldly examines them by destructuring the whole. He then reconstructs the figures as robot-dancers, stressing their animal-like aspects, disorderly gracefulness, precarious balance and unseemly attitudes. A manifest erosion of form takes place, yet the painter's genius nonchalantly conceals all action; a new gracefulness springs from the consummate skill of the artist, who like a man of the Renaissance exercises complete control over lines, colours, light and space. The contrast between the butterfly-dancer, the swan, an embodiment of divine gracefulness, and the new modern mode of representation – which is more life-like, almost brutal – emerges in certain pastels with an aggressive, ruthless rendering. Degas brings out the awkwardness of the girls and their physical limits, while creating images of unearthly beauty. This is the artist's very own "human comedy", in line

with the principles of life-likeness defined by Honoré de Balzac. From the *Dancer at the Photographer's Studio* (c. 1875, Moscow, Pushkin Museum) one would expect a certain aplomb, a balanced and graceful position; instead, Degas portrays a stiff ballerina, like a cross between an ape and the figurine one might find on the top of a musical box. The artist has completely forgotten the languidly reclining beautiful women, the gorgeous Venuses, the odalisques. He is not inspired to paint "well" by the presence of flesh, be it golden, white or carmine; instead, he is determined "to reconstruct the specialized female animal as the slave of the dance, the laundry, or the streets. Those more or less distorted bodies" – the author of *Degas danse dessin* writes – "whose articulated structure he always arranges in very precarious attitudes (trying a ballet shoe, or pressing iron down on linen with both fists) make the whole mechanical system of a living being seem to grimace like a face".[15] This taste for mimicry reflects Degas's Neapolitan roots:

"Mimicry comes from Naples, where there is no word without gesture, no tale without imitation, no person without a multitude of characters – always possible and always ready", Valéry notes.[16]

A few years earlier, in 1894, a giant of sculpture, Auguste Rodin, had created *Iris, Messenger of the Gods*. This headless female body, with one arm missing, has been brutally destructured and given a pose halfway between sexual provocation and highly contemporary dance. It is as though the woman portrayed were suffering from a cramp: with one leg outstretched, she is massaging the sole of her foot. Her pose breaks the rules of academic decorum, so that the human figure proves more beastly than divine. Titian's *Leda*, with her perfect proportions, is a very distant memory indeed. Like Degas, Rodin brings the female figure

Edgar Degas, *Dancers
at the Barre*, c. 1900
Oil on canvas, 130.1 x 97.7 cm
Washington, DC, The Phillips Collection.

down to the horizontal level, with no academic or idealistic filters. With crude materialism, he emphasizes the apotropaic and feminine aspects of the body, its sexual power, from which modern female gracefulness rises, like a phoenix from its ashes. An admirer of Phidias and Michelangelo, the sculptor found inspiration in Khmer dance, in the small, muscular, asexual women dancers of Cambodia. He devoted a series of extraordinarily beautiful sketches and watercolours to them. According to Rainer Maria Rilke, who worked as Rodin's assistant for some time, this was his finest work. When he first witnessed a performance by Khmer dancers, Rodin felt that he had finally grasped the origin of movement he had been searching for his whole life.

Works such as those of Rodin and Degas are already part of the twentieth-century avant-garde: they constitute its necessary prerequisite. By working on unbalance, anatomical deconstruction and the new aesthetics engendered by a cacophony of poses, the modern artist reverted to a geometry of broken lines, to inorganic, formless matter, to the world of insects and invertebrates. Buried within this framework of genealogical regression was a new form of beauty. As an example of terpsichorean perfection, Paul Valéry mentions the Medusa: "not wom-

en but beings of an incomparable substance, translucent and sensible, madly irritable flesh of glass, domes of floating silk, hyaline crowns, long living strips cut across by rapid waves, fringes and gathers that they fold and unfold … No ground, no solids for these absolute dancers: no stage, but a centre on which to rest, on all points, which give way wherever one wishes. No solids in their crystalline elastic bodies, no bones, no unchanging joints, no countable segments".[17]

Something close to an elastic body, a human being with neither bones nor sinews, is to be found in Francis Bacon's *Painting 1978*. Here a naked man is portrayed in the acrobatic pose of a mountaineer testing his joints, destructuring his body, and turning it into rubber. The figure, standing in precarious balance on its left leg, extends the other leg towards a door, in order to turn a key with the toes of its right foot. Its gesture is like that of a chimpanzee trying to escape from the lab in which it is imprisoned. Bacon thus blurs the difference between caged animal and acrobat. The British artist's oeuvre is replete with scenes of this sort: consider, for instance, *Three Studies from the Human Body* from 1967 or *Man Walking Down the Stairs* from 1972. In the latter work, an elegant gentleman performs the same action – but via a different, or indeed opposite route – as the figure depicted in Alessandro Allori's *Death*, an anatomical drawing executed in the latter half of the sixteenth century. In other words, what we find is always the risk of falling: "At every instant, the dancer can slip from the signifying function attributed to him without his knowledge, and which strips him of his body, so to speak; he will once again come to coincide with his own material reality".[18] And this, not because the dancer will be heavily pulled to the ground and fall on the floor, but because he will be "stripped of his ideal significance and reduced to the dismal evidence of his presence in the flesh".[19]

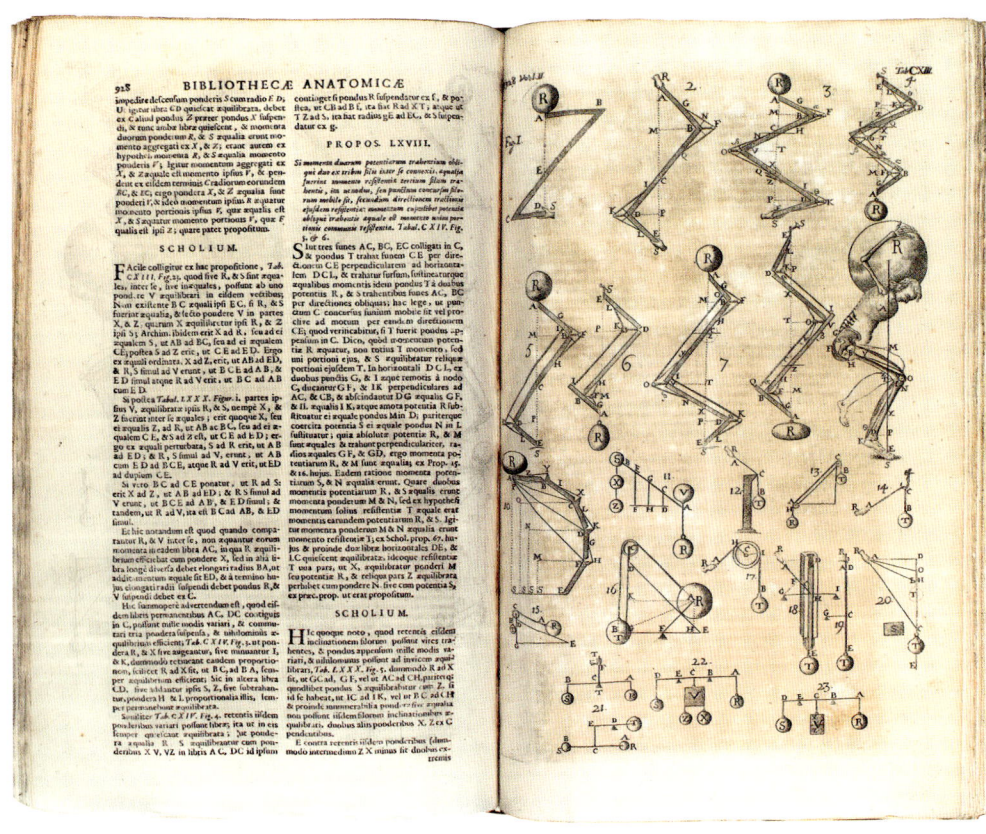

Daniel Leclerc and Jean-Jacques Manget, *Bibliotheca anatomica sive recens in anatomia inventorum thesaurus…* (Geneva: J. A. Chouët, 1699), vol. 2, 37 x 24 x 8.5 cm Florence, Biblioteca Nazionale Centrale di Firenze.

DANCING SKELETONS

In his *Discussions on the Rules of Drawing*, Alessandro Allori (Florence, 1535–1607) stresses the importance of anatomical dissection for painters, as a means for the acquisition of self-knowledge through a journey within the human "model", on a quest to discover the deepest and darkest recesses of the body. Artistic anatomy was thus approached as a planned, educational endeavour, whereby drawings were not seen as mere preparatory sketches, but rather acquired a value of their own. In some studies, the representation of skeletons and skinned corpses reflects the increased interest in necromantic, macabre aspects, producing a sense of estrangement that transcends the naturalistic starting point – the specific study of human anatomy. From as early as the mid-sixteenth century, "Deaths" (skeletons) started serving an allegorical function, devoid of any religious significance. In order to exorcise fear and dread, they embraced the grotesque: in some illustrations by Juan Valverde, for instance, and in Jean-Jacques Manget's *Theatrum anatomicum* the corpses play the part of leading actors, or improvise some dance steps. Allori's skeletons stroll about, make their way up steps, and try to keep their balance. They look like automata just crafted by a stage designer, who for the moment only wishes to test their joints and hinges; the artist studies their first steps, footing, and joints. The skeleton seems to be brought to life, like a puppet, through some invisible strings that make it stand without falling. This group of drawings – homogeneous works in terms of quality and underlying spirit – were probably executed in preparation for some of the decorations set up for Michelangelo's funeral, held on 14 July 1564 in the Basilica of San Lorenzo in Florence.

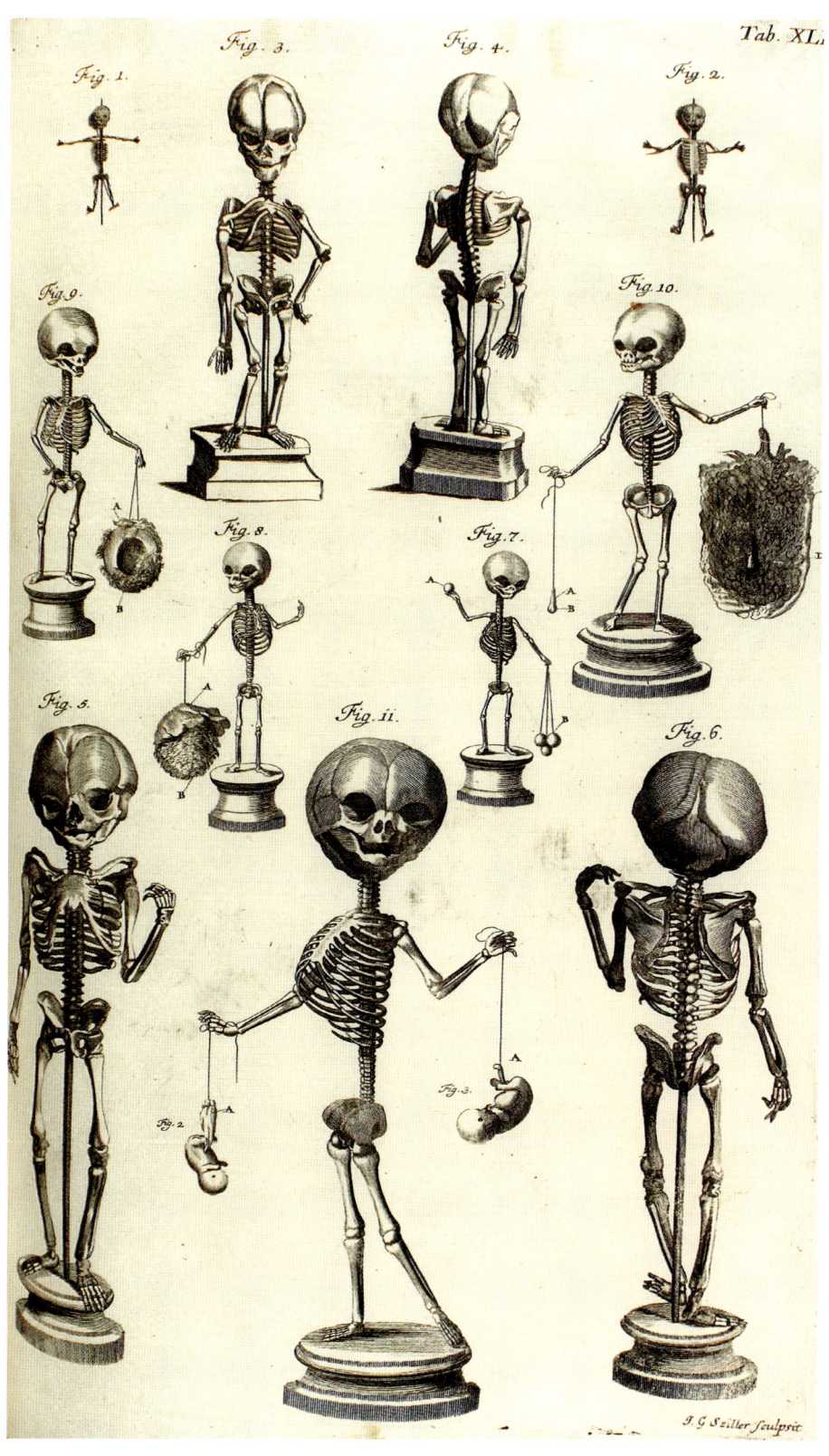

Jean-Jacques Manget, *Theatrum anatomicum…* (Geneva, 1716–17), 2 vols., 44 x 29 x 8.3 cm Florence, Biblioteca Nazionale Centrale di Firenze.

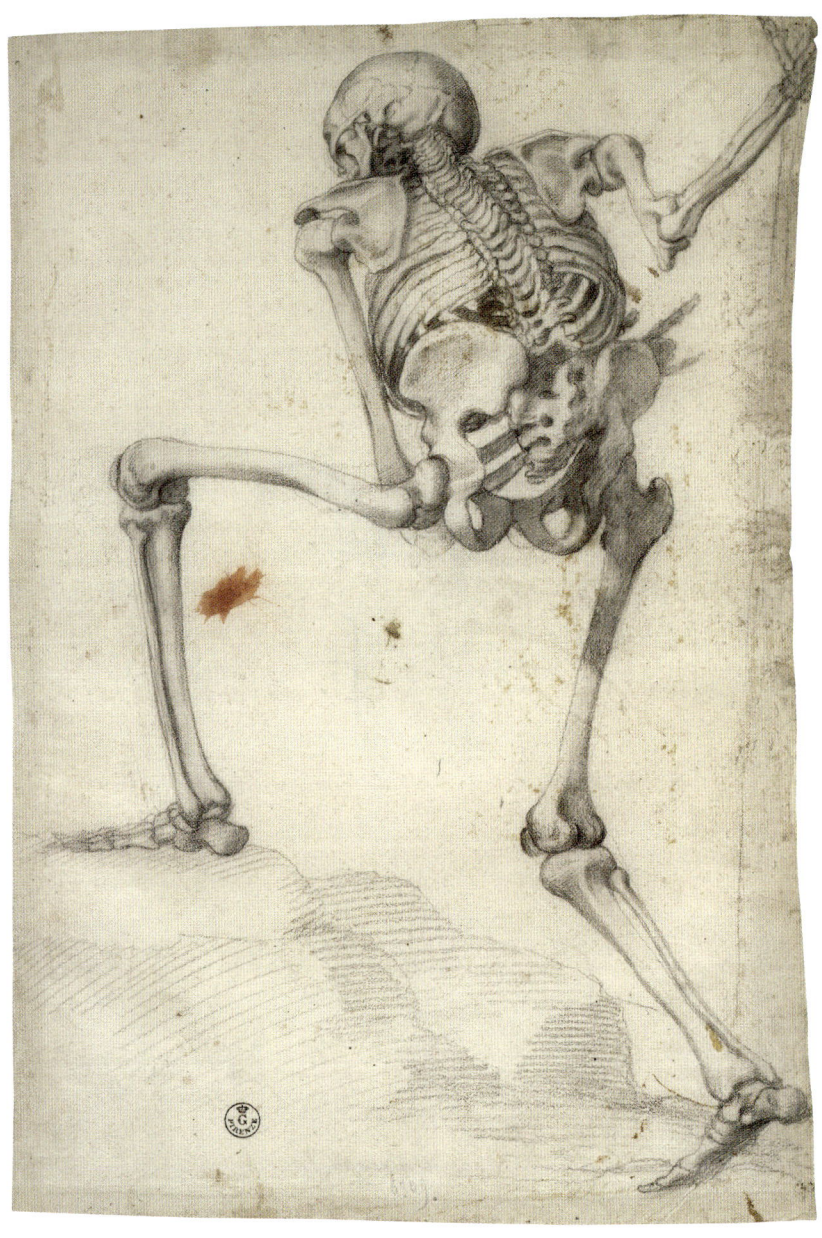 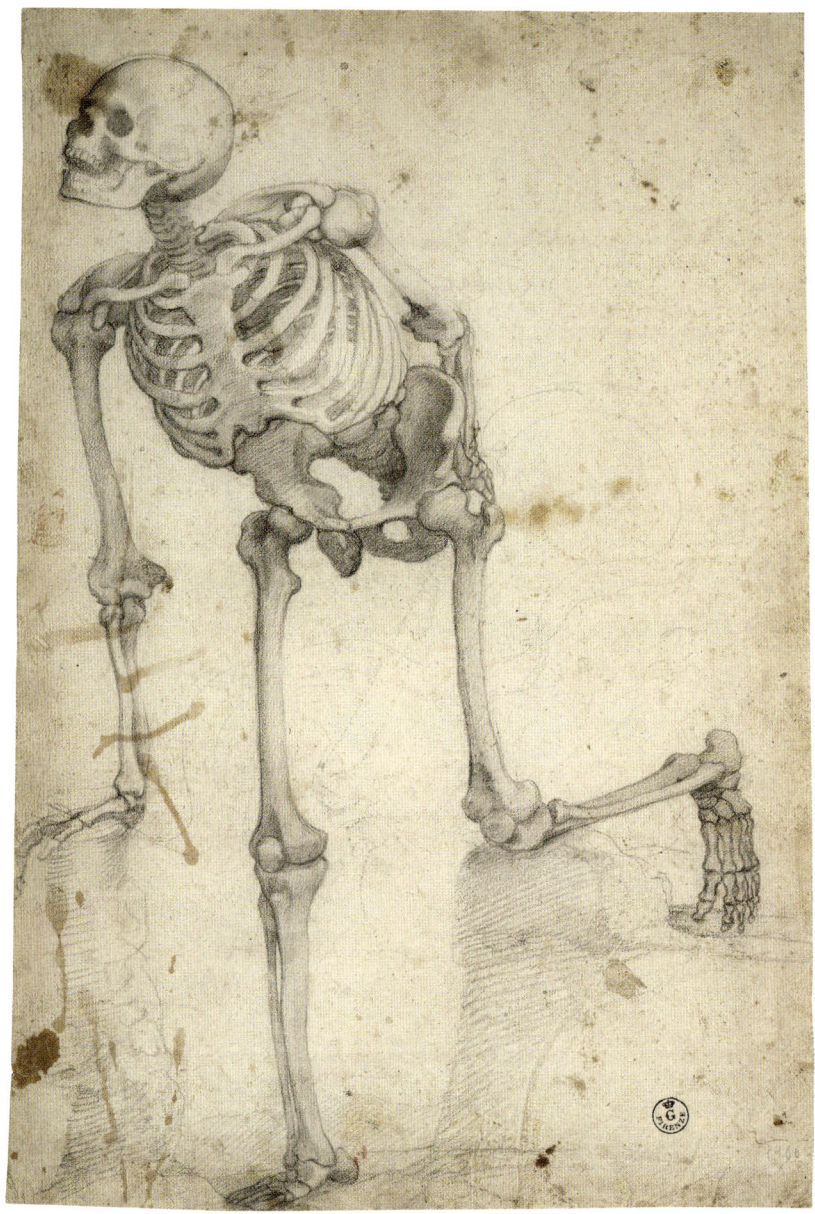

Alessandro Allori, *Animated Skeleton*,
c. 1564–65
Black pencil on white paper,
42 x 28 cm
Florence, Gabinetto Disegni e Stampe
degli Uffizi.

Alessandro Allori, *Animated Skeleton*,
c. 1564–65
Black pencil on white paper,
43 x 28.7 cm
Florence, Gabinetto Disegni e Stampe
degli Uffizi.

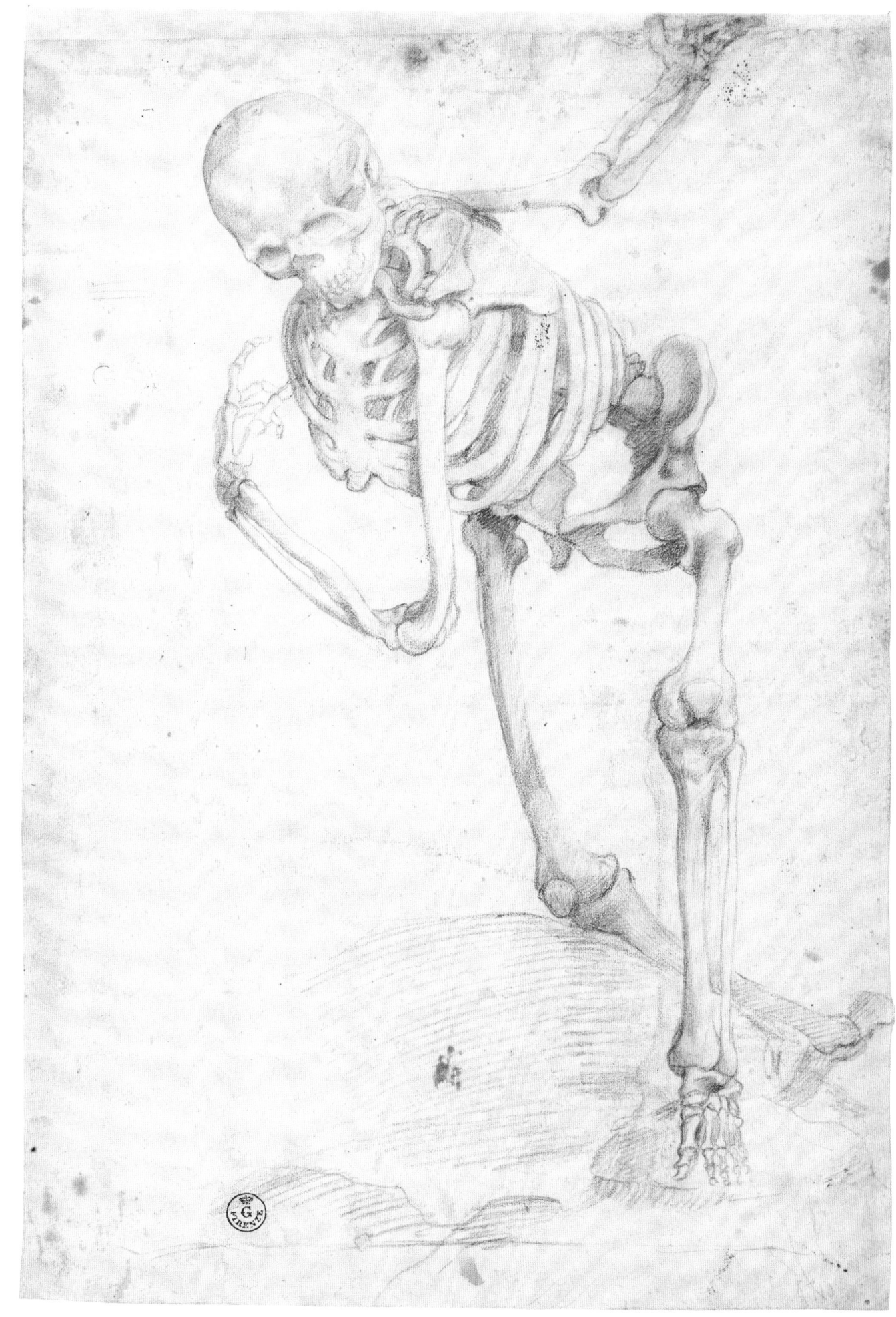

Alessandro Allori, *Animated Skeleton*,
c. 1564–65
Black pencil on white paper,
41.5 x 27.8 cm
Florence, Gabinetto Disegni e Stampe
degli Uffizi.

WALKING ON ALL FOURS

Unlike Allori, Bacon forgot about the inner structure and led the beautiful Renaissance body to the slaughter, creating what Gilles Deleuze describes as "a zone of indiscernibility or undecidability between man and animal".[20] Bones acquire the function of "gym equipment" used by the flesh for its acrobatics. With Bacon, beauty and horror are gently mixed on the dissection table or gym floor, and "the athleticism of the body is naturally prolonged in the acrobatics of the flesh".[21] As the joints collapse, the rigid structures of the skeleton are loosened up, yet a hint of nervousness, or almost angst, endures in the body. In *Francis Bacon. The Logic of Sensation*, Deleuze reminds us of the Irish painter's interest in Degas, and especially in the young woman in *Après le bain*, "whose suspended spinal column seems to protrude from her flesh, making it seem much more vulnerable and lithe, acrobatic".[22] Bacon identifies a special pictorial tension between flesh and bones (those of the foot as much as of the trunk or head). In *Woman Emptying a Bowl of Water / Paralytic Child Walking on All Fours* mankind is portrayed in an animal-like form, as an alien monster or corpse. The child seems to be fleeing from a horrifying creature; both are walking in precarious balance on a circular metal frame. The chase and flight might continue forever, as the geometry of the structure suggests. The viewer gazes at the scene as though perched on a branch or at the circus – as though he or she had rushed to witnessed the slaughter. We are flung back to the primordial age, to life in its bare form, to a state of utter instinctual violence, where the flesh becomes an object of bestial appetites. By an odd evolutionary twist, art here reverts to its very first steps by offering us a hybrid humanity – halfway between beast and God.

Francis Bacon, *Woman Emptying a Bowl of Water / Paralytic Child Walking on All Fours*, 1965 (from *The Human Figure in Motion* by Eadweard Muybridge) Oil on canvas, 198 x 147.5 cm Amsterdam, Stedelijk Museum.

ACROBATICS

Jean Starobinski has devoted some extraordinary pages to the topic of balance, vertigo, and elation, starting from those of *Portrait de l'artiste en saltimbanque* (1970).[23] According to the Geneva-born philosopher, psychiatrist and literary critic, the Parisian Théâtre des Funambules maintains some links with primitive life. There one can still breathe the mysterious and disturbing air of archaic games and rituals. The feats of tight-rope walkers and acrobats, their choreographies and agility tests, stir certain layers of the collective unconscious: in the form of gymnastic exercises, strange connections resurface with the Earth's first inhabitants, with the time in which mankind – as we know it – was still at the stage of "assembly". In the theatre on Boulevard du Temple, a venue also praised by Edmond de Goncourt, the infancy of man is brought to life once more. Starobinski writes that "acrobats know the password for the superhuman world of divinity and the subhuman world of animal life".[24] In other words, today as much as in the past, the acrobat performs by keeping in contact with the open side of animal and natural life, with all the risks and adventures which this form of gymnastics entails. By walking in mid-air, he can sense the divine presence in the heavens, without feeling dizzy at the void beneath him. Nietzsche, a philologist-philosopher and an expert on the Apollonian and the Dionysian, must have had tight-rope walkers in mind when in *Thus Spoke Zarathustra* (1885) he wrote that man is a rope stretched between the beast and the superman. From that day onwards, the two poles became the distinguishing feature of the artist, who sees the tight-rope walker as his double – as the source of his schizophrenia. Both possess lightness and agility, qualities which according to Théodore de Banville, the author of *Odes funambulesques* (1857), may be found in cats, fawns, monkeys and panthers. The tight-rope walker triumphs on the air, experiences the vertigo of height, fixes his gaze on the chasm of the heavens, and soars over the heads of astonished fellow-humans, who – slaves to their own weight – remain anchored to the ground: "Whether he be / a sublime or grotesque hero, / O muse, whether he chases vultures / Or comes down to perform some tricks / On the tight-rope, / A tribune, prophet or clown, / He will always contemptuously flee / the paths beaten by the crowd, / He walks on the proud peaks / Or on the vile rope; yet / Far above the faces of the crowd".[25] Every tight-rope walker yearns for the greatest heights and walks on the verge of an abyss, aiming for the blue which he sometimes almost touches with the tip of his nose: "In end, from his wretched stage, / The clown jumped upwards, so high that / He broke through the canvas ceiling / At the roll of the drums, at the sound of the horn / And his heart devoured by love / Performed somersaults among the stars".[26]

METAMORPHOSES AND HIEROGLYPHS

According to Starobinski, dancing and acrobatics no doubt "only refer to the body itself, its gracefulness, its vigour, its erotic appeal". However, something special happens to the dancer: "however scantily dressed she may be, she always acquires an illusionary role, becoming someone else".[27] The dancer can embody a flower, a bird, an imaginary deity, a sylph, Titania, or Péri. But this "representational estrangement … is entirely sustained by the dancer's body and the presence of the flesh is invariably predominant, down to the positive outcome of those allusive embodiments. … The appeal exercised by the dancer largely consists precisely in this restrained overcoming, which envelops her in an aura of brief significations, which are constantly taken up and recreated by the being in flesh and bones".[28] In his short story *La Fanfarlo*, Baudelaire perfectly grasps the significance of the charm exercised by the dancer: "She appeared in an agreeable succession of metamorphoses … *La Fanfarlo* was, in

turn, respectable, fairy-like, mad, mirthful: she was sublime in her art, as much an actress with her legs as a dancer with her eyes … Dancing can reveal all the mystery hidden in music, and it has, moreover, the merit of being human and palpable. Dancing is poetry with arms and legs: it is matter, graceful and terrible, beautified by movement".[29] The story told by Baudelaire does not have a happy ending: the dancer's lover, captivated by the abstractions she embodies, despises her fall into reality – the appearance of a real woman in place of the dancer, an idol unavoidably abandoned. Farfarlo thus forever loses the prestige associated with her role and "allows herself to be utterly pervaded by the inertia of the flesh".[30] Her lightness, her ability to rise up on the tips of her toes, to soar up in the air, vanishes in an instant and all that is left is an excessive body: "As for her, she grows fatter by the day; she has become a plump, clean, gleaming, and cunning beauty, a kind of ministerial call girl".[31] According to nineteenth-century aesthetics, the body was still something evil – a dark power. At the circus and the theatre, people wanted to admire bodies "gloriously, vainly seeking redemption through movement". Mallarmé was aware of the problem but dodged it, by seeking refuge in idealism and reintroducing the notions of ritual and worship. The French poet found the graceful interweaving of movements quite enthralling, as long as the signifier pointed to a mysterious poetic revelation: "Dancing for him was a text sprung from a silent speech of the body, but within which the body is abolished. And ballet turns into a *hieroglyph*".[32] Acting ahead of the avant-garde, Mallarmé envisaged dancing as a corporeal form of writing: "The dancer *is not a woman*, but a metaphor – sword, goblet, flower, etc. – that encapsulates one of the basic aspects of our form; and she *does not dance*, but rather suggests – by virtue of her prodigious delays or leaps, through corporeal writing – what it would take whole paragraphs of dialogue and descriptions to express in letters: a poem free from the limitations imposed by the scribe's tools".

MOBILES

In certain cases, the tight-rope walker is like a crawling insect. Théophile Gautier admired the androgyny, gracefulness and strength of Auriol, a famous acrobat of his day: "Monkeys seem clumsy and lame compared to Auriol; the laws of gravity seem foreign to him; he climbs, fly-like, up the varnished surface of a tall column; he could walk on a ceiling, if he wished to".[33] Acrobats' performances brought amazement and mirth. They would switch between the perfect geometries of tight-rope walking and precipitous tumbles; they would display absurd, unbelievable bodily poses, evoking fanciful new natures and potentials. The human figure here is no longer recognizable as such, but acquires a geometrical, organic, Cubist appearance, as a deconstructed anatomical structure. Ultimately – Gautier explains – "we feel humiliated ... at having to walk on our feet: we would like to return home by making cartwheels".[34] Human, semi-divine and animal-like, the acrobat is a breed apart. In him, spiritual virtue prevails over physical gravity, which he overcomes by growing disjointed, like a puppet. The body of Jean-Gaspard Deburau, a French Pierrot, was indeed an anatomical miracle. Deburau would play with his skeletal frame, tendons and muscles, creating impossible or unlikely balances. He would move on his head, balance a ladder on the tip of his nose, dance on stalks, perform somersaults on a rope. His was a "broken, open and deboned" body, as Starobinski notes.[35] Like some of Léger's figures, which remind one of rag dolls abandoned in children's playrooms. The consequence of this anthropomorphic destructuring is even more radically perceivable in Julio

Julio González,
Dancer with Daisy, c. 1937
Bronze, 46 x 30.5 x 9.5 cm
Madrid, Museo Nacional Centro
de Arte Reina Sofía.

Julio González,
Dishevelled Dancer, 1935
Forged and welded iron,
53.5 x 37 x 20 cm
Nantes, Musée des Beaux-Arts.

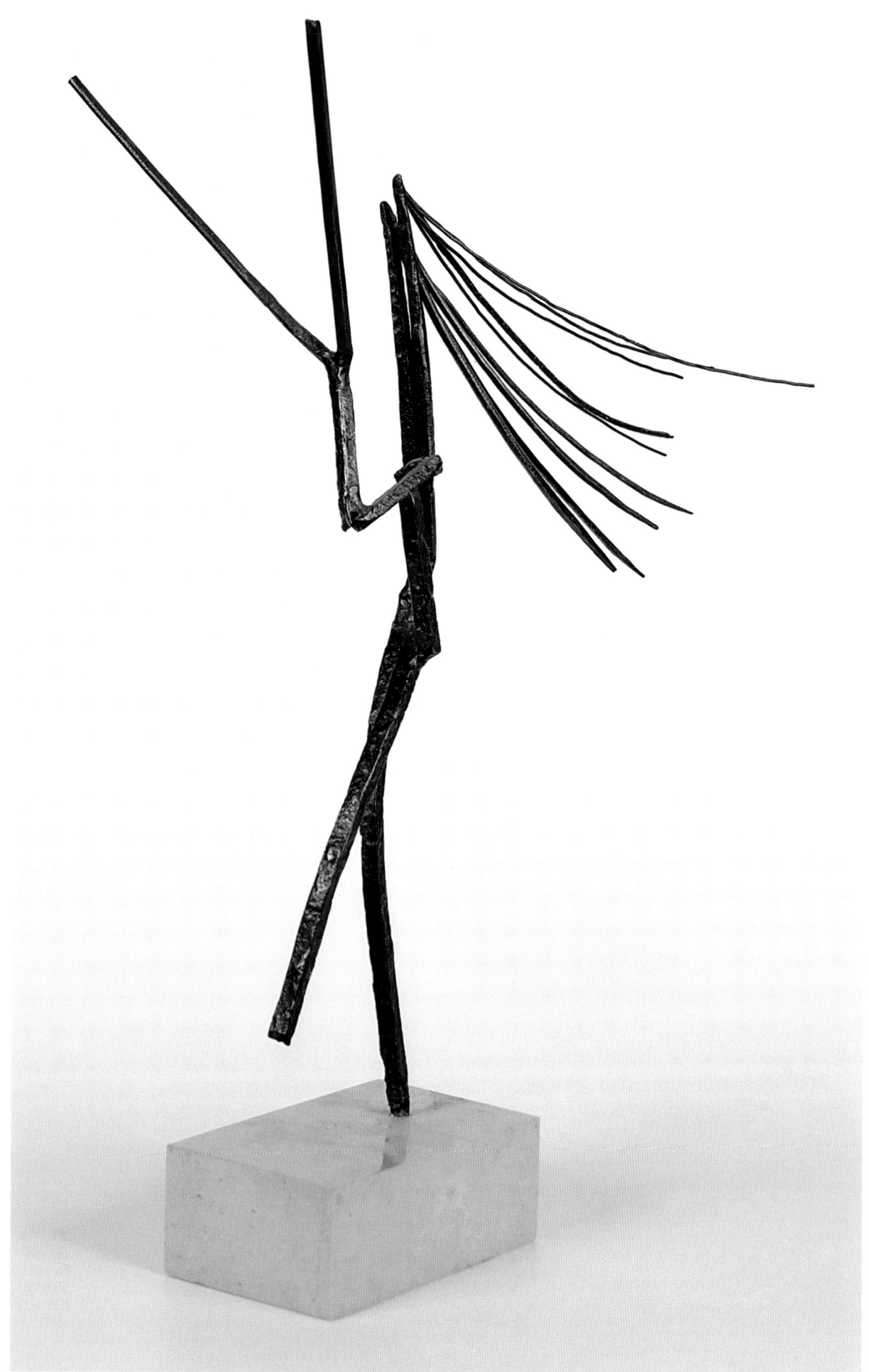

González's dancers, based on a process of subtraction and reduction that leaves only some essential broken lines. Having reached this point, the paths of art and nature meet once more and these creatures become insects – a grasshopper or mantis. Similar works include Fausto Melotti's slender sculptures and the precariously balanced *Mobiles* by Alexander Calder, who spent much of his time gazing at and depicting circus heroes. Calder is famous for constructions suspended in mid-air, or highly colourful ones, in balance. What are less well-known are his bronze sculptures of balancing acrobats and stout jugglers, with roughly sketched anatomies. These are like the merry amusements of a child, who pictures the world as a circus. In more recent years, the artist moved beyond the organic realm, delving deep into that of molecules, where the deepest connections between man and nature, anatomy and cosmic laws are established. The biped figure of Antony Gormley gives the impression of dissolving into the immaterial; his skeleton is reduced to simple segments – almost trajectories traced by energy fields; yet the overall structure is perfectly balanced, like a constellation, so that its primal origin, or primeval outline, is visible.

SILENT RITUALS AND ADMIRABLE AGILITY

By walking on two legs, man has had to learn how to stand in balance, without falling. From the still ape-like yet familiar silhouette of Ardi – a female ancestor 4.4 million years older than us, a hybrid biped with long arms suited for swinging on the highest branches of trees to escape from saber-toothed felines – to Marino Marini's *Juggler* and George Segal's *Red Woman Acrobat Hanging from a Rope*, the history of humanity is also that of its quest to stand in balance on the ground or on branches, against emptiness and the force of gravity. Standing, on a floor or rope, branch or ball, means not falling – assigning primary importance to verticality. It means identifying balanced verticality with the stability of the mankind's fate. Proof of this may be found in the representation of Fortune/Opportunity, which has often been pictured in precarious balance on a sphere or wheel, with a lock of hair ruffled by the wind – as in the case of the figure in the Palazzo Ducale in Mantua, an early sixteenth-century work in the school of Mantegna. Albrecht Dürer may be credited with one of the most famous versions of this theme. His *Great Fortune* (c. 1501–02) is a figurative archetype that circulated in artists' studios through printed reproductions, resurfacing with Pablo Picasso and then Gino Severini. In Sir George Sitwell's Montefugoni Castle and in Léonce Rosenberg's apartment in Paris, Severini created figures of female dancers on a sphere, Pulcinellas balancing on columns like modern stylites, masked music players with picturesque poses, set against sublime ruins and ancient remains – the solemn beauty of the Tuscan landscape. For Dürer, the woman on the sphere is also a symbolic embodiment of Justice and Nemesis: through her wings, she can fly over the world and govern the lives of men from above – of the rich and powerful as much as the humble and simple. In this respect, a fall can be both a tragedy and a comedy. In Joyce's mock-heroic prose poem, Tim Finnegan dies after falling down the stairs, drunk; Little Tramp instead ends up with his legs in the air, in a pose that is comical without being joyful, because of the sadness that all forms of social humiliation exude. Kings or queens who fall expose themselves to ridicule: they express frailty, instability, fleetingness. On all fours, lying on the stomach, or on the back, human beings, be they male or female, make their body – and especially their orifices – liable to the most basic instincts. This is the animal-like position in which the boys in Pier Paolo Pasolini's *Salò* walk, or in which they are exposed to the light of a torch for a tragic competition between backsides. Like wild beasts, some of the damned in Michelangelo's *Last Judgement* obscenely twist upon themselves, showing those body parts that for centuries had been frowned upon and which Freud was finally to vindicate in order to explain childhood traumas and social frustrations.

Antony Gormley, *Domain LXVIII*, 2009
Welded stainless steel rods,
188 x 64 x 29.5 cm
Florence, private collection.

George Segal, *Red Woman Acrobat Hanging from a Rope*, 1996
Bronze with red patina,
223.5 x 104.1 x 49 cm
New York, The George and Helen Segal Foundation.

Marino Marini, *Juggler*, 1939
Polychrome bronze,
161.5 x 63.4 x 13.6 cm
Florence, Museo Marino Marini.

Albrecht Dürer, *The Great Fortune*,
c. 1501–02
Burin, 30.5 x 23.2 cm
Pavia, Musei Civici.

Albrecht Dürer, *The Small Fortune*,
c. 1495–96
Burin, 10.6 x 5.7 cm
Pavia, Musei Civici.

Gino Severini, *The Acrobat*
(or *Masks and Ruins*), 1928
Oil on canvas, 160 x 145.5 cm
Siena, Collection of Banca Monte
dei Paschi di Siena.

Pablo Picasso, *The Acrobats*, 1905
Drypoint, 22.8 x 32.6 cm
Albenga, Collection of Galleria
d'Arte Moderna.

Pablo Picasso, *Salome*, 1905
Drypoint, 40 x 34.8 cm
Albenga, Collection of Galleria
d'Arte Moderna.

AVERTING DEATH

The acrobat pursues, strives for and almost attains the impossible. This is a picture Théodore de Banville found moving: "Between the adjectives 'possible' and 'impossible', the pantomimist and acrobat has made his choice: he has chosen the 'impossible'. And it is precisely the impossible that he lives by, the impossible that he accomplishes".[36] According to Starobinski, modern artists choose the path of the impossible, risking their all. Picasso, for example, offers unforgettable images of acrobats, street performers, dismal masked figures and small, sinuous Salomes dancing before clowns and winged horses. A sort of mysterious solemnity pervades his circus; Apollinaire was well aware of this and in 1905 undertook to describe some of the *Saltimbanques* exhibited by the Spanish artist. The rendition was a rather imaginative one: the poet and refined critic superimposed a mythical universe on that presented by the painter. The protagonists are the same acrobats, monkeys, white horses and bear-like dogs. But the show also features some "teenage sisters with feet balanced on large circus balls" and who "bestow on those spheres the radiant movement of worlds. These prepubescent teenagers have the restlessness of adolescents, and animals teach them about religious mystery. … These acrobats cannot be mistaken for stage actors. Before them, the spectator must display a pious demeanour, for they celebrate rituals with demanding agility".[37] Apollinaire is an Alexandrian who combines "the mysteries of different religious traditions: miraculous fecundity, hermaphrodites, sacramental silence, nativity... The poet's interpretation, superimposed upon the painter's imaginative composition, turns the spectacle that is taking place on stage into a gnostic ceremony. This is no gratuitous game:

it is a rite, the unveiling of secret wisdom".[38] Agile acrobatic feats may be associated with burial ceremonies: "The acrobat's leap, the contortionist's agility, the tight-rope walker's daring ought to avert death by mimicking the unbounded moment in which life springs forth".[39]

Another poet was also drawn by Picasso's imagery. For Rilke, the world of street performers is always reminiscent of a secret passage, a passage which nonetheless does not ensure any liberation, transcendence, or redemption. Still, it is a transition from awkwardness to nimbleness, from imperfect performances to the successful creation of a pyramid, leading from "sheer insufficiency" to "empty superabundance".[40]

Horses and knights, jugglers and dancers are among the favourite subjects of a Tuscan artist who is much loved and sought after among overseas collectors. Marino Marini descended from the Etruscans and based his training on the plastic values of Masaccio and Donatello. From the very beginning of his career, he was well attuned to the innovations introduced by Picasso, Giacometti, Lipchitz and Moore. Through these encounters Marini began his voyage as a timeless artist: through a formal synthesis between the avant-garde and tradition, he attained a classicism devoid of any affectation – a rather archaic and archetypal form of art. He perceived the mysterious ritual quality of the circus world, the enchanted atmosphere it retained, and the disturbing solitude of its protagonists, always grappling with the impossible and the void. Marini's creative hive gave birth to unique figures and embodiments of myths and archetypes: the tight-rope walker, the acrobat, the juggler, the clown; and the dancer, who encapsulates all possible choreographies, conveying the idea of "vital impetus" through the tip of a toe.

Far away from the more archaic Europe, George Segal adopted a different approach. Yet, with his real life casts he was one of the few artists capable of bestowing a mythical aura on everyday life: for instance, on the resurrected figures that seem to be rising from the ashes of metropolitan stories – as in the case of his famous circle of *Dancers*.

George Segal, *The Dancers*, 1971
Plaster, 182 x 182 x 274 cm
New York, The George and Helen
Segal Foundation.

Marino Marini, *Dancer*, 1953
Polychrome plaster,
171.5 x 44.5 x 30 cm
Florence, Museo Marino Marini.

Marino Marini, *Dancer*, 1953
Polychrome plaster,
149 x 65.5 x 36 cm
Florence, Museo Marino Marini.

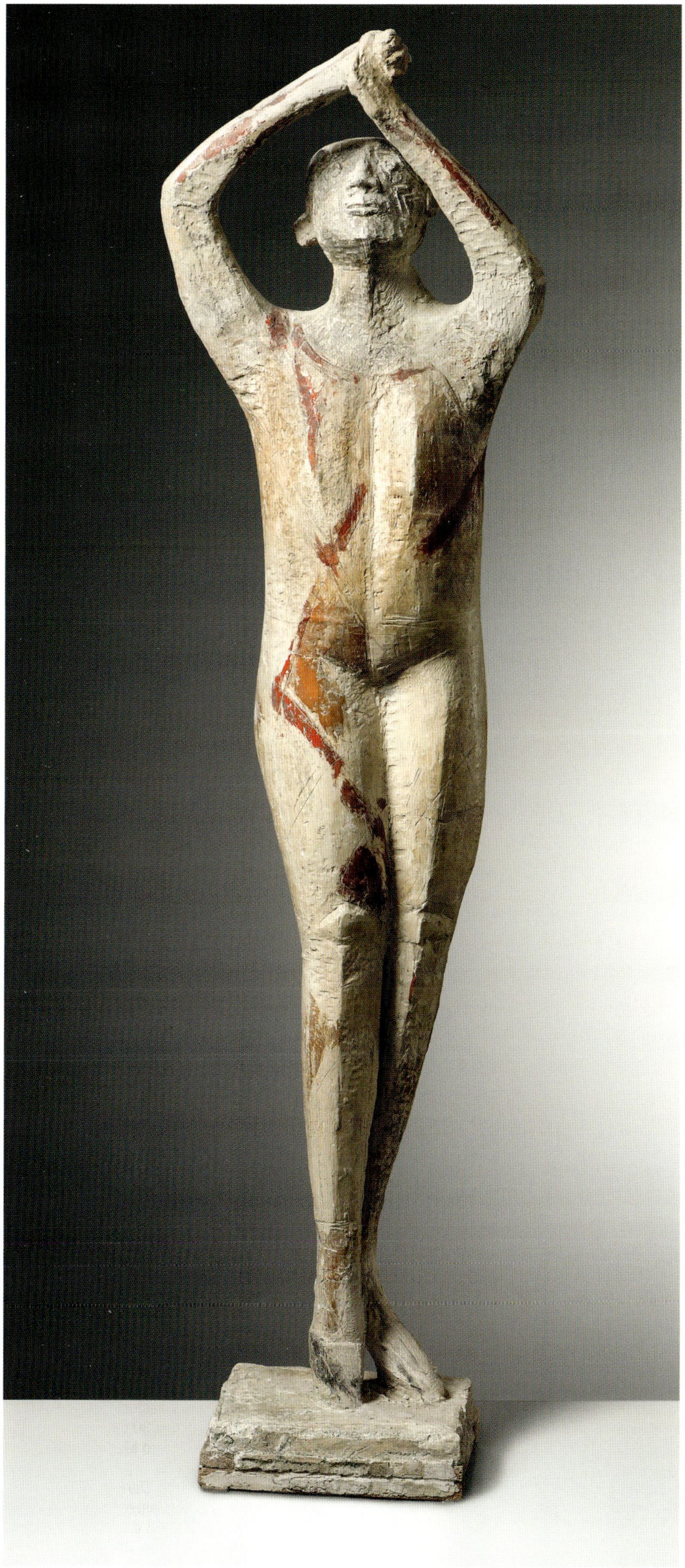

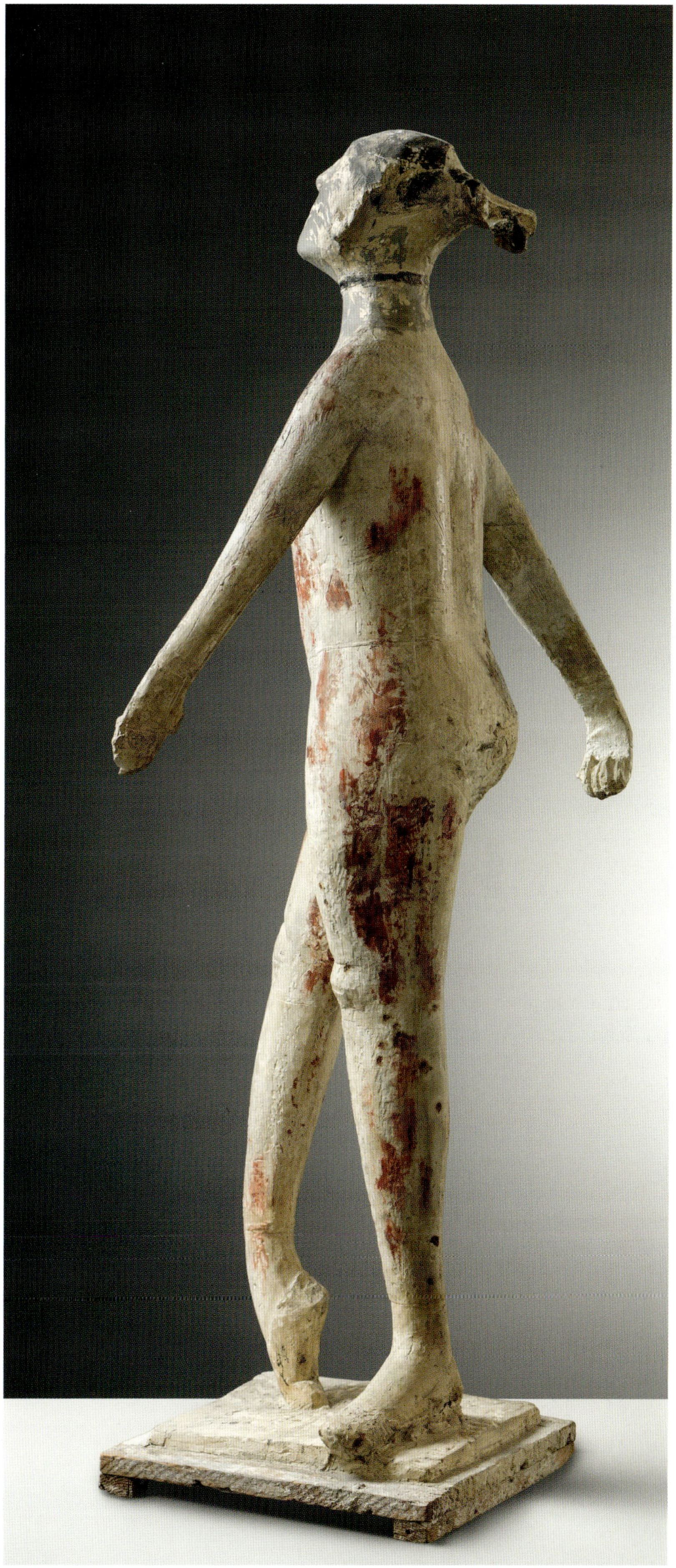

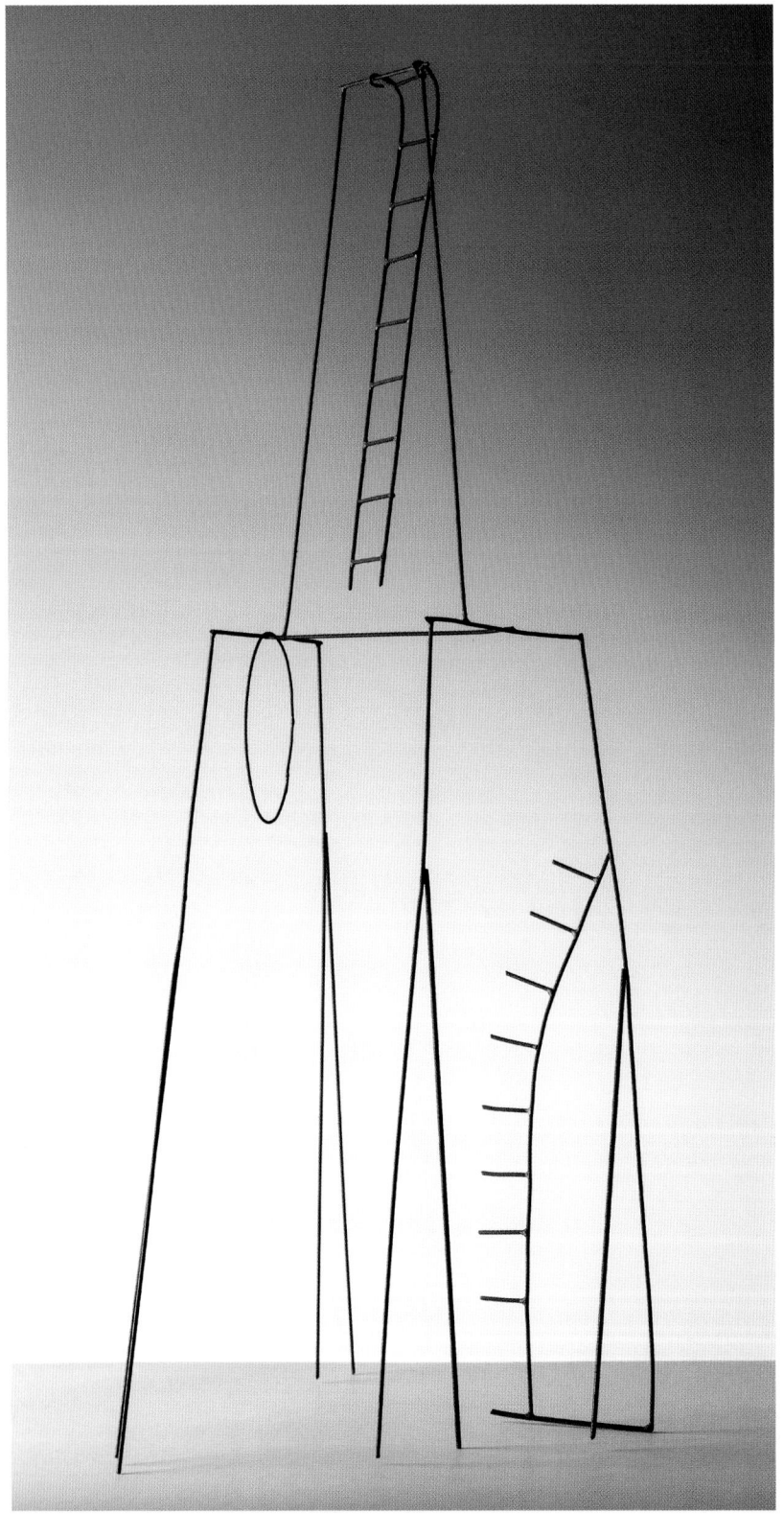

THE INVISIBLE ACROBAT

"Where was she going and where did she come from?" Wilhelm Jensen wondered in 1903 with regard to the Gradiva, the Hellenistic bas-relief that became the subject of his most famous novel. This text was much admired and studied by Sigmund Freud, who in the summer of 1906 wrote an important essay, published the following year under the title of *Der Wahn und die Träume in W. Jensens Gradiva*. In his novel, Jensen writes that the name given to the ancient image reminds one of the German *Bertgang*, meaning "she who shines in walking". In Fausto Melotti's atelier in Milan, the Gradiva archetype took shape on paper and then through sculpture in 1933: the artist was already about to cross the threshold of Metaphysical-Surrealist figurative art and attain a leading role in the world of abstract sculpture. In his plans for a statue, Melotti interpreted Eurydice as a dancer advancing with a miserable attitude. He may have associated the photographic memory of nymphs and maenads from ancient bas-reliefs with the more modern images of Isadora Duncan. In all likelihood, Melotti was influenced by the Gradiva's renown, as embodied by some of the female figures by Ghirlandaio and Botticelli, after coming across the writings of Freud and Warburg, which he could read in the original language. With their particular poses, these figures appear to be suspended between gracefulness and melancholy, between a serene immobility and visible anxiety. Almost dancing in everyday life settings, they elicit a strong emotional reaction from those who admire them as frescoes, paintings, free-standing sculptures or bas-reliefs. With its lines, eurhythmics and relation to time, dancing frequently provided a source of inspiration for Melotti: he would make lines dance to create real flourishes, as in certain drawings and sketches marked by a sort of visual frenzy. Still, Melotti never lost control, since he followed the principle of dynamic balance even when dealing with a dissonant harmony.

Fausto Melotti, *The Invisible Acrobat*, 1980
Brass, 66 x 28 x 22 cm
Florence, private collection.

Wassily Kandinsky, *Circle of Friends of the Bauhaus*, 1932
Drypoint, 20 x 24 cm
Albenga, Collection of Galleria d'Arte Moderna.

Melotti was an eclectic artist, who could not stand barriers and remained loyal to the rules of musical counterpoint even as a sculptor. He freely embarked on a creative quest by employing both clay and thin metal or cloth threads in order to create vibratile objects, striking sculptures that play with the air and balance. Germano Celant was to present Melotti as an acrobat, interpreting his 1980 *Invisible Acrobat* as a self-portrait: "Each stage of his work is experienced as a leap. Bodily dynamics, like those of a trampoline jumper, are illustrated by *The Invisible Acrobat*, who – in line with an unpredictable physicality – leaps and dances from one level to another, from one elevation to another".[41] Melotti, "while adopting a repertoire of geometrical objects reminiscent of de Chirico, wishes to point to a more concealed and lively surface in which poetic perception rests and which brings plastic construction to life from within, through a subtle vibration. With Melotti the play of metaphors is extended and not just architecture but music too is brought into play. Music means a rhythmic and harmonious execution that expands across space: a vibration of the filaments, shapes and ramification of metal or plaster constitute the figurative echo or implication, extending beyond the limit set by the objects chosen".[42] Music means time, and Melotti is conscious of the achievements made by the avant-garde, particularly by Kandinsky, whose works he had appreciated at the Galleria del Milione in Milan, where Melotti's own work had also been put on display in 1934.

FIRST STEPS IN CONQUERING THE WORLD — WITHOUT FALLING

A child soon gets tired of moving on all fours, of exploring the world of adults – the living room or garden – in the position of a cat or turtle. He will then stand up and totter about on unsteady feet, doing his best to walk around with an erect posture, his hands stretched forward in search of a suspended resting point. Deep down, the child is a little afraid of falling. It is interesting to note how the idea of "first steps" played a primary role in Florence and Paris. At a certain point in the history of art, at the turn of the twentieth century, figures of intrepid tottering children cropped up alongside the silhouettes of dancers, tight-rope walkers and acrobats. This repertoire is clearly to be found in Rainer Maria Rilke's *Elegies*: animals, angels, lovers, children and street performers are all featured there. It is as though the adventure of the "wee one" were perceived as a genuine act of heroism, like that of Prometheus. Adriano Cecioni, Van Gogh and Brancusi, for instance, focused on the little biped with great attention and interest; they studied his expression and sought to capture a fleeting moment: the beginning of the world as represented by the initial stages of human life.

Adriano Cecioni (1836–1886) executed a free-standing plaster sculpture of a child embarking on his *First Steps* (c. 1869). Depicted here with a certain degree of nonchalance is the "launching" stage – the moment in which the toddler, having acquired an erect posture, walks on his own legs. For some time, this human puppy will look like a puppet or chimpanzee. Our cute little athlete has just left the reassuring embrace of an adult and chosen to venture into the world, on a quest to discover his own fate. Cecioni, an insightful interpreter of child psychology, recorded the emotional reactions of the toddler – the dawn of reason and feelings within him. In *First Steps* the Florentine artist illustrates the child's unstable balance with sober lifelikeness; he

First Steps, 1946
Light brown calfskin derby, 12.5 cm.
The shoes belonged to Ferruccio
Ferragamo. For this specific model
Salvatore created special anti-slip
soles and patented them.
Florence, Museo Salvatore
Ferragamo.

Adriano Cecioni, *First Steps*,
c. 1869
Plaster, 71 x 30 x 27 cm
Florence, Galleria d'Arte Moderna
di Palazzo Pitti.

does away with superfluous impressions and is deeply moved by his little hero. He grasps his naivety and fragility, along with his courage in facing potential risks. The child's first steps also coincide with an investigation of moral pride, a spiritual tone that marks each person's character right from the start. The representation of movement is a very important theme for Cecioni: "What represents nature in motion is a great subject; what expresses motionless nature is far less great".[43] A toddler's first steps, then, also constitute a test of skill for the artist, who must assign a fixed pose to a restless physical structure, a being that moves unsteadily, with little coordination, staggering along. The eyes of the plaster figure are like those of one possessed: caught between fear and amazement, they suggest that the child is about to lose his balance. He is about to fall. Or perhaps we are mistaken and his efforts will soon be crowned with success. A spark of marvel and suspicion fills his gaze, adding an expressionist touch to this almost photographic depiction.

First Steps is also the title of a famous painting by Vincent van Gogh, who in this case drew inspiration from a work by François Millet. A child takes his first steps to meet his father, who has just returned from the fields. The mother is leaning over her baby son and holding his hands, to offer him protection and support. Behind the silhouette of the child a bouquet of blood-red flowers stands out. Van Gogh here sought to allude to a likely fall – and the injury and crying that would ensue. But that colour might also symbolize a far more tragic event in store for the family: the death of the father, here greeted by his son for the first and last time.

Constantin Brancusi's wooden sculpture *First Steps* was only put on public display once, in 1914, during the first solo exhibition of the Romanian sculptor in New York. Soon afterwards, the work was destroyed by the artist, who nonetheless decided to keep the head of the statue, like a symbolic embodiment of the decapitation of the world of childhood. Among the many reasons that led Brancusi to destroy the work and preserve only a "relic" of it, one was formal: for many years, the artist repudiated his former interest in African art, which had also attracted the interest of figures such as Picasso, Derain and Matisse – his direct rivals in the definition of modern art.

In a photograph now in the Centre Georges Pompidou (see here on p. 93), we see Brancusi holding a child. The artist helps the baby stand erect and take his first steps, resting his feet on the ground in such a way as to find his balance. Brancusi was very familiar with Ovid's *Metamorphoses*, which he turned to for his *Danae*. *First Steps* may have sprung from an association between mythology and sacred knowledge. In the *Metamorphoses*, Ovid describes the moment in which the human being, after having been fashioned by Prometheus, stops walking on all fours and dares lift his head to gaze at the sky. Brancusi follows Ovid's text to the letter, fully grasping its philosophical, poetic and spiritual implications. Man was fashioned because "the Earth still lacked a more sacred animal with a lofty intellect, a Lord to the other animals". Thus the Earth, "still rough and formless, shaped the hitherto unknown forms of man".

THE WALKING MAN

According to Merleau-Ponty, there is a substantial difference between art and photography. When gazing at a man who is walking, what the artist searches for is not the outer appearance of motion, but rather its hidden quality. According to the French philosopher, Rodin had a clear view of the matter: "It is the artist who is truthful, while the photograph is mendacious; for, in reality, time never stops".[44] Art does not freeze motion, rather it shows it in its overall dislocation and inner discordance. Marey's photographs, for instance, or those of Muybridge show "a rigid body as if it were a piece of armour going through its motions; it is here and it is there, magically, but it does not go from here to there. Cinema portrays movement, but how? Is it, as

Masaccio, *Expulsion from Paradise*, 1424–25
Fresco, 260 x 88 cm
Florence, church of Santa Maria del Carmine, Brancacci Chapel.

Auguste Rodin,
The Walking Man, c. 1905
Bronze, 213 x 161 x 72 cm
Paris, Musée d'Orsay.

we are inclined to believe, by copying more closely the changes of place? We may presume not, since slow motion shows a body floating between objects like seaweed, but not moving itself. Movement is given, says Rodin, by an image in which the arms, the legs, the trunk and the head are each taken at a different instant, an image which therefore portrays the body in an attitude which it never at any instant really held and which imposes fictive linkages between the parts, as if this mutual confrontation of incompossibles could – and alone could – make transition and duration to well up in bronze and on canvas".[45]

Rodin must have had Masaccio's *Expulsion from Paradise* in mind when modelling his *Walking Man*, a sculpture of crucial importance for twentieth-century art. The underlying conception is very sophisticated. According to some scholars, the work is an assemblage made after 1887 by combining a *Torso* from 1879 with the legs of *John the Baptist* (see here on p. 125), a sculpture fashioned between 1878 and 1879, and modelled after a peasant from Abruzzo by the name of Pignatelli. According to other scholars, the work might be a preliminary study for *John the Baptist* that later came to be appreciated on its own right. Ultimately, Rodin must have established a mental connection between two different characters: Adam, making his way out of the gate of Paradise; and St John, the man who roamed the desert and was beheaded at the bequest of the dancer Salome, whose bacchant-like moves had

Etruscan art, *Statuette
of Laran (Etruscan Mars)*,
fourth century BC
Bronze on wooden base,
29 x 8.5 x 7 cm
Soprintendenza per i Beni
Archeologici della Toscana –
Florence, Museo Archeologico
Nazionale.

seduced King Herod. Masaccio's Adam has just betrayed his God after having been spurred by Eva, who in turn has been hypnotized by the snake, an animal who moves by crawling on the ground – the only living creature who turns its body into one big foot, as Jünger suggests: "If we regard as the distinguishing feature of the foot the fact that it touches the ground, then we may well argue that, with respect to its bodily surface, the snake is almost all foot, and hence as far as it could get from the human posture. In this regard, among all creatures the snake is the most animal-like of all – the animal quality is most evident within it. This explains the sudden disgust and horror man feels before it".[46] The snake, then, derives its intelligence from below, from the ground: it is dear to Asclepius and has therefore been regarded as demonic, ever since it offered an "earthly meal" to our ancestors by the Tree of Knowledge. But let us now make our way into the church of Santa Maria del Carmine, in Florence: on the left as we enter the Brancacci Chapel we see Adam, about to cross the threshold of Eden, which he will be forbidden to return to until the end of time. Weighed down by guilt and put to shame, he takes his first steps as a mortal by resting his feet on the earth he is henceforth called to inhabit and till by the sweat of his brow. The shadow of his body recalls the passing of time, the duration of the seasons. Before him the whole history of humanity from a Judeo-Christian perspective unfolds. Henceforth, Adam and Eve will never stop walking. Walking will become the distinguishing feature of human creatures; bipedism will mark the difference between us and the other inhabitants of the Earth. Rodin stretches out the limbs of his *Walking Man* like a compass: the man's body is bent forward, with an almost excessive weight; it almost shifts the air as it moves and an invisible countering force seems to pull it downwards. Both feet are planted on the ground and yet the figure perfectly conveys the effort and inner yearning

to walk, to move about. Rodin's man wishes to conquer the planet, to move from one point of the Earth's surface to another. He wishes to get away from paradise, responding to an innate sense of freedom.

THE CAPSIZING MAN

Giacometti has left us a famous statement concerning a distinctively human capacity. It is not easy to stand on two legs, to walk without falling: "I always have the impression or feeling of the frailty of living beings, as if at any moment it took a formidable amount of energy for them to remain standing".[47] This acknowledgement helps explain many of the artist's creations, such as *The Walking Man* and *The Capsizing Man*. Having forgone the idea of finding inspiration in the grand sculpture of the Renaissance, like many of his colleagues Giacometti had turned to non-European art: "I find that certain idols from Oceania or Africa have far more truthfulness to them, and are much closer to what we are searching for, than any sculpture by Michelangelo,

Alberto Giacometti,
The Walking Man, 1960
Bronze, 182.2 x 26.6 x 96.5 cm
Buffalo, Albright-Knox Art Gallery.

Alberto Giacometti,
The Capsizing Man, 1950
Bronze, 60 x 14 x 22 cm
Zurich, Kunsthaus Zürich, Vereinigung
Zürcher Kunstfreunde.

were the shouts of the crowd to stir her, it would be a disaster. This tight-rope walker, of ash and smoke, advances towards a non-place. She's like a cross between Hermes and Harlequin.

WITH SLOW AND SCANT STEPS

The three major Italian poets – Dante, Petrarch and Leopardi – all took the experience of walking, the walker's effort, quite seriously. The *Divine Comedy* is a long journey made on foot in the netherworld, among the damned and the blessed. It all starts in a gloomy wood. In the midway of a life. Dante has gone astray from the path of Christian perfection and must roam hell and purgatory – just as Odysseus must sail the Mediterranean – before reaching paradise. During his journey, the poet falls and picks himself up again; he is dazzled and faints; he slips into darkness and loses his memory. By a boiling stream, the Phlegethon, he advances along a landslide and suddenly realizes that the boiling water before his eyes is actually blood – he sees something that no one would like to see. Dante and Virgil take slow, scanted steps in the twentieth canto of *Purgatory*. They make their way up a narrow, difficult mountain pass. They are about to meet Ugo Capeto, whose role within the economy of the poem is to provide examples of avarice and prodigality. Dante often counts the steps he has taken and those he must take: both in order to give a clear idea of the distance between one place and another and in order to remind us that he is alive – that he is truly walking in that world, and that everything is real. He wants us to sense the strain of the sinner walking to achieve salvation. At the summit of purgatory Dante finds himself walking through the Garden of Eden, in what is arguably one of the happiest moments in the poem. This place is reminiscent of certain medieval depictions of the *hortus conclusus*. Through a journey

Donatello or any other sculptor from the traditional art of the West".[48] In an exchange he had with Pierre Schneider in the halls of the Louvre, the artist described his interest in Chaldaean sculptures and the portraits from Fayum: "What I love from the past is that which is closest to what I can see, to my own vision of things. Chaldaean sculptures, for instance. And I am thousand times fonder of Byzantine painting than Western painting".[49]

In a little-known 1943 sketch, Giacometti fixes his psychopomp's gaze on a female *Tight-rope Walker*, celebrating her feats with phantasmal strokes. The figure balancing on a cable looks like a sleep-walker;

Alberto Giacometti,
Tight-rope Walker, 1943
Pencil and charcoal on paper,
36.8 x 28.6 cm
New York, Yoshii Gallery.

à rebours the Florentine poet has reached the site where everything began, at the dawn of creation, in a day that is still unfolding, evergreen – in the perfection of the primal splendour. Dante is about to lose touch with the ground. He will be dipped into the River Lethe up to his neck; once purged of his sins, he will be able to rise up to the dazzling light of the God of love. After so much walking, he lifts his feet from the ground and soars up. He soars up filled with ardour.

Dante finally seems rested and serene; the journey has been a purificatory one, circle after circle. Faced with evil and abjection, the poet has grasped the spiralling logic behind the punishments he has witnessed. He now happily strolls, enjoying the beauty, colours, light and peace which the place exudes. He makes his way into the garden of paradise, into a grove; he meets a stream and moves even slower. Dante is contemplating nature. He meets Matelda, an embodiment of female perfection, as in John Flaxman's representation: "a solitary woman moving, / singing, and gathering up flower on flower – / the flowers that colored all of her pathway". Dante addresses her because he wants to know what she is singing. Matelda is so nimble, so graceful, that she seems to be dancing. "Even as a woman, dancing, turns around / with feet close to the ground and to each other, / and scarcely places foot in front of foot, / she turned upon the red and yellow flowers / in my direction, no differently than would / a virgin lowering her modest eyes".[50] The poet describes a genuine dance step, the frame of a courtly choreography, evoking the image of a dance performed on toe tips in the hall or garden of some Florentine palace.

Alone and broody, or even melancholic, Francesco Petrarca (Petrarch) advances with "tarrying and slow" steps. It is the autumn of 1337. The poet wishes to spend some time alone and eschews human contacts: he wanders across the fields, far from the city, avoiding all human traces. He is living under the influence of Saturn. He trails along, weighed down by gloom and exhausted by his heartache. He feels like Homer's Bellerophon. This is how unrequited lovers walk, those saddened by the loss of their loved one. They move slowly, with their head hanging low, scorning social interaction and friendship. In a later poem, Petrarch admits that he has always pursued a "solitary life" – as even rivers "and the fields and woods" know. The poet's most celebrated walk, however, is an ascent: his difficult climb to the summit of Mont Ventoux together with his brother Gherardo. At one point, Francesco wished to call it off, "yet nature does not yield to human will, nor can something corporeal ever attain a height by descending".[51] Again and again, Petrarch sought to find a level path, but then he reasoned within himself: "What you have repeatedly experienced today in the ascent of this mountain, happens to you, as to many, in the journey towards the blessed life. If this is not so readily perceived by men, it's only because the motions of the body are obvious and external while those of the soul are invisible and hidden. The life which we call blessed is to be sought for on a high eminence, and narrow is the way that leads to it. Many, also, are the hills that lie between, and we must ascend, by a glorious stairway, from strength to strength. At the top is at once the end of our struggles and the goal for which we are bound. All wish to reach this goal, but, as Ovid says, 'To wish is little; we must long with the utmost eagerness to gain our end'".[52] Having reached the peak, Petrarch reads a passage from St Augustin's *Confessions*: "And men go abroad to admire the heights of mountains, the mighty waves of the sea, the broad tides of rivers, the compass of the ocean, and the circuits of the stars, yet pass over the mystery of themselves without a thought". In the *Canzoniere* we also read the tale of the "ancient graybeard" who sets off in the direction of Rome "by years broken, spent by the long road". The old pilgrim has reached the end of his life and has chosen to go and pray before the true face of Jesus in the Basilica of St Peter. He yearns "to look upon the painted face of Him" whom he hopes to meet in Heaven. These are his last steps, those necessary to cross the threshold.

TO THE WORLD'S END

Similitude (*Ein Gleiches*) by Goethe is as much of a cornerstone of Romantic poetry as is Caspar David Friedrick's *The Wayfarer*. In both cases, a contemplative figure, fuelled by melancholy or exaltation, ascends a mountain in order to broaden his horizon, touch the sky, and commune with the sublime and boundless. These Romantic walks sometimes take place in the moonlight. According to Goethe's biographers, *Similitude* was written in pencil on the walls of a hunting lodge near Weimar, on 7 September 1780. It is said that the poet retraced his steps back to the lodge on 26 August 1831 and was moved by his youthful verses, particularly the closing ones: "Über allen Gipfeln / Ist Ruh; / In allen Wipfeln / Spürest du / Kaum einen Hauch; / Die Vögelein schweigen im Walde. / Warte nur, balde / Ruhest du auch" ("Over all the hilltops / is calm; / in all the treetops / you feel / hardly a breath of air; / the little birds fall silent in the woods. / Just wait, soon / you'll also be at rest"). By retracing his steps, Goethe realized the inevitability of human fate: the fact that after so much walking eternal rest awaits us.

Another night song was put to verse by Leopardi in Recanati, between 1829 and 1830. The poet's sublime reflection here unfolds as he contemplatively turn his gaze upwards. The solitary wayfarer in this case is a wandering shepherd from Asia, who converses with the Moon: "Why are you there, Moon, in the sky? Tell me why you are there, silent Moon. / You rise at night, and go contemplating deserts: then you set. / Are you not sated yet with riding eternal roads?"[53] The shepherd conjured up by Leopardi suffers from cosmic pessimism, as the poet philosophically ponders the vanity of all, the futility of human fate, and the pointlessness of his constant wandering in search of meaning: "Like an old man, white-haired, infirm, / barefoot and half-naked, / with a heavy load on his shoulders, / running onwards, panting, / over mountains, through the valleys, / on sharp stones, in sand and thickets, / wind and storm, when the days burn / and when they freeze, / through torrents and marshes, / falling, rising, running

Dante Alighieri, *La Divina Commedia.* Commentary by Cristoforo Landino, illustrations attributed to Sandro Botticelli (Florence: Nicolò di Lorenzo della Magna, 1481), 40 x 28 x 8.7 cm Florence, Biblioteca Nazionale Centrale di Firenze.

faster, faster, / without rest or pause, / torn, bleeding; till he halts / where all his efforts, / all the roads, have led: / a dreadful, vast abyss / into which he falls, headlong, forgetting all. / Virgin Moon, such / is the life of man".[54] These verses start with a slow rhythm, which grows more and more frantic. The poet leads us to discover a nature that is ungenerous and cruel. He takes us back to the dawn of humanity, when man found himself facing the savannah for the first time. And suddenly realized that he was destined to die.

[1] J. Clair, "Parade et palingénésie. Du Cirque chez Picasso et quelques autres", in *La Grande Parade*, exhibition catalogue, edited by J. Clair (Paris: Gallimard, 2004), p. 21.

[2] E. Jünger, "Testa e piede", in *Il contemplatore solitario*, edited by H. Plard, afterword by Q. Principe (Parma: Guanda Editore, 1995), p. 59.

[3] Ibid., pp. 59–60.

[4] Ibid., p. 69.

[5] Ibid., p. 70.

[6] P. Valéry, "Degas Danza Disegno", in *Paul Valéry. Scritti sull'arte* (Milan: Guanda, 1984), p. 13.

[7] Ibid.

[8] Ibid.

[9] R. Calasso, *La follia che viene dalle ninfe* (Milan: Adelphi, 2005), pp. 37–41.

[10] G. Didi-Huberman, *Ninfa moderna. Saggio sul panneggio caduto* (Milan: Il Saggiatore, 2004), p. 13.

[11] Ibid.

[12] Ibid.

[13] Jünger 1995, p. 59.

[14] E. Degas in *Degas. La vita e l'arte. I capolavori* (Milan: Rizzoli-Skira, 2003), p. 128.

[15] Valéry 1984, p. 39.

[16] Ibid.

[17] Ibid., p. 15.

[18] J. Starobinski, *Ritratto dell'artista da saltimbanco* (Turin: Bollati Boringhieri, 1984), pp. 87–88.

[19] Ibid., p. 88.

[20] G. Deleuze, *Francis Bacon. The Logic of Sensation* (London and New York: Continuum, 2003), p. 21.

[21] Ibid., p. 23.

[22] Ibid., p. 22.

[23] Starobinski 1984

[24] Ibid., p. 139.

[25] Ibid., p. 62.

[26] Ibid., p. 26.

[27] Ibid., p. 81.

[28] Ibid., pp. 81–82.

[29] Ibid., p. 82.

[30] Ibid., p. 84.

[31] Ibid., pp. 84–85.

[32] Ibid., p. 87.

[33] Ibid., p. 56.

[34] Ibid., p. 58.

[35] Ibid.

[36] Ibid., p. 136.

[37] Ibid., p. 138.

[38] Ibid., pp. 138–39.

[39] Ibid., p. 139.

[40] Ibid., p. 145.

[41] G. Celant cit. by S. Risaliti, "Fausto Melotti. An 'espace Melotti' has come into being", in S. Risaliti, *Melotti. Catalogo generale della grafica* (Milan: Electa, 2008), vol. I, note 45, p. 33.

[42] Ibid., p. 28.

[43] A. Cecioni, *Scritti e Ricordi con lettere di Giosuè Carducci, Ferdinando Martini ecc.*, foreword and notes by G. Uzielli (Florence: Tipografia Domenicana, 1905), p. 233.

[44] M. Merleau-Ponty, *The Merleau-Ponty Reader* (Evanston: Northwestern University Press, 2007), p. 375.

[45] Ibid., p. 374.

[46] Jünger 1995, p. 60.

[47] "Conversazione con Brassaï. La mia ultima visita a Giacometti", in *Alberto Giacometti. Il mio lungo cammino*, edited by E. Grazioli, with an introduction by M. Pozzati (Cernusco Lombardo: Hestia Edizioni, 1998), p. 85.

[48] Ibid.

[49] "Conversazione con Pierre Schneider. Al Louvre con Giacometti", in *Alberto Giacometti* 1998, p. 58.

[50] Dante, *Divine Comedy*, Purgatory, XXVIII, verses 52–57.

[51] F. Petrarca, letter of 26 April 1336 to Dionigi de' Roberti da Borgo San Sepolcro, in *Rerum Famliarium. Libri*, vol. IV, n. 1.

[52] Ibid.

[53] G. Leopardi, *Night Song of A Wandering Shepherd In Asia*, verses 1–6.

[54] Ibid., verses 21–38.

WALKING FOR SIX MILLION YEARS

TELMO PIEVANI

The intricate history of our sub-family began in Africa around six million years ago, when hominins broke away from the common ancestor they shared with chimpanzees. Six million years is not much on the geological timescale but quite long enough to accumulate a fair number of changes in behaviour, morphology, posture and gait. Scientists have pondered the question of the crucial innovation introduced by our earliest ancestors for decades, focusing attention obviously on the admirable human head. The growth of the brain and hence of intelligence was regarded as the prime mover of the human saga. Not so. We were looking in the wrong place. For more than two-thirds of their history, the twenty-plus hominin species that followed one another in our past were not in the slightest concerned about having a big brain. Their secret lay hidden elsewhere, at the opposite end of the human corporeal architecture.

TO EACH HIS OWN GAIT

The growth of the encephalon began only two million years ago and only with the *Homo* genus. The innovation that caused us to diverge from all the other great apes started long before with the feet and their mechanics. To be more precise, it started with the two-footed upright posture, a formidable invention but also an imperfect one as is often the case in evolution, which is a succession of compromises rather than an ascent to the optimum. The abandonment of the four-footed gait entailed a

Traces of footprints left in wet volcanic ash. The hominins crossed the plain of Laetoli in Tanzania 3.6 million years ago.

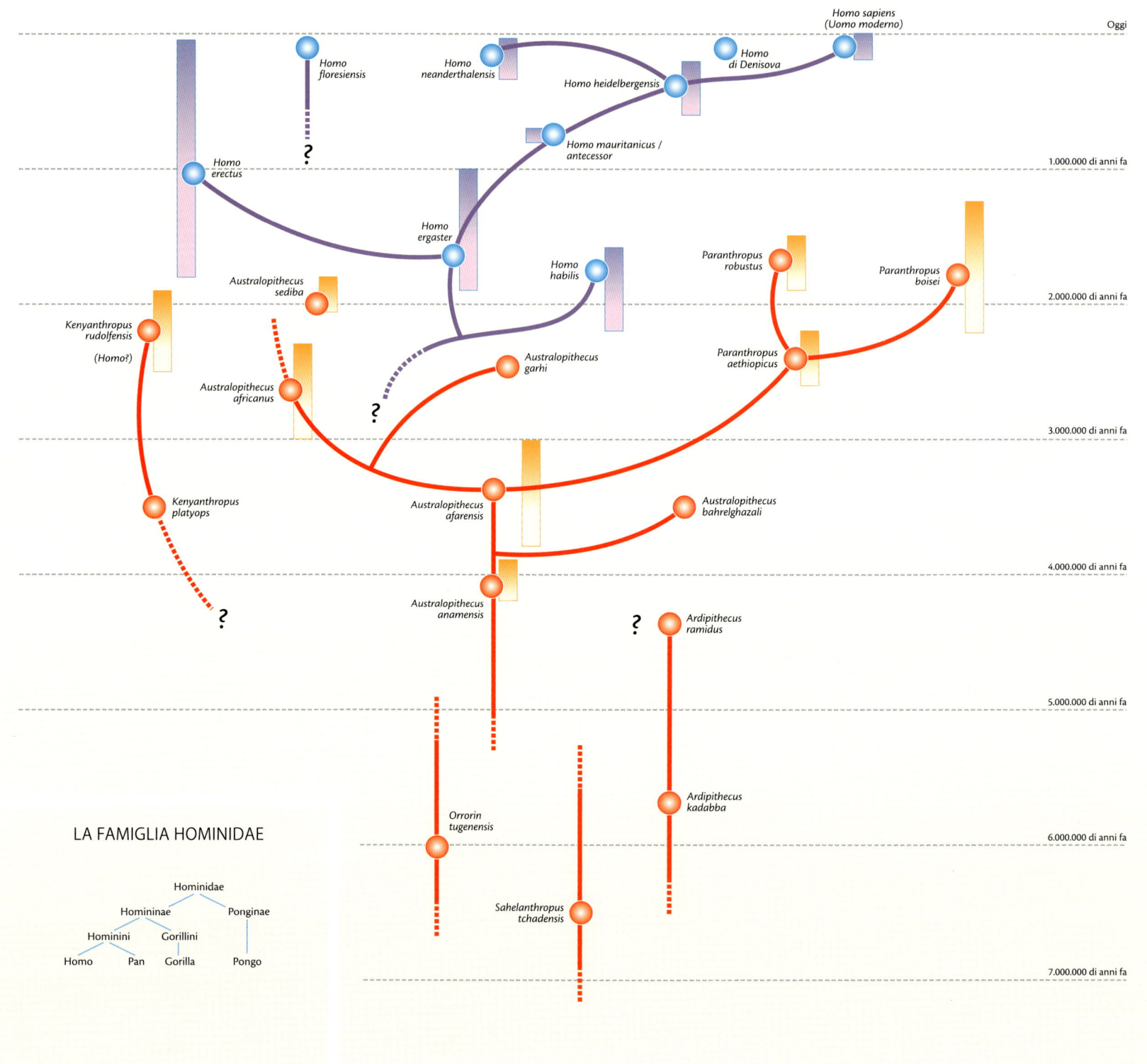

Oggi

1.000.000 di anni fa

2.000.000 di anni fa

3.000.000 di anni fa

4.000.000 di anni fa

5.000.000 di anni fa

6.000.000 di anni fa

7.000.000 di anni fa

Homo sapiens
(Uomo moderno)

Homo floresiensis

Homo neanderthalensis

Homo di Denisova

Homo heidelbergensis

Homo mauritanicus / antecessor

Homo erectus

?

Homo ergaster

Homo habilis

Australopithecus sediba

Paranthropus robustus

Paranthropus boisei

Kenyanthropus rudolfensis

(Homo?)

Australopithecus africanus

Australopithecus garhi

Paranthropus aethiopicus

?

Kenyanthropus platyops

Australopithecus afarensis

Australopithecus bahrelghazali

?

Australopithecus anamensis

?

Ardipithecus ramidus

Ardipithecus kadabba

Orrorin tugenensis

Sahelanthropus tchadensis

LA FAMIGLIA HOMINIDAE

Hominidae

Homininae — Ponginae

Hominini — Gorillini

Homo — Pan — Gorilla — Pongo

Phylogenetic tree of the hominins
(from T. Pievani, *Homo Sapiens.*
Il cammino dell'umanità, Novara:
Istituto Geografico De Agostini, 2012).

costly reorganization of our anatomy as a whole. Our body is still not completely suited to the upright position even today, as those who suffer from slipped discs and backache unfortunately know to their cost. Some compromise is necessary, however, in order to survive. For anthropomorphic African apes blocked off to the east and obliged to move with ever-increasing frequency over red-hot clearings, reducing the body surface exposed to the sun was a good idea, as was standing upright to keep watch over the grassy expanses. Whatever its initial function may have been, bipedalism gave us children of the Rift Valley and the subsequent formation of the dry savannah such invaluable gifts as the ability to run long distances and free use of the hands, not to mention the fact that a biped can also climb a tree or swim if necessary. This is the gift of behavioural flexibility. At the cost of some aches and pains, our success as explorers of the planet stems precisely from this unfinished anatomical revolution and its cultural effects.

Therefore, as the great anthropologist and archaeologist André Leroi-Gourhan wrote in 1964, "the history of mankind begins with the feet". It should not be thought, however, that this great leap towards the future was the triumphal march that we still often find illustrated in newspapers and magazines: the ape on all fours that begins to stand upright, loses its bodily hair, develops tools, displays its enlarged skull, lengthens its stride and turns as though by magic into a fully-fledged *Homo sapiens* – the peak of progress, the goal long pursued. The zigzag course of our history is instead made up of diversity and evolutionary experiments, most of which proved unsuccessful.[1] Since primordial times, ours has been a story of the coexistence of various species, each endowed with a set of peculiar adaptations, in an area with its heart in the valleys and plateaus of East Africa and the Horn of Africa but stretching as far as Lake Chad at one end and South Africa at the other. While walking upright on two feet is the common feature of these species, examination of the earlier forms on our evolutionary tree reveals that it is possible to be bipedal primates in a whole variety of ways. With their occasional and oscillating bipedalism, chimpanzees also seek to maintain uncertain balance on the soles of their feet as best they can.

The anatomical diversity of ancestral gaits is full of oddities and unique cases. The first hominins were adaptive hybrids, good climbers when necessary (especially at night) and walkers at the same time. The elongated femur of *Orrorin tugenensis*, who lived in East Africa about 6 million years ago and was described in complete morphometric detail in December 2013, is a mosaic of features already typical of a facultative biped.[2] *Ardipithecus ramidus*, whose skeleton was described in 2009, is another very early ancestor of the period between 6 and 4.4 million years ago, preceding the australopithecine and living in what is now northern Ethiopia. "Ardi" presents a unique mixture of arboreal ape-like characteristics (long upper limbs stretching to below the knee and a divaricated, prehensile hallux or big toe) and forest-dwelling bipedalism. With broad feet and huge hands, it literally walked on the branches. Another biped in its own way, it was probably adapted to a hybrid environment combining trees and open spaces.

Some time later, 3.66 million years ago to be precise, another hominin cousin performed a memorable feat immortalized in stone due to a lucky turn of events. This took place at Laetoli in northern Tanzania, when the Sadiman volcano erupted light ash that mixed with rain. Amongst the fleeing animals were some strange bipeds. Mankind's earliest footprints were left in the volcanic tuff like a fossil snapshot – a great stroke of luck for palaeontologists. Seventy prints in a line displaying great haste. What happened? In their flight, two australopithecines of the species *A. afarensis*, perhaps a male and a (much smaller) female or a mother with a child, probably together with a third individual (again very small), left their footprints in the wet volcanic ash of Laetoli, which then fossilized to preserve

a record of the earliest escape of our ancestors discovered so far.

Their locomotion was unquestionably bipedal, as the discoverer Mary D. Leakey deduced in 1979 from the prints of the arch of the foot and the big toe, which is parallel to the others even though slightly separated from them. The two creatures belong to the same species as the celebrated "Lucy", whose skeleton was discovered in 1974 (40% complete and 3.2 million years old). Recent analyses published in 2011[3] have confirmed that the arch of their foot, similar to ours, was already typical of a markedly bipedal species. Computer simulations carried out on the prints show that during movement, their weight of barely 40 kilos rested on three points of support in sequence, from the heel to the hallux along the medial arch and the lateral arch: nearly the same as a complete biped.

ON FOOT OUT OF AFRICA

It is thus by walking that we conquered the world, crossed the most disparate ecosystems and put our adaptive flexibility to the test of environmental instability. The first specimens of our genus, descendants of the australopithecines, made their appearance in Africa around 2.4 million years ago. They coexisted for a long time with the last australopithecines (like the South African *A. sediba*) and the *Paranthropus* genus of robust individuals, which were to survive for another million years: an authentic jumble of forms. *Homo ergaster* stands out in this human bush as an unprecedented model of hominin: slim and completely bipedal with lighter bones, increasing cranial capacity (from 600 to 800 cc), the long stride of the prairie dweller and the stone hand tools of the Oldowan industry, the earliest known technology. A new way of being human thus emerged in eastern and southern Africa, made increasingly barren by the beginning of a new ice age and now covered with grassland as far as the eye could see. Walking humans, bipedal apes of the grassland, lived around the Olduvai Gorge, hunting small animals and fleeing from potential predators.

The articulation of the hip and the angle of 180° between the femur and the vertebral column now show a bipedal gait in the modern sense. The dispersal of heat is maximized all the way along the long-limbed body. Development is slower with respect to the australopithecines with prolongation of the infantile period: this is a crucial change in human evolution. The walking species began to expand its territories of settlement in search of food and space or in flight from drought. It had long limbs, an omnivorous diet (beginning to compete with predators for meat) and more organized encampments, and had perhaps already mastered fire. It moved in small groups of 25–30 individuals and spread over the African valleys and plateaus with a preference for coastal and riverside areas. Traces of its completely bipedal stride dating back 1.5 million years have been found in the Kenyan sites of Koobi Fora and Ileret. The hallux is aligned and the three arches of the foot are very pronounced.

Preliminary study for the creation
of a model of two australopithecines
(*A. afarensis*) on the move.
Drawing by Lorenzo Possenti.

It was in this phase that hominins moving upright with the typically human gait crossed the borders of Africa for the first time in history. Very early remains dating back just under 2 million years have been found in the valleys of the Lesser Caucasus in Georgia (in the flourishing site of Dmanisi, dating back 1.85 million years, the earliest ever documented so far outside Africa), at Ubeidiya in the Middle East, along the coasts of Asia and at Riwat in present-day Pakistan. It was at Renzidong and then Zhoukoudian in China and at Sangiran on the island of Java that the eastern branch of *Homo erectus* began 1.5 million years ago. A process of fragmentation known by evolutionists as "adaptive radiation" also commenced. Over a period of tens and hundreds of thousands of years, the first specimens of the *Homo* genus, who started out from a valley in the Horn of Africa, reached the Pacific and divided into subspecies. Walking humans also arrived for the first time in Europe, where *Homo antecessor* was born.

A second long march began, again from Africa, over a million years later and a new species, again apparently scattered all over the Old World (Africa and Eurasia) and differing markedly from *Homo ergaster*, appeared about 780,000 years ago. The species, characterized by considerable cranial expansion (up to 1,200 cc or just over) and the Acheulean industry of hand axes or biface flake tools, is known as *Homo heidelbergensis*, whose remains have been found in Europe, Africa and China. Scholars suggest that *Homo heidelbergensis* played the leading role in a second great wave of human expansion that set out from Africa and then, due to geographical separation, began to divide into a number of variants, some of which took the place of previously established species.[4]

Already present in Africa about 600,000 years ago and possibly including finds of 350,000 years ago with pre-sapiens features in the skull and palate, *Homo heidelbergensis* was geographically close around 200,000 years ago to *Homo erectus* in the Far East, where the two species deriving from two different waves of expansion may

have coexisted. It was a branch of *Homo heidelbergensis* in Europe that gave birth to the *Homo neanderthalensis* species, our closest cousin and best-known evolutionary alter ego.

This period saw a repetition of the fortunate dynamics of sedimentation that occurred at Laetoli. Fossilized footprints of human beings (five adults and children, possibly constituting a family group) left in fluvial mud 850,000 years ago have recently been found near Happisburgh on the Norfolk coast in England. The earliest ever discovered outside Africa,[5] they could indicate the presence of descendants of *Homo antecessor*, the European pioneers of the first wave of migration out of Africa. Southern Italy can, however, boast some of the world's earliest tracks of individuals of the *Homo* genus, dating from a period between 385,000 and 325,000 years ago. It has proved impossible to determine the species they belong to (possibly *Homo heidelbergensis*, but of a slightly smaller size). The prints were left in the fresh ashes of the Roccamonfina volcanic complex in the northwest of the Campania region. These mysterious tracks, known by the

Detail of footsteps at Laetoli.

Australopithecines on the move. Model based on a series of footprints found at Laetoli in Tanzania (Africa) on a volcanic layer deposited 3.6 million years ago. Model by Lorenzo Possenti.

local inhabitants as "the devil's footprints", suggest a scene of panic. Together with other terrified animals, three unquestionably bipedal individuals rushed down the slope of the volcano in warm mud mixed with ashes during the eruption. They left 56 footprints. Losing their balance, slipping and searching for footholds, they also touched the ground with their hands occasionally during their wild flight, which was imprinted in stone forever.

THE FIRST AFRICAN MARATHON RUNNERS

The most recent and certainly most versatile walker has a familiar name. When our species, *Homo sapiens*, was born in Africa, about 200,000 years ago from an African form of *Homo heidelbergensis*, movement appears to have been one of its first activities. As the Old World was already crowded with species from the previous waves of migration, our ancestors – who may also have spread out from Africa in a succession of waves – met up with their earlier cousins in the different regions and coexisted with them in the same territories, which were sparsely populated at that time. At a certain point in the comparatively recent past, for reasons that are still unclear, we were then left as the only representative of our kind in the world – with our flat faces, long legs, highly developed frontal and parietal lobes of the brain, new lithic technologies and prolonged infancy. Like the Neanderthals, *Homo sapiens* was now an excellent hunter. According to the biologists and anthropologists Daniel Lieberman and Dennis Bramble,[6] the locomotory specialization of the *Homo* genus was probably running at a regular pace over long distances so as to tire the prey out rather than capture it with a sprint; in short, we are bipedal apes that run the marathon rather than the one hundred metres.

The South African sites of Langebaan and Nahoon have yielded fossil tracks of anatomically modern humans that lived in the region

around 125,000 years ago. They are like ours. The *Homo sapiens* species spread through Africa and then left the continent, probably in successive waves. The first, between 120,000 and 100,000 years ago, went directly from the Horn of Africa to the coasts of the Arabian peninsula through the strait of Bab el-Mandeb and perhaps already via a more northerly route along the Red Sea and the Nile corridor to the Mediterranean and then the Levant. The second, between 85,000 and 70,000 years ago, drove into Asia. The third took place with more permanent results, not least due to favourable climatic conditions, between 60,000 and 50,000 years ago. About 50–45,000 years ago, from the east and perhaps the southwest *Homo sapiens* made their entrance into Europe, where they gave birth to the Cro-Magnon population characterized by highly advanced behaviour. They were also to be found during the same phase in the innermost areas of Asia, on the edges of the northern steppes and in the Far East at Zhoukoudian in China, where they arrived 67,000 years ago. The settlement of the Old World, from South Africa to France, from Spain to China now regarded various latitudes and took place at unprecedented speed. The *Homo sapiens* hunter-gatherers penetrated Europe repeatedly during the subsequent periods, sometimes from Central Asia.

These population movements should not be conceived as caravans of human beings in search of better land or mass exoduses from inhospitable regions but rather as a slow advance, from generation to generation, of groups similar to the still existing tribes of hunter-gatherers. If we assume an advance of ten kilometres every century, the hominin tribes of the first and second waves could have covered the distance between Ethiopia and all the most distant regions of the Old World in 200,000 years, even allowing for the geographical barriers. Due to cultural adaptations, while remaining a gradual sequence of tribal movements, the expansions of *Homo sapiens* could have been slightly faster, taking 15,000 years to reach the Far East from Africa and 25,000 to complete settlement all the way to South America.

MANKIND ON THE MARCH

The last epic journeys of the human race are no less enthralling and brought our ancestors also to the new worlds of Australia and the Americas.[7] With his sunburnt skin, calloused feet and watchful eyes, *Homo sapiens* now found the Old World too small. While the islands of the Indonesian archipelago formed an uninterrupted bridge as far as Bali between 60,000 and 50,000 years ago, the Australian super-continent (Australia, New Guinea and Tasmania were joined during the ice ages) could only be reached by crossing a channel of 70–100 km, a distance at which it is hard to see the other side, from Timor or Sulawesi. Some tribes of *Homo sapiens* looked beyond the sea and accomplished this feat almost certainly earlier than 55–50,000 years ago, as they were to be found a few thousand years later both in the coastal site of Bobongara in eastern New Guinea and in the Australian settlements on Lake Mungo in New South Wales. The Australian pioneers of our species left a total of 450 footprints documenting 22 walks in the clay at Willandra Lakes more than 20,000 years ago. The ancestors of the aborigines spread from New Guinea to Tasmania along the coasts and then into the interior, altering the environment through the use of fire. They had advanced technologies and marked aesthetic awareness. As skilful hunters, they made a crucial contribution, either alone or perhaps with the aid of the climate, to the disappearance of the Australian megafauna made up of large marsupials and huge flightless birds.

Were we alone while we expanded in this way? Not exactly. Neanderthal man became extinct in western Europe only 30,000 years ago, a comparatively very recent period, and not before coexisting with us for a long time in Eurasia. A print of the large, sturdy foot of a hunter used to difficult conditions was left in calcareous mud in the cave of Vârtop in Romania 97–62,000 years ago. Meanwhile, another and very unexpected human form still survived in oceanic recesses in the Far East: a pigmy human species with long, flat feet hidden in

Reconstruction of *Homo floresiensis*.
Model by Lorenzo Possenti.

the tropical forest. Discovered in 2003 on the island of Flores in Indonesia, *Homo floresiensis* today enjoys a place of honour as the most curious representative of the diversity of the *Homo* genus. Though little over one metre tall, and therefore with a proportionally small brain, he possessed advanced technology and was an excellent hunter. The nine individuals studied on Flores are similar to *Homo erectus nani* but possess some characteristics so primitive (above all in the shape of the skull and the long feet of the archaic type) that it was suggested in a study of 2009 that they may be the descendants of a much earlier African form that reached the farthest point of its territory of expansion and was trapped on the island a long time before.[8]

Some tools dated in 2010 indicate that the first settlement of Flores, in the cave of Mata Menge, was about 900,000 years ago. There would thus have been sufficient time for an archaic human form to develop "insular dwarfism", an adaptation typical of large-sized species forced to live on islands. With scarce resources and in the absence of predators, it is more efficient on islands to become small. The dating tells us that this extraordinary pigmy species of ancient origin lived on Flores until as recently as 15–12,000 years ago. In other words, they existed till a few thousand years before the invention of agriculture and the birth of the first civilizations. We do not know why they became extinct and there is no evidence of any meeting with *Homo sapiens*. Given that the first representatives of our species reached Australia long before 12,000 years ago and that the chain of the Sunda islands was practically the only possible route, it is probable that there were close encounters between the two species.

In the meantime, perhaps already starting about 25,000 years ago, *Homo sapiens* Siberian hunters occupied the emerged land of the Bering Strait and then followed the herds of mammoth and caribou across it down into North America along the Canadian corridor of Saint Lawrence and along the jagged coasts of the northern Pacific teeming with fish. A further cooling of the climate between 22,000 and 18,000 years ago then slowed down their expansion and cut off the pioneers that had already reached the great prairies. About 16–15,000 years ago the ancestors of the Amerindians again began to move south in small groups, occupying the Mississippi valley, Florida and California. They also arrived in South America, as attested by fossilized footprints of *Homo sapiens* dated from between 10,550 and 7,250 years ago found in the Chihuanan desert in Mexico.

Subsequent arrivals, again of Asian populations, from Beringia and the Pacific ridge then gave birth to the north-western branch of the Na-Dene cultures, which include the Haida, Navajo and Apache. Other Siberian peoples settled still more recently in the northern lands and formed the Inuit and Aleutian groups.

Nor did it end there. *Homo sapiens* hunters spread out from New Guinea over the islands of the Pacific and split up into countless cultural strands. They arrived in the great north in pursuit of the woolly mammoth. The first human footprints were left in the most distant lands and even in extreme environments like Iceland. The Arctic region from northern Canada to Greenland was inhabited beginning 4,500 years ago by fishermen of Mongol stock from Alaska. The capacity of *Homo sapiens* for settlement and dispersal in search of food and who knows what else was now strengthened by cultural adaptations (like weaving garments and making footwear) to cope with any terrestrial environment. In the meantime, human beings reached the tip of South America along the valleys of the Andes and the Atlantic coast, arriving in Tierra del Fuego and giving birth to the Yamana culture around 9–8,000 years ago.

An incredibly long and breathtaking march from the Cape region of South Africa to the southernmost tip of South America, just to see what lay beyond the next hill. Human evolution and globalization meant putting our best foot forward. Mankind has been on the march for six million years.

[1] T. Pievani, *La vita inaspettata* (Milan: Raffaello Cortina Editore, 2011).
[2] S. Almécija et al., "The femur of Orrorin tugenensis exhibits morphometric affinities with both Miocene apes and later hominins", in *Nature Communications*, no. 4, art. 2888, December 2013 (DOI: 10.1038/ncomms3888).
[3] C. V. Ward, W. H. Kimbel and D. C. Johanson, "Complete Fourth Metatarsal and Arches in the Foot of *Australopithecus afarensis*", in *Science*, no. 331 (6018), 2011, pp. 750–53.
[4] For an overview of these recent studies, see G. Manzi, *Il grande racconto dell'evoluzione umana* (Bologna: il Mulino, 2013).
[5] N. Ashton et al., "Hominin Footprints from Early Pleistocene Deposits at Happisburgh, UK", in *PLOS One*, February 2014 (DOI: 10.1371/journal.pone.0088329).
[6] D. M. Bramble and D. E. Lieberman, "Endurance running and the evolution of Homo", in *Nature*, no. 432 (7015), 2004, pp. 345–52.
[7] For an overall reconstruction of the paths of human population in the world, see L. L. Cavalli Sforza and T. Pievani, *Homo sapiens. La grande storia della diversità umana* (Turin: Codice Edizioni, 2011).
[8] W. L. Jungers et al., "The Foot of *Homo floresiensis*", in *Nature*, no. 459, 2009, pp. 81–84.

Buzz Aldrin on the Moon
photographed by Neil Armstrong,
20 July 1969.

WALKING. A PHILOSOPHY AND LIFESTYLE

DUCCIO DEMETRIO

WHEN A WORD IS MORE THAN JUST ITSELF

Is there anyone who is unaware of the meaning and function of walking? With some confidence, for once, we can answer in the negative. The word is all too comprehensible, since it describes an action we cannot do without, regardless of our culture or the vocabulary we may employ to describe its various aspects. Everyone – men and women of all ages, whatever their language or place of origin – knows its meaning, value and purpose. We have picked it up by listening to those who taught us to take our first steps – and then to run, get up, and measure our breaths and efforts. We have come to grasp it as a fact as well as a word, invariably by experiencing it in person, each in his or her own particular way, season after season; we now employ it in everyday life and as a rule for improvement. We make the most of it (or eagerly look forward to it) in spare moments during the day or in times of leisure – and freely so, since walking is as cheap as it gets. Sometimes we walk to forget everything; at other times, to collect our memories and thoughts, to ponder and grow. We also walk to train and purify our body, conscious that this lifestyle choice will influence our behaviour. Choosing to walk rather than drive, hop on a bus or take the tube is not simply a costless and non-profit sabbatical whim. Those who never walk across cities and plains, who don't use their legs to reach high peaks and to walk the roads of the world; or who see walking as an unconscious, automatic movement which does not deserve much attention – all of them miss out on some of the beauty sur-

Each man's life represents a road to himself, an attempt at such a road, the intimation of a path.
Hermann Hesse[1]

A life without a journey cannot be regarded as successful.
Sébastien Jallade[2]

Mahatma Gandhi during the Salt March, India, 1930.

rounding them. To walk is to continuously learn how to admire and observe new things, to gather memories, feelings and charm along a path that, as we all know, is not already there but is rather created step after step. Those who do not like walking will experience beauty distractedly, or even fail to detect it – for this is a kind of beauty that is not buried in museums, but must rather be grasped in the open. They will fail to enjoy the moments of elation, freedom and wisdom which travelling on foot brings to those who have come to see it as a constant and exciting exposure to unpredictable events, coincidences and encounters.

Consequently, the more we will choose to walk, as far as this is possible, the more genuine we will be as men and women. We must heed this call, which can make us more humanly conscious, whatever may be the odds against us – so as to uphold the values of dignity, pride, courage and mercy. We must prove to ourselves that this is how we prefer to travel: by panting, sweating, and arriving late – not to any appointment, but to a renewed encounter with our awareness of being alive. And this, even when faster means of travel are available. Not out of infirmity, weakness or laziness, but to stress that we were born to go out and meet someone, and then return, relying on our own energy alone. Choosing not to walk due to bodily discomfort is something that damages our image – and not just our physical image either; it is tantamount to witnessing a profound transformation in our own being. What suffers from it is our personal identity. So unless we are affected by

some congenital inability, by old age or a debilitating condition, we must hold true to what constitutes a display of vitality, particularly in our own eyes. What is at stake is not simply the enduring performance of an act of motion: the word in question conveys many other messages. While apparently being one of the most common of all actions, it always brings some unexpected discoveries. For this reason, we must be ready to examine it again and again, as it may well reveal things that still lie concealed, which escape ordinary assumptions. The word ultimately carries such implications as to require a few additional considerations regarding its importance from a cultural as well as practical perspective – concerning its history, its place in our lives, and the philosophical interpretations and notions surrounding it.

PHILOSOPHERS OF OUR STEPS AND OF WHAT WE WEAR

Since our earliest childhood we have learned what nimbleness, agility and sharp reflexes this pleasure and power requires, since from one moment to the next it can turn into a heavy burden. Whatever steps we may take, by walking we convey our state of health, mood and ease or difficulty of living to others – as well as to ourselves. Each walker – that is, anyone with the power to walk – is a potential philosopher and psychologist in the process of examining himself and capable, to various degrees, of analysing what his steps reveal. We

Queen Elizabeth II out walking at Balmoral, Scotland, 1967.

François Mitterrand walking in the country, photographed by Richard Melloul, 1978.

grasp the aesthetic aspects of this – for the comeliness of a walking human body knows no match – as well as its connection to distinctive traits of the personality, character, qualities and defects we attribute to ourselves. We choose our gait, seek to correct it if it is not to our liking, or when we realize that we must change our rhythm to suit a new place, climate, or weather. Often these changes are imposed through education, or shaped by the land we inhabit, which influences body and mind. To a large extent, feet and the attitudes they express may be regarded as an additional source of meaning, a further tactile quality of our physical appearance. Soles and toes react to the shoes encasing them – like a blindfold, nose or ear plugs, or gloves which prevent us from sensing the skin and surface of living or inanimate things. Feet communicate with us, they warn us. Through them we perceive what kind of terrain we are moving across. This is particularly the case if we move barefoot and only slip shoes on when the ground gets rough, making thick leather, rubber or wood, or high heels, create an unnatural distance between our flesh and the earth. Some people, however, go for footwear that allows them to perceive the inertia of the surfaces they tread

upon. All footwear ought to become an ally of our skin, so as to cause us as little discomfort as possible – and let us forget about it. This is why we have given footwear various different names, so as to stress its contrasting qualities and reflect the various meanings of the word "walking": slippers, clogs, moccasins, flippers, sandals, boots... Through the vocabulary related to walking we find ourselves shuffling, echoing, creaking, nimbly sliding, swimming, clicking our heels or marching.

Any discomfort felt along the path – or before taking it – with or without thick, comfortable, reliable and supple protective layers, will prevent us from experiencing the world as we would like to in order to meet people, attend unmissable events, and enjoy the pleasure afforded by direct contact with earth, water, sand, grass and mud. Each moment of our life is conditioned by the wellbeing of our legs and feet – like every organ of perception. Moreover, philosophy is said to have been developed on the move, since it was the feet of its inquisitive and often unshod founders that led them to raise their far from otiose or pointless questions towards the sky, the trees and the other things surrounding them, and towards birth as well as death.

President John F. Kennedy photographed by James Atherton while leaving a conference, Washington, DC, 11 March 1963.

FEET HAVE EYES, AND MORE

Feet seem to listen, see, touch, and perceive things. Each time we hobble, shuffle, sway, use a stick or prop of any kind, as well as each time we move with confident speed, leaping with joy, this delivers far from obscure information. We convey these messages to those who are looking at us, to those patiently walking by our side, despite our cursing – be it justified or unjustified. At times, by contrast, the message we deliver is a subtle one that conveys our desire to return, give in, and find our dear, tattered house slippers. We prove to others that we are independent, reliable and happy if we show ourselves to have fully mastered those basic rituals that we must perform whenever we walk: getting up, standing in balance, adopting a suitably erect posture, advancing. We move forward to reach a different place from that in which the motion – however brief – began or to return to it after a few metres. Walking is a crucial activity that allows us to live at our best as human beings. Still, it is not as crucial to our sheer survival as breathing, eating or sleeping: for without these we would die. Clearly, if someone walks to provide care and food for us, and prostheses of various sort, and if this person then ceases to perform this selfless, merciful and generous duty, it affects all other activities in our life. Yet – even if all we do is go around the block – choosing to stop walking, to forever rest that haversack which has been with us for so many kilometres, is like murdering our soul by clouding our mind and shutting it off from everything and everyone, making it blind and depriving it even of imaginary journeys.

WE ARE NOT THE ONLY ONES TO WALK

Walking, as a mere exercise, is a universal action: an experience that is always voluntary (be it freely performed or under constriction) and common to any human being with no debilitating condition affecting his or her lower limbs. It may also be a mourned for, longed for and missed experience. Yet it is not a bodily exercise exclusively reserved to our biped *sapiens sapiens* species. Man is what he is thanks to his ancestors' tendency to move so as to avoid threats, ultimately in order to adapt to ever-new environmental conditions that required the development of cerebral capacities unknown and unattainable by other hominids. These included: the adoption of forms of language (not limited to vocal expression); the ability to turn local resources into tools, items, clothing and primitive footwear; and the bold idea of exploring new latitudes, with the thrill of reaching new horizons, gazing beyond mountain ridges, and yearning for unexplored or promised lands.

Human intelligence would never have evolved if communities had remained sedentary like quite a few peoples that are now extinct – among other reasons, because of the impossibility of travelling due to cowardice, lack of enterprise or a spirit of self-preservation. Our cells would not have multiplied their neuronal networks without the need to solve concrete problems. This led to the rise of abstraction, hypothetico-deductive thinking, and plans to move beyond the physical surface. The starting point, however, was always a matter of questions and answers that could only arise by way of leaving a given place and moving into a less hostile one, a better hunting and fishing ground, more suited for reproduction. Through lucky coincidences and repeated attempts, this led man on the move. Yet human beings – it is worth noting – are not the only ones to have made the most of this genetic advantage. All sorts of animals walk, each in its own

Muslims on a pilgrimage to Mecca, 1963.

way: insects crawl and fly, fish use fins as their feet, and birds have feathers as well as feet. There are even some plants that have this power. Trees walk by soaring up towards the sky. Countless plants expand by winding their way across the surface of the ground, while others creep up trunks and walls. There is no plant which does not "walk" by plunging its rhizomes, bulbs and roots into the soil, so as to search for nutrients and water, avoid the sun, flee from predators, and produce fruits in darkness. Finally, not just living beings, but even inanimate and inorganic things (smoke, vapours, clouds, rolling pebbles or any other projectile, leaves or seeds in the wind, smells…), whether by virtue of their intrinsic properties or for other reasons, fall under the laws of motion and gravity. Using colourful expressions, we often say, for instance, that "time runs on", "fortune comes and goes", and "thoughts roam freely". Having reached this point, our word turns into metaphor, analogy and anaphora, shedding its literal meaning to acquire one that merges with any implication of motion: at times reassuring, at times – like the wandering, restless mind – exposed to the unknown variable any becoming will bring. Its charm evokes ancestral echoes, myths and archetypes, since it is bound to conjure up images of the great wayfarers from the most time-honoured legends: Gilgamesh, Cain, Moses, Odysseus, Aeneas, Orpheus, the Argonauts, the searchers of the Holy Grail, and pilgrims headed for Santiago, Jerusalem, Mecca, or the Ganges. And all the aimless, faceless wanderers.

SPACE AND TIME: UNAVOIDABLE CONVERGENCES

The term "walking" thus encompasses many others, evoking images, symbols and fantasies. Each of these refers to the general motif of shifting from one point in space to another, along linear, concentric or maze-like trajectories, which bring into play the other intrinsic aspect of going, travelling and moving across: the variable of time. This existential time reminds us – and proves to us – that every life, every manifestation of life, is subject to a limited duration, just as it must always rest on a circumscribed surface, be it big or small. The space we must depart from, cross, reach or leave behind inevitably engenders and multiplies temporal situations. Step after step, mile after mile, walking punctuates and measures the duration required to reach a destination. Every centimetre of ground that is passed corresponds to a second that has ended for good; through steps, our personal, bodily "indicators", we choose whether to defer an arrival, better enjoy the journey and heighten the sense of expectation, or whether to shorten it instead. Speed and slowness are carefully measured according to the ordinary or exceptional actions we must perform: once again, we cannot escape our fate in having to find food, mate, avoid possible threats, and – alas – occupy others' land through violence, or relinquish our own. We have always been the most nomad species on earth. Our constitutive and endemic migrating turns into a restless affair because of our thirst for knowledge and remote horizons. However, it is not identical for every human being. Established practices, planned or casual changes, cultural and ethnic difference, environmental influences and distinctly individual inclinations or interests (as in the case of hiking or sports, for instance), aside from people's more or less sedentary professional roles, are all factors that influence people's approach to walking. In other words, while all human beings walk, they do so in different ways, using their feet to serve different purposes.

Tibetan caravan,
India, 1959.

BASIC NEEDS, AND MORE

Those that have been recalled so far may be regarded as the basic needs which walking enables us to meet. The loss of the ability to walk properly will prevent us from fulfilling the so-called secondary needs, which are no less important for our physical and psychological wellbeing. To carry our investigation further, it is possible to argue that the choice to walk a lot or only little, nimbly or awkwardly – but always determined to go ahead – represents an indicator of the quality of life we adopt. And there is no need here for marathons, races and prizes, which are best left to other pursuits. Competitiveness, brawls, conflict and the humiliating of losers, or the weak, are age-old practices. Walking without the ambition of becoming a winner is something more than a usual, all too common motor skill: it is a need that reflects other intellectual and spiritual urges. In literature, as well as in real life, these gave rise to human types which differed from those of the men who approached walking as a "profession of arms" or – more simply – through farming, shepherding, geographical exploration and trade. While these people continued to stand for the first current in the conceptualization of walking, in the wilderness of deserts, forests and mountains, out of monasteries and hermitages, lovers and masters of contemplation, meditation, retreat and preaching strode forth: representatives of the second current. We experience this latter achievement whenever in our globalized contemporary society we "embody" this – be it secular or religious – approach by walking with no anxiety, restlessness or excessive eagerness to reach our destination. This coincides with the Romantic *Wanderung* spirit, Goethe's philosophy of wandering, with the American idea of walking in the wilderness exemplified by Thoreau, and with the contemplative *marche* of Rousseau: three philosophical lifestyles and ways of walking. Each of them only embraces walking when it is suffused with spiritual values. This research on society's mores and personal, as well as collective, modes of conduct has led to an acknowledgement of the significance of walking as a long overlooked intellectual concept – on a par with food, sexuality, the senses, thought and other such themes. Indeed, walking as a theme had previously been taken into consideration only by philosophical and religious cultures. In such a way, this humble, obvious and all too common practice of choosing to move on foot instead of by other means – a "pedestrian" practice, to use a disparaging expression precisely related to walking – has finally become an integral part of the "history of ideas" and human behaviour.

THE LAST CURRENT: THE RELUCTANT

Anthropologist David Le Breton has identified yet another current in the various approaches to the practice of walking. This is a typically contemporary current, embodied by all the people who are "reluctant", if not openly opposed, to taking steps as a lifestyle choice. Most prominent in their ranks is the "anti-pedestrian party". The French scholar paints a vivid picture of its many followers. For these people, he argues: "feet are mostly for driving cars, or for supporting the pedestrian when he needs to go up an escalator. Their owners are largely reduced to the rank of invalids whose body serves no purpose other than mortifying existence. ... To waste time by walking is seen as something *passé* in a world governed by haste".[3]

By contrast, there is a minority of people – a growing counter-trend – who are eager to quit all means of transport on wheels in order to devote themselves to walking as a cure and culture. This attitude almost invariably springs from an encouragement to keep up a habit inherited from one's family, from one's place of birth, or from an early love for life in the open air, for the countryside or mountains, or even

farming. The anti-reluctant never draw upon a merely practical culture of walking as a sport or health practice. Their choices represent a challenge directed not against others but towards their own selves: a means to test their lungs and muscles – not a hedonistic pursuit. Their recurrent desire is to establish a regular, regenerating relationship with movement, with the heartbeat, with less stressful everyday life habits; with spiritual traditions – usually Eastern ones – on a quest for the kind of solitude which leads to self-engagement; with those sublime moments that frenzied tourism can never offer, for all its mystifying and spectacular fictions. Far from any form of advertising manipulation, the thrill is felt of having an inner life that we can draw upon by walking, not least in order to discover the importance of silence,[4] as an antidote to the racket, bothersome bustle and distracting noises of urban life. In other cases, we may enjoy walking down dirt roads, paths and alleys with fellow travellers who are also on a quest for what every eager pilgrim and wayfarer has always searched for, with

Plinio Nomellini, *The Strike*, 1889
Pen drawing on paper,
9 x 13.3 cm
Florence, private collection.

Plinio Nomellini, *The Strike*, 1889
Oil on canvas, 29.5 x 40.5 cm
Tortona, "il Divisionismo" Pinacoteca
Fondazione C.R.

the ambition of tracing new itineraries or following the legendary ones leading to places of great secular or spiritual significance. We may wish to pay homage to a tragic event whose commemoration is our civic duty; and if we subscribe to one of the many world faiths, we may find ourselves in sites made memorable by their powerful sacred and mystic pull. In these cases, Le Breton notes, "the act of walking represents the triumph of the body and the senses. ... It favours the development of a basic life philosophy grounded on a series of small things; at least for a moment, it leads the wayfarer to engage in critical self-reflection. ... One walks for no apparent purposes other than the pleasure of enjoying the passing of time, of making a detour in order to more easily find one's way at the end of the journey, to discover new places and faces".[5]

THE WAYFARER'S WISDOM

In the nineteenth century, the Danish philosopher Søren Kierkegaard, a precursor of the current of Christian Existentialism, prayed to God that he might "not lose the desire to walk. Every day, I walk myself into a state of wellbeing and walk away from every illness. I have walked myself into my best thoughts [as Jean-Jacques Rousseau had already claimed before him, and as Henry David Thoreau was to claim shortly afterwards with his famous saying that walking "is itself the enterprise and adventure of the day"[6]] and I know of no thought so burdensome that one cannot walk away from it".[7]

In these circumstances, we walk in search of fuller, more intense experiences, of something which may increase our knowledge, by enabling us to make the most of a span of time – be it long or short – which admits no distractions but requires recollection for prayers, meditation, and the choice of less empty words. Through steps that

grow slower and slower as marvels are rediscovered, we come to enjoy landscapes that we would neither see nor observe by rushing ahead – humble plants and animals that would otherwise be invisible; or pause to chat with the inhabitants of a countryside or village that otherwise we would never have met.

In a page from his rural diary, Hermann Hesse evokes similar moments and states of grace in the following terms: "The old pleasant feelings that accompanied my wandering passed over my soul, as changeable and varied as the shadows of a cloud: a feeling of sadness for things lost, for the brevity of life, and for the variety of the world ... a feeling of detachment from space and time. ... We wanderers are used to cultivating desires of love precisely because they cannot be fulfilled. ... We wanderers free love from the object, love itself is enough, as in our roaming we do not look for the destination, but only for the enjoyment of wandering itself, to be on the way".[8]

Travelling on foot or deciding to include moments devoted to walking in every journey is a choice that, aside from providing forms of wellbeing unknown to people who do not enjoy walking (the reluctant), also transforms our life habits to the point of deeply transforming ourselves. It discloses a different way of approaching the present and future, whose implications we are bound to perceive in our lives, which are too subject to routines and too geared towards the pursuit of superficial and fleeting, if not trifle, goals. As we shall see shortly, walking also stirs our memories. When walking, we pay silent visits to different moments of our past – including the ones that have made of us wayfarers, wanderers and tireless searchers of what might (still) amaze us.

EACH NEW LIFE ALSO BRINGS ONE'S FIRST STEPS

Walking is always a journey through one's memory. So it is almost impossible not to tell of the memories we have from when we were still young walkers. "The kid's learned how to walk!" is one of the most memorable phrases ever uttered by our parents, on a momentous day of which we will inevitable remember little or nothing at all. This verb, then, was among the first words we picked up as we toddled about, encouraged to take the first steps in our life – tottering, yet confident that someone was there to support us or greet us with arms out-stretched after a few metres' journey. When we finally did away with all supports, without too much fear, we could still enjoy the pleasure of having someone encouraging us with a smile to reach him or her, a few metres away. By calling our name and reassuring us, this person showed us that the word in question was a valued one, a cause of praise and reward. Crawling thus became part of our developmental past, as a crucial chapter in our psychophysical development came to an end. This episode, a real act of release from the law of gravity, showed us that – whether on our own or with the help of others – we had undergone the first initiation in our life. Despite the constant fall-ing and standing up, we were proud of learning how to go about the world. Yet we still ignored just to what degree that moving "up and down" from the floor would mark our own life – as it indeed marks everyone's life. The good omen of our need and desire to grow was being fulfilled as we experienced a surface harder than that of a soft, familiar body. At the time we had no idea that things would no longer be the same, that our fate as little men or women was already sealed. So much so that those first momentous steps enabled us to enrich our vocabulary: the shifting of the horizon achieved with bare feet – or with shoes especially donned to mark the occasion – revealed new words as well as new things to touch. By wandering about the house, loitering on a balcony or in a garden, inquisitively exploring previously unknown ground, we witnessed our body extending towards each new marvel by its own will. Hands, feet and mouth soon became crucial tools enabling us to access the beautiful and ugly sides of life, with an increasing degree of freedom and awareness of the risks faced and the respecting or breaking of rules. Yet, inevitably, it also brought unexpected tumbles and falls. By the time we had grown more confident in our stride and managed to turn a handle to move from one room to another, we were already well into the "game of life". Even though we were still unaware of this increasing linguistic transformation, the word "walking" started acquiring new meanings for us – meanings that all boil down to the contrast between stillness and motion, change and fixity, progress and involution. The knowl-edge we acquired at home, in a garden, or on a carpet started open-ing up new horizons for us; in falling and getting up again, we were already experiencing one of the inevitable rules of existence destined to accompany us throughout life. It is hardly a coincidence that it is customary to give newborn babies their first, tiny shoes, in expecta-tion of what the future will bring. These shoes of wool, soft leather or rubber symbolically signify that the newborn may now be regarded as a rightful member of the human species.

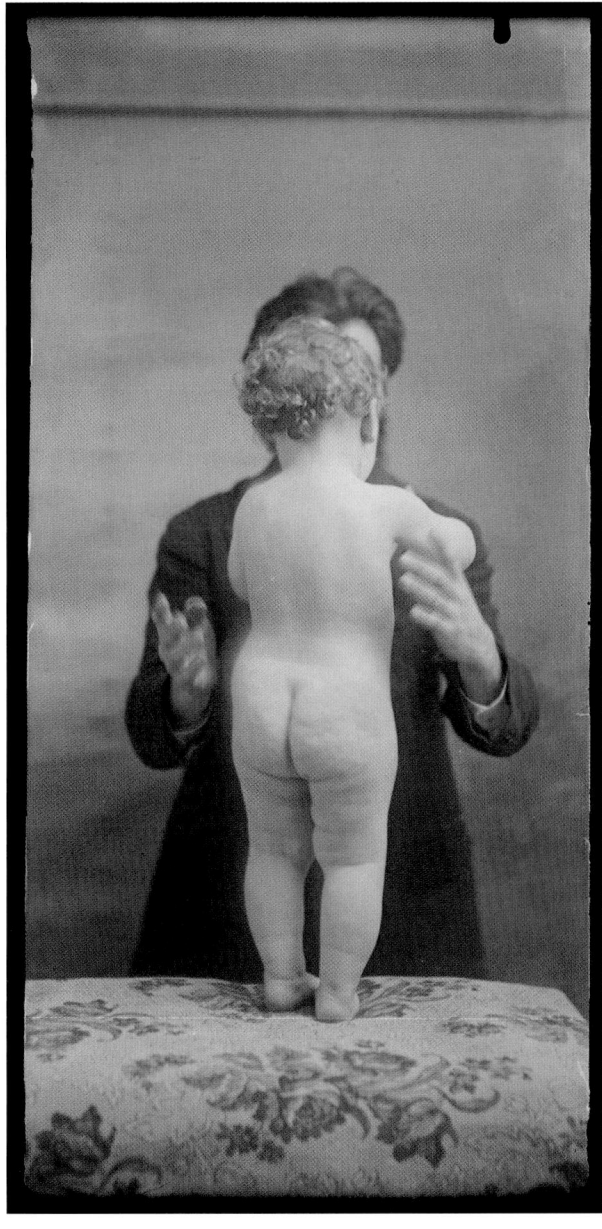 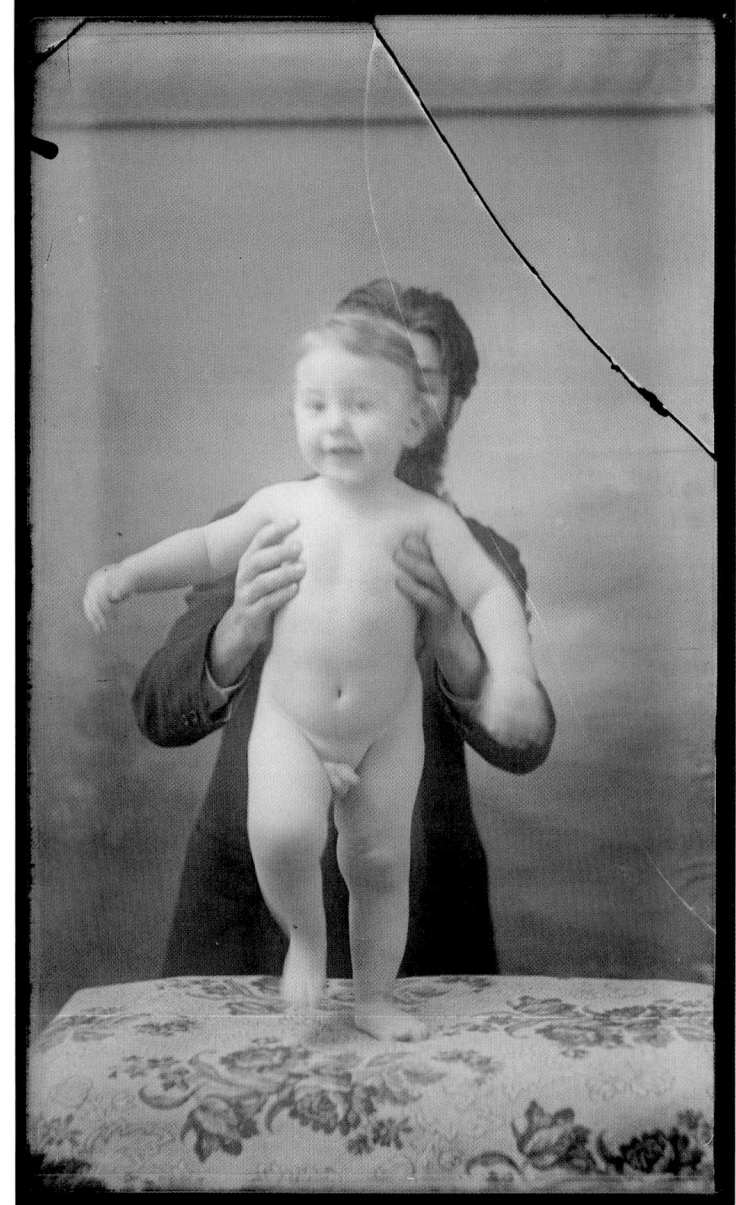

Constantin Brancusi, *Portrait of Baby George Farquhar*, 1911
Gelatin silver print on glass plate, 17.8 x 8.2 cm each
Paris, Musée national d'art moderne / Centre de création industrielle – Centre Georges Pompidou.

Italian troops shouldering weapons
and skis as they march upwards
during the First World War.

OUR STORY IS WRITTEN BY THE PATHS WE HAVE TRAVELLED

One question we should ask at this stage is: "Tell me how you walk, how you have been walking, and I'll tell you who you are". In its directness and adaptability to other prerogatives of ours, this indirect question enables us to shed light on different psychological, cultural, emotional, social, civil and even religious attitudes. For it may be argued that the way in which we take our steps reflects our individual "autobiography as walkers". Our life story is also the story of how we have used our feet, of how much and where we have walked – either independently or because compelled to do so. Walking makes us eager to tell our own personal story, notebook in hand; to put down in writing the places we have visited and adventures we have experienced. We will thus find ourselves gazing differently at urban settings, secluded groves, side roads, alleys and many other places that would not ordinarily draw our attention. We will discover that if we are able to do so, if we have the right shoes, we will want to climb up among the trees. And so our feet will become the protagonists of stories set in silence – intimate tales. The feet on which we rest will turn into an extra pair of hands through the sheer pleasure of relying on them, partly to enjoy a general sense of wellbeing, of the sort increasingly recommended by doctors, and partly because of an ancestral impulse that stirs within us, one inherited from our ancestors, who were forced to walk as a matter of life or death.

Benito Mussolini leads Fascist officials in the goose step for a parade of the militia in 1928 in Rome.

Adolf Hitler walks down the steps of the Zeppelinfeld during the Reich Party Day ceremonies in Nuremberg, Germany, 1935.

Chinese officials and Tibetan Lamas inspecting troops, Lhasa, Tibet, 1956.

Pope John Paul II walks along the pilgrim route near Santiago de Compostela, 25 July 1989.

WALKING IN SYMBOLS: TRAVELLING

Moving on foot brings into play another far from trifle story – a more conceptual, philosophical, religious and symbolic one. Here walking takes on more interesting and complex meanings than those we usually refer to. Writer Erri De Luca paints an evocative picture of this through his recent considerations about his own walking along some memorable paths: "On ancient paths I feel and see myself following in the trail of my ancestors. Not in a museum or by an ancient monument, but on paths do I sense that I am treading in the footsteps of the countless men and women who came before me. On paths, more than in book pages, do I sense that I am indebted and heir to works of genius. … I stand in a current formed by paths that meet and branch out, as dense as the wrinkles on our skin or furrows on our hands".[9]

Walking, then, means conjuring up those who travelled the same road before us; it means pausing to consider the different verbal expressions of the word and reinterpreting them in the light of the many synonyms we have hastily quoted and which now require more in-depth examination. These include: strolling, moving forward, going, marching, advancing – not to forget the other predicative word implied: travelling.

Walking, then, constitutes al alternative to the contemporary mode of travel, as Sébastien Jallade has written: "In the modern world, travel is increasingly less based on disorientation, to the point that we must acknowledge that we are experiencing the end of exoticism … our childhood was nourished by artificial images from television programmes … through the Internet each of us is constantly informed about the world he or she lives in".[10]

It is no coincidence, perhaps, that in Italian, just as in French and Spanish, the Latin etymology of the word is related to the idea of "passing through", of the movement of a physical object down a corridor or passageway. These images are bound to evoke that of smoke spiralling up the narrow passage of a "chimney" (etymologically related to the Romance word for "walking"). For genuine travel, particularly when made on foot, always presents some tight spots, past which we can take some deep breaths.

Even though we can travel with our mind – and nowadays through the Internet too – without moving an inch, travel in the literal and most common sense of the word is inconceivable without the use of feet. When travelling, we make use of these lower appendages each time we walk down stairs, choose what path to take, and hop in a car or aboard any other means of transport to depart from the place where we had temporarily stopped.

Our word is also the source of other exemplary metaphors. Can we not compare our life to a journey? To a travel, route or itinerary whose final destination is often unknown, as are sometimes the reasons for which we embarked on it? Each difficulty we face evokes images of roads with no exit, of steep climbs and ascents – unwanted yet inevitable. But we may also start making our way, almost blindly, down winding paths that teach us bravery and resoluteness. We walk and travel for as long as we are granted to live, if necessary replacing feet with thoughts, the imagination, and an eagerness to discover new things even when in a condition of forced immobility. Nowadays we walk and travel online, visiting websites that enable us to explore faraway places from a desk or anywhere else, putting us in touch with sources of knowledge from the past or present – with what is simultaneously taking place across the planet. We walk and travel for real, raising a cloud of dust along the path, leaving our shadow behind us; but now we also walk by creating virtual worlds, without necessarily having to be science-fiction writers, mystics or visionaries. So:

– we walk along the lines of books, following their protagonists' progress, as well as in film scenes, leading from one tale to another, and wander across exhibition halls;

– we walk because individual or collective life is nothing but a journey with a beginning and an end, made up of stages and stops creating the illusion that it is possible to freeze time: we continue to develop and grow older as we sleep, and conform to the needs and changes of our body regardless of our wishes. Each moment of quietude, of rest or relaxation, is merely a pause in our usual toil;

– we walk to pass the torch on to others – a testimony, our material or spiritual heritage;

– we walk to put ourselves to the test, to face the unexpected in unknown territories, driven on by a yearning for novelty, change, and alternative routes.

Besides, is there anyone who has never found himself at a crossroads in the course of his life? Who has never had to speed up or slow down? Who has never found himself not knowing what the best path to take and follow was? Who has not rejoiced when life enabled him to move on smoothly, through the help of some fellow traveller? Who has never come to enjoy moments of profound happiness, free from all danger, because of the absence of obstacles along his path?

WALKING IS A RITUAL, SACRED OR SECULAR

Such questions should lead us to greet the rising and setting of each new day, and to celebrate events in our lives with a meditative, wandering walk: not so much in order to reach a place – regardless of whether our walk is contemplative, refreshingly brisk, or of variable speed – but rather to celebrate the fact that we are still alive and capable of appreciating life, of sensing its breath. The fullness of life can only be experienced, through a feeling of incomparable joy, if we are granted the freedom to reach a person or a place out of the pure pleasure of the discovery or of the meeting. Whatever our religious beliefs, this morning ritual makes us secular witnesses to the return of darkness and light – in line with the Ptomelaic picture of a sun running

its course from east to west. The longing for its return is familiar to monks, mountaineers, wayfarers, poets, nature lovers, and artists – precisely the kind of people who have always been keen on walking, conceived as an endless quest. Sébastien Jallade reminds us that: "In each of our personal stories we seek to access a hidden, ancestral knowledge that stems from the mysterious source of life. ... The itineraries traced by each one of us are – at different levels – unconscious fragments of our quest for the miraculous, for self-transformation".[11]

[1] H. Hesse, "Demian", in *Romanzi*, edited by C. Magris and M. P. C. Palin (Milan: Mondadori, 1977), p. 306.
[2] S. Jallade, *Il richiamo della strada. Piccola mistica del viaggiatore* (Portrogruaro: Ediciclo, 2011), p. 21.
[3] D. Le Breton, "Camminatori e cammini", in *Pensieri viandanti. Antropologia ed estetica del camminare*, edited by I. Testa (Reggio Emilia: Diabasis, 2007), p. 22.
[4] Concerning this experience, see D. Demetrio, *I sensi del silenzio. Quando la scrittura si fa dimora* (Milan: Mimesis, 2012).
[5] Le Breton 2007, pp. 23–24.
[6] Quoted from H. D. Thoreau, *Camminare* [1862], edited by M. Jevolella (Milan: Oscar Mondadori, 2009), p. 22.
[7] T. Espedal, *Camminare. Dappertutto (anche in città)* (Milan: Ponte alle Grazie, 2009), p. 31.
[8] H. Hesse, *Vagabondaggio* (Rome: Newton Compton, 1992), pp. 66 and 46.
[9] E. De Luca in *Le parole che sono importanti* (Milan: Feltrinelli, 2014), p. 153.
[10] Jallade 2011, p. 54.
[11] Ibid., p. 17.

NEOCLASSICAL BALANCE. CANOVA AND THE RHYTHMS OF GRACE

CARLO SISI

In search of models to substantiate his concept of ideal beauty, Winckelmann repeatedly visited the excavations of Herculaneum and other archaeological sites in the Campania region, receiving impressions not always in line with his own celebrated description of the works of the antiquity as endowed with "noble simplicity and tranquil grandeur". On his return from a stay in Naples, he thus avowed his unconditional admiration for the depictions of dancers discovered on the walls of what was known as Cicero's Villa at Civita, pictorial versions of Greek models that

went beyond the formal conventions of contemporary figuration to display the drive for escapism and departure from reality underlying the Pompeian third style. As he wrote to Count Bruhl in 1762, "The most beautiful of all are the figures of dancers and centaurs, measuring about a hand's breadth on a black background, which are evidently the work of a great master, being as fluid as thought and as beautiful as though fashioned by the Graces". These encaustic works were described in a section on painting of his *Geschichte der Kunst des Alterthums* as the work "of a learned and forthright artist … as though created with a single brushstroke".[1]

Different kinds of balance therefore from those recognized by Winckelmann in the "grace" of ancient works characterized by *ponderatio*, simplicity and peace of

E' della stessa grandezza dell'originale

Antonio Canova, *Dancer Holding Veil*, 1799
Tempera, 29 x 25 cm
Possagno, Museo e Gipsoteca Antonio Canova.

Dancer, from *Le Antichità di Ercolano esposte*, vol. 3, plate XXIX (Naples, 1757)
Florence, Biblioteca Nazionale Centrale di Firenze.

mind undisturbed by undue warmth or violent passion. Painted in a condensed, dynamic style, the dancers of Herculaneum instead manifested the "dawning of delight" admitted by the writer as in keeping with the kind of subject represented. As he wrote in the *Geschichte*, "The pose of the figure encloses the rationale of its action". Depictions of bacchants must therefore reflect their wild nature, just as dancing figures will be more frenzied and contorted than those of gods and heroes designed for places where stillness and its emotive counterparts must instead predominate.[2]

A vital breeze therefore stirred the Apollonian imagination of the fathers of Neoclassicism. Even Gotthold Ephraim Lessing acknowledged in the *Laocoön* that grace is "beauty in motion",[3] hard to capture in painting but suited to the resources of poetry as fleeting enchantment. Examples include Ariosto's description of Alcina and the elusive charms of her eyes ("Which ever softly beam and slowly move") and bosom ("Where, fresh and firm, two ivory apples grow, / Which rise and fall, as, to the margin pressed / By pleasant breeze, the billows come and go"), and Anacreon's advice to the painter intent on portraying his beloved ("Let the Graces hover around her delicate chin and marble neck").[4] It was Foscolo that heeded the call of these eighteenth-century theorists at the beginning of the nineteenth and introduced into his poetry fleeting images of beauty that were to engage the figurative culture of his era in dialogue with undisputed authority. From the odes *A Luigia Pallavicini caduta da cavallo* ("And you are called by the dance / Where zephyrs wafted / Unusual fragrance, / When, impatient of restraint, / Your hair to your rosy arm / Was a sweet encumbrance") and *Alla amica risanata* ("… or when / You dance and, / Your agile body to zephyrs entrusting, / Unknown charms escape / From your mantle and the ruffled veil / forgotten on your swelling breast") to the many verses on the dance in *Grazie*, the poem he devoted to the sculpture of Antonio Canova and an ideal of beauty that celebrated the foretold harmony between the sister arts.

Radiant Panathenaeas alternate with twilit recesses where rites to Pan are celebrated; views of plains, mountains and the sea with celestial epiphanies; sumptuous allegories with fleeting visions that pursue the impalpable, elusive grace which appeared with unexpected freshness to the devotees of ideal beauty on the walls of Herculaneum ("But when she dances, / See! All the harmony of sound / Slides from her beauteous body and the smile / Of her mouth; and a movement, an act, a gesture / Sends sudden grace to the eye. / And who can paint her? While to portray her / I gaze industriously, she escapes me, / and the circles she slowly describes / Are quickened till she flies / Over the flowers and I can barely glimpse / the gleam of her veil amid the myrtles").[5]

The search for felicitous continuity between the eras of myth and everyday contingency, for possible osmosis between ancient and modern sensibility, gave rise to literary transfigurations like those of Foscolo but also to practical experiments that made use of archaeological finds or historical sources as tools for daring and imaginative disguises in an extraordinary theatre of the passions. It was in Naples in 1787 that Canova beheld Emma Hamilton's *attitudes*, performances of dance and mime inspired by the encaustics recently discovered at Herculaneum and by the supple figures on the Greek urns collected by her husband Sir William and made famous by the engravings in the great volume of *Antiquités Etrusques, Grecques et Romaines tirées du cabinet de M. Hamilton* published by Baron d'Hancarville. In this series of evocative *tableaux vivants* the beautiful lady adopted a whole variety of poses drawn from ancient models and alluding to scenes from Greek tragedies. Goethe, who was also among the audience, noted her ability "to choose and change the hang of a veil" and arrange "her hair in a different hundred ways" against a background of minimal scenery. This was limited to a chair, some Greek vases and a few extras but above all a sort of black box edged with gold and illuminated by a single source of light, which emphasized the poses struck and facilitated comparison in both thematic and spatial terms with the vases and encaustics "studied" by the eccentric lady of the house.[6]

The success of these private performances is demonstrated by the numerous portraits of Lady Emma by the leading artists of the time, including George Romney, Angelica Kauffmann and Elisabeth Vigée-Lebrun, as well as the scenes engraved by Tommaso Piroli and Pietro Antonio Novelli. In the latter, she is captured in poses that were adopted at the same time in public theatres, which also saw a corresponding simplification of scenery, the elimination of masks, the replacement of heeled shoes with sandals or dancing pumps and the adoption of light garments more suited to the movements of the body.[7] The principle of Neoclassical analogy and synthesis was therefore extended to dance and drama, previously subjected to the canons of Baroque theatre, and in this way new fronts were opened up in the dialogue between the arts, to which Canova himself was not extraneous. As stated above, he was a spectator at Lady Hamilton's *attitudes* and indeed good-naturedly scolded by Sir William in a letter of 21 March 1795, which spoke of his wife's wrath on identifying some of them – which "you stole when you were in Naples"[8] – in the engravings of Homeric reliefs made the year before. This confidentiality gives an idea of Canova's close relations with a host enamoured of antiquity and in possession of one of the largest collections of urns, which were made available for study to scholars and to artists as stimuli for compositions and details of costume. Examples include Wilhelm Tischbein's engraving of a dance scene with musicians playing pipes and tambourine, which Hamilton connected with the tarantella and had reproduced in modern guise in a landscape by Philipp Hackert.[9] In imagining Corinne swaying to the rhythm of the dance, Madame De Staël described her as well ac-

quainted with "all the attitudes represented by ancient sculptors and painters": moving with agile skill and beating a tambourine above her head, "she recalled the dancers of Herculaneum and later gave rise to a whole series of new ideas for drawing and painting".[10]

Those who have studied Canova's travel notebooks know how interested he was in theatrical performances and especially in music and dance. These arts saw major changes in the late eighteenth and early nineteenth century both in theoretical works and on the stage, where Salvatore Viganò's *coreodramma* came to rival opera, accentuating the dialogue and osmosis between gesture and the figurative

Elisabeth Vigée-Lebrun, *Lady Hamilton as a Bacchante*, 1790–91
Oil on canvas, 159 x 135 cm
Liverpool, Lady Lever Art Gallery.

arts on a par with Lady Hamilton's *attitudes*. A performance at the San Samuele theatre was described as follows in the *Gazzetta Urbana Veneta* on 18 January 1792: "How then to describe the lightning quick movements and eloquent groupings that constantly recall the most felicitous inventions of painting and sculpture? ... a school for people of every kind. Painters go to reap an abundant harvest of drawings for their paintings and poets to enrich their imaginations with the choicest images of beauty".[11]

While Foscolo's lofty example is sufficient for poets, the passion of Alessandro Verri for scenery inspired by ancient monuments is also significant. As he wrote of an experience as actor and director submitted to the appraisal of painters: "Everything forming part of the production was carefully studied. The costumes were magnificent and truly authentic, being drawn from ancient reliefs, statues and especially Trajan's column. I was dressed as an ancient Roman, just like the statues of the Caesars and with an ancient helmet. My poses and every action were also governed by those of painting and I always endeavoured to conduct myself, even in the most violent passions, after the model of the best painters, as though my every act were to be depicted. According to the best artists in Rome, I was successful in this".[12] Among the artists, Canova's interest in the subject of dance filled his workshop with countless sheets of paper where the space and time of choreographic action were captured immediately in quick sketches. Antonio D'Este sheds some light on this in his memoirs, where he states that the sculptor "was greatly attracted by dancing and would have learned its principles had he not been held back by his love of art. I remember how we would sometimes walk through the Monti and Trastevere districts during festivities to see the girls of the populace dance in all their innocence. He delighted in this and always made some observations on the natural movements of the dancers for the benefit of his art".[13] This testimony is interesting because it supplements the biographical data with further information on how

Antonio Canova, *Two Dancers on Tiptoe*, n.d.
Pencil on paper, 20.4 x 13.5 cm
Bassano del Grappa, Museo Biblioteca Archivio.

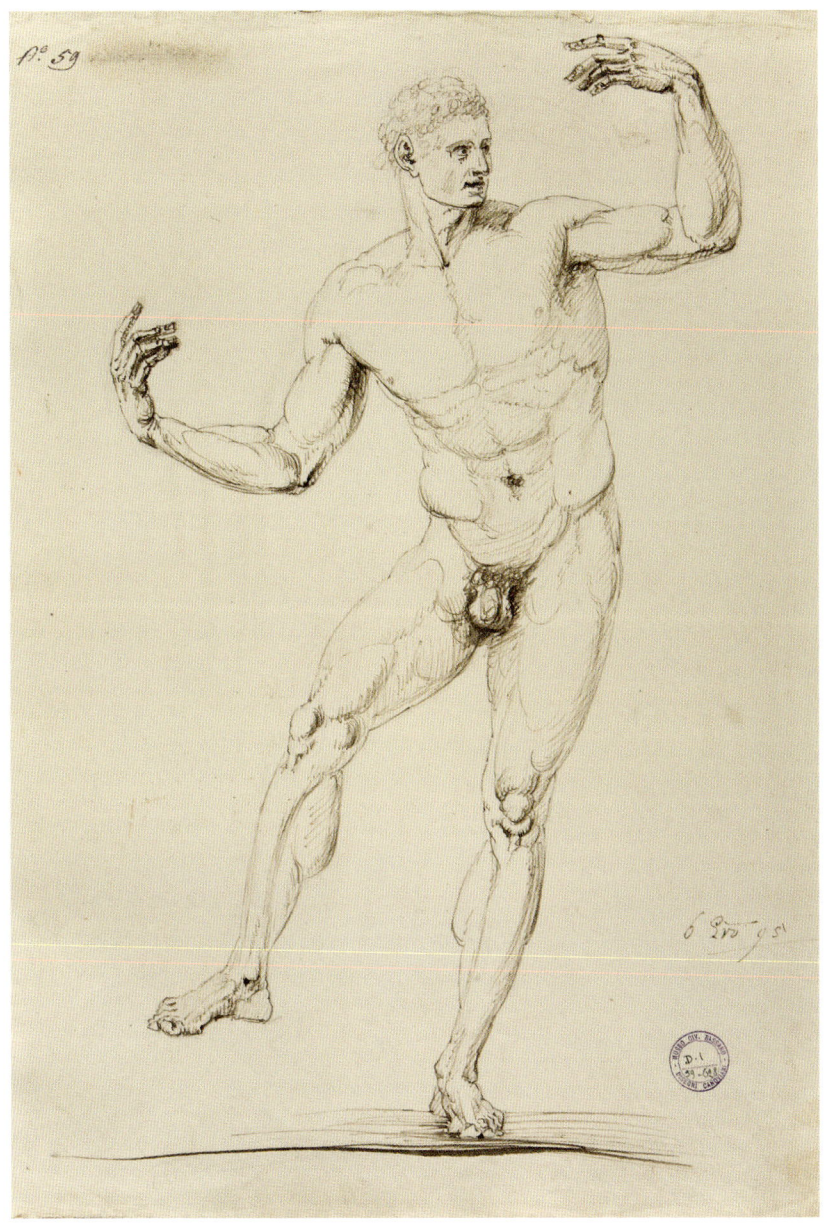

Canova observed natural appearances and assimilated them within the process of derivation from antiquity, understood in turn as a place of "natural gesture", a simple, orderly, organic form and at the same time a sublime concentration of the passions. In contrast to the narratively sterile "dishonest leaping" of the Baroque, Francesco Algarotti had already described dance as "an imitation that, through musical movements of the body, is made up of the qualities and inclinations of the soul", thus assigning this art the task of expressing the passions through the organic discipline of attitudes: "it is required to speak continuously to the eye and to paint with gesture. And dance must also have its exposition, its entanglement and its unravelling. It must be the extraordinarily rich compendium of an action".[14]

Some of Canova's drawings in the Bassano sketchbooks are of athletic male figures evidently based on ancient models but captured in natural poses or rather in the balanced gestures of dance, with precise reference to the *port des bras*, the *arrondissement* of the arms, and the *saut de chat*, all figures that document the importance

Antonio Canova, *Full-face Nude with Arms Raised*, n.d.
Pen on ivory paper, 45.3 x 32 cm
Bassano del Grappa, Museo Biblioteca Archivio.

Antonio Canova, *Full-face Male Bust*, n.d.
Pen on ivory paper, 45.5 x 31.5 cm
Bassano del Grappa, Museo Biblioteca Archivio.

attributed to the upper part of the body in dance in the late eighteenth century, when the raising of arms was required to accompany the striving for elevation and for speed in the execution of the new steps at the same time. According to the choreographer Carlo Blasis, the correct positioning of the cubital fossa or elbow pit was one of the most difficult things for the male dancer, requiring perfect harmony of movement so as to create a continuous undulating effect like Hogarth's line of beauty, "a sort of imitation of what we see in Giambologna's celebrated Mercury" (Blasis).[15]

Canova often recorded his impressions of performances in his notebooks, admiring the "prowess of a young woman and splendid leaps of a man aged sixty but of such speed and precision that it seemed impossible" and citing innovators of contemporary ballet such as Onorato and Salvatore Viganò, Giacomo Tantini and Charles Lepicq with all the critical awareness of someone in close contact with the leading figures in theatrical debate, such as Francesco Milizia and Ranieri de' Calzabigi. In opposition to the conventional *danse d'école* that had reigned supreme with its formal, textbook rules in the first half of the eighteenth century, the period of Canova's debut on the national art scene also saw the development of a "historiographical" conception of ballet based on examination of the classical sources and hence in line with the return to antiquity championed by Neoclassicism. The result was a revaluation of classical pantomime and a wholly enlightened concept of the naturalness of song and dance, born with mankind and developed with its emancipation. This gave rise to a bitter dispute about the autonomy of dance that was to contribute to the invention of the *ballet d'action*, where the union of music and dance would permit narrative solutions of great emotional involvement similar to those adopted by the history painting of David.[16]

The success of the new kind of dance caused Metastasio to write that opera had been relegated to the intermissions between ballets, which had "usurped the art of representing human emotions and ac-tions" to the point of capturing "the hearts and minds of the spectators".[17] This aesthetic and physical domination was to manifest itself with great impact in the *coreodramma* of Salvatore Viganò and the demanding methodological work of Carlo Blasis. Deeply admired by Stendhal, Viganò's productions aimed at the fusion of pure dance, pantomime dance and character dance in accordance with an expressive code that highlighted individual gesture within the overall framework of magnificent settings, often designed by Alessandro Sanquirico, and the composition of ensembles peculiar to this choreographer.[18] Carlo Blasis instead replaced tragedy with "grace", accentuating the lyrical character of dance with constant references to ancient and modern art in an effort to attain the aerial transfiguration of the body, beyond all fatigue, that Canova was also soon to pursue in sculpting his female figures. This parallel is supported by the fact that Blasis chose to use an engraving of the sculptor's *Terpsichore* as the frontispiece for his *Manuel complet de la danse*.[19]

Canova demonstrates his knowledge of the latest developments in ballet in his relief *Dance of the Children of Alcinous*, where the two figures in the centre of the scene are executing *sauts de chat*,[20] their speed and agility suggested by the device of fluttering veils. As Faustino Tadini observed, "It was a great feat for him to portray them naked and in mid-air with a long veil held in the left hand and curving over their heads like a rainbow, almost as though about to take on new shapes: a playful invention that imbues the painting with charm and the dancers with suppleness."[21]

While this nudity is related to Winckelmann's ideas as to the habits of young Greeks in antiquity, the formal lightness corresponds to the ideals of grace, the triumph over gravity sought by modern dancers and even more by Canova, whose experimentation with bodily balance during the 1790s was to culminate early in the nineteenth century with the extraordinary conception of the flying *Hebe*. First drawings and highly original sketches in tempera on canvas that form

Antonio Canova, *Dance of the Children of Alcinous*, 1790
Plaster, 120 x 266 cm
Venice, Museo Correr.

rhythmic sequences of studied harmony where memories of the Pompeian dancers, "as fluid as thought", are informed with the modern invention of grace and blossom anew; then the sculptures executed between 1806 and 1811, serene embodiments of youthful beauty that amaze by virtue of their lightness and the intimate artifice whereby they struck the patrons and the public of the time as instilled with natural life.

Importance attaches here to the immediate impressions of some of those who witnessed the sublime flowering of Canova's genius. Isabella Teotochi Albrizzi wrote as follows on the relief *The Three Graces and Venus Dancing Before Mars*: "three delightful figures that barely touch the ground with one foot while the other soars with such animated gracefulness that you would not believe it's the work of marble and chisel alone but rather pliant limbs nourished with celestial ambrosia and divine nectar".[22] The wonder aroused by the *Dancers* through the wholly unprecedented handling of the subject and the extraordinary balance of the figures is instead recorded in Melchior Mis-

sirini's life of Canova: "The first of the dancers … holds a corner of her long garments, placing her hands on her hips and displaying all the strength of vigorous youth, the elasticity of her sinews enabling her to rise swiftly on tiptoe. The second is of a different nature, touching her chin with a finger and presenting in the charm of her motion the graceful waving line upon which Hogarth based his theory of beauty. The third belongs to the category of bacchants, and no figure could be captured in the free impetus of dance with greater simplicity and decency. She is shown in the act of leaping gracefully to the clashing cymbals".[23]

As we can see, the analysis of the works focuses above all on the emotion aroused by the perception of movement and life, the precarious balance that works through the representation of dance to endow marble with the illusion of naturalness and lithe motion, elements that were to triumph in the mobile epiphany of *Hebe* gliding down amongst mankind from the heavens of a reformed mythology. While Ippolito Pindemonte wondered in his celebrated sonnet at the

Antonio Canova, *Five Dancers Holding Hands*, 1799
Tempera, 35 x 80 cm
Possagno, Museo e Gipsoteca Antonio Canova.

pp. 108–09
Antonio Canova, *Dancer with Her Arms about Her Head*, 1799
Tempera, 29 x 25 cm
Possagno, Museo e Gipsoteca Antonio Canova.

Antonio Canova, *Dancer with Cymbals*, 1799
Tempera, 29 x 25 cm
Possagno, Museo e Gipsoteca Antonio Canova.

soft, living marble and the hand that sculpted its steps, it was the authoritative critic Cicognara that deciphered the admirable artifice of the sculpture: "The fact of leaning forward and cleaving the air at a certain speed produces the very natural effect that the garments are blown back and display the naked body beneath with no affectation whatsoever. The raising of an arm to pour the liquid from the vase involves the entire outline of the figure so deftly that even though the eye finds it draped with the utmost decency, the avid gaze discerns its every contour, displaying nothing other than the budding freshness of its forms".[24]

This perfect ekphrasis introduces the theme of drapery into the stylistic and emotive reading of the work. This was by no means a secondary consideration in the compositional conception of figures, as demonstrated both by Canova's studies of ancient works and models posing in his studio and by the theoretical writings of Neoclassicism, where Winckelmann and others regarded drapery as an essential element of beauty. The aim was to capture the natural contours of apparel and the lightness and transparency of veils and materials by means of formal devices not unlike those of Foscolo's odes in poetry. Dancers like Maria Medina Viganò and Maria Del Caro appeared instead on the stage free of all constriction, wearing simple tunics to enhance the grace and agility of their movements. This is precisely what was seen at the same time in Canova's sculptures in accordance with the classicistic recommendations formulated by Cicognara: "the clothing of sculpture is usually thin and moulded to the forms, soft and clinging as though damp".[25]

The search for naturalness in the representation of figures caught in precarious balance, developed on ancient models and then translated into the modern vocabulary of grace, thus proves to have been a key issue in aesthetic debate in the late eighteenth and early nineteenth century, when the dialogue between the "sister arts" was at its liveliest and artists and theorists in their laboratories mixed the

Antonio Canova, *Hebe*, 1816
Marble, h 175 cm
Forlì, Pinacoteca Civica.

Antonio Canova, *Dancer
with Hands on Hips*, 1809
Plaster, 179 x 75 x 67 cm
Possagno, Museo e Gipsoteca
Antonio Canova.

Antonio Canova, *Dancer
with Cymbals*, 1812
Plaster, 197 x 80 x 55 cm
Possagno, Museo e Gipsoteca
Antonio Canova.

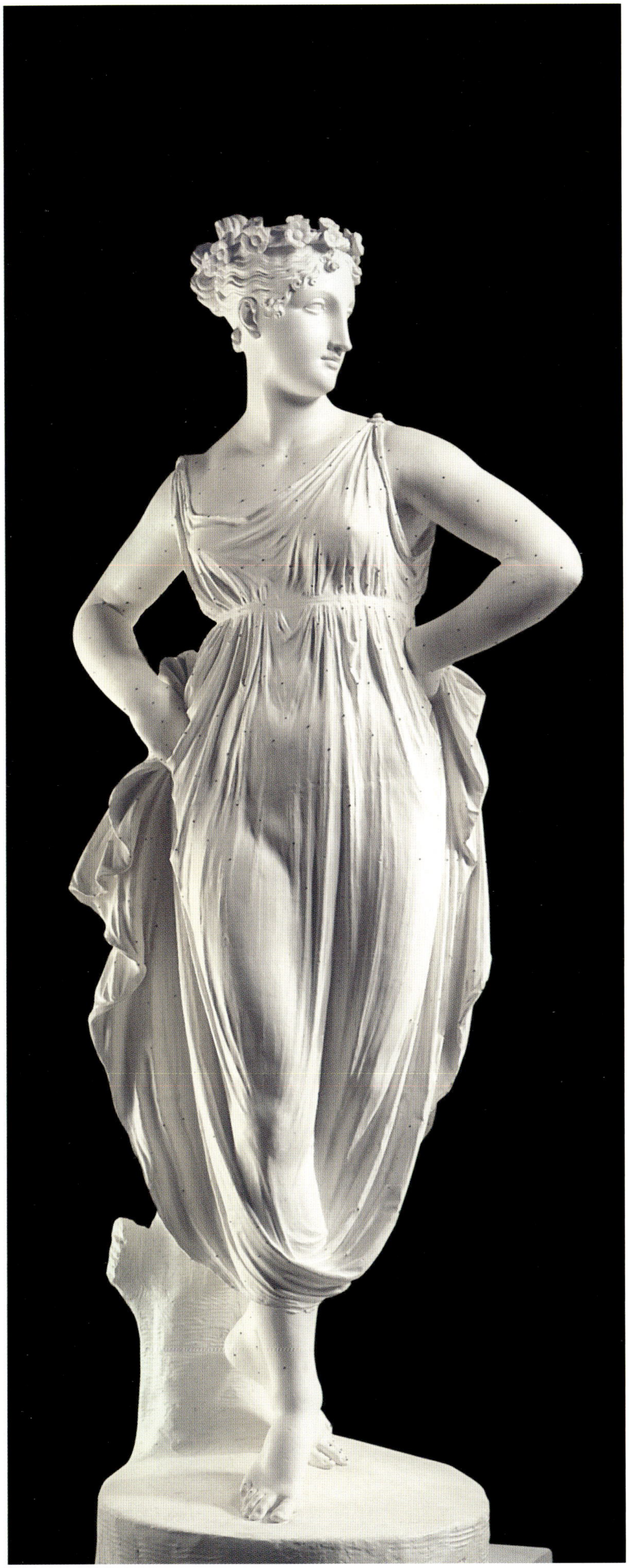
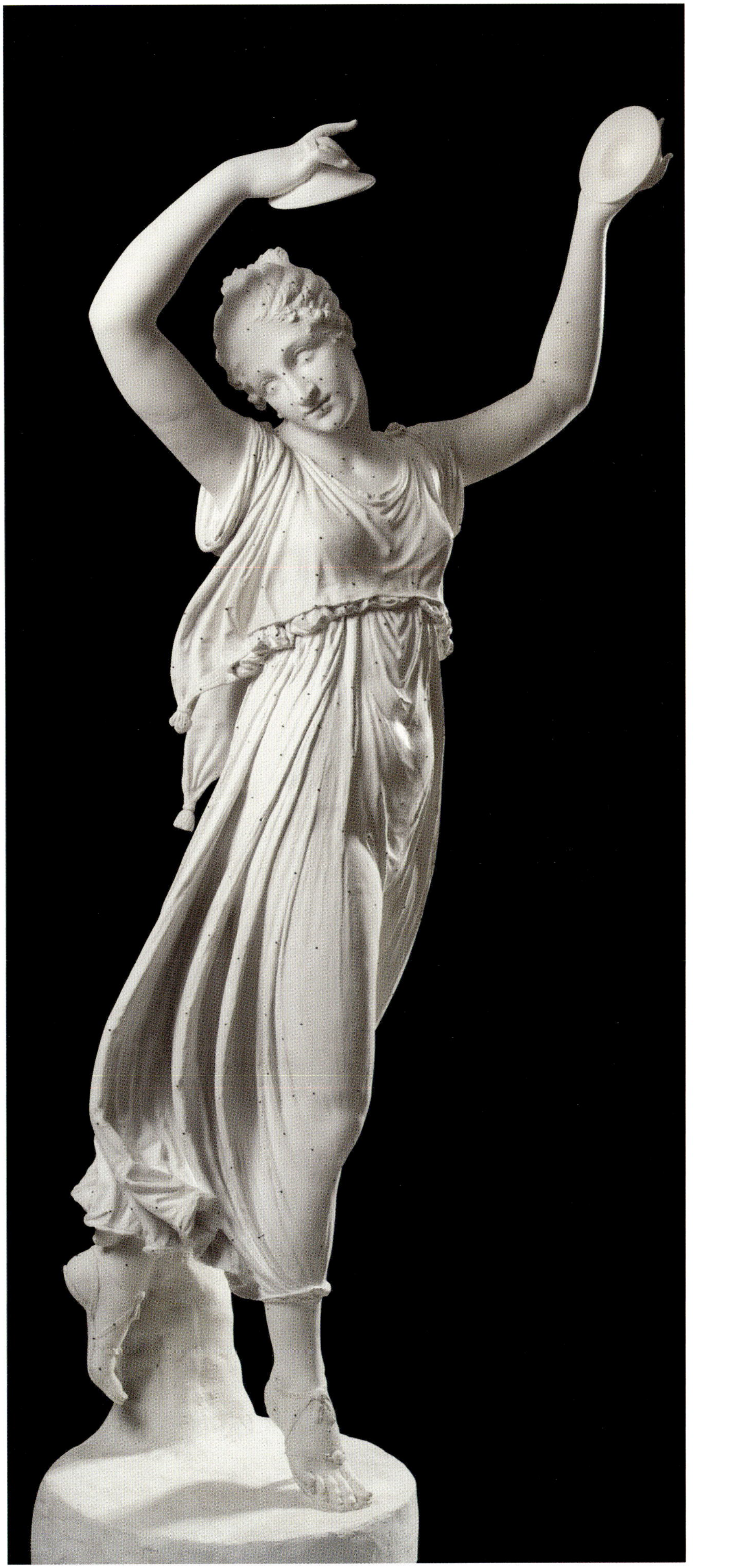

treasures of the classical sources, poetry and the manifold resources of gesture offered by observation of natural life. The episode described in Wilhelm Tischbein's autobiography of Charlotte Campbell being frightened by a carriage during a royal hunt in Naples provides insight into the ability of a mind educated in the processes of analogy to transpose everyday events into the realm of ideal beauty: "It offered a splendid vision to anyone interested in the quick, lithe motion of a beautiful figure. Enchanting movements can be seen in some dance steps, but what is comparable to this natural running, this agitation and turning, resolute and irresolute at the same time? Every movement was expressive and clearly showed her inner emotion and slim, youthful figure, with her clothes blown back against her body in her flight. Everything I had particularly admired in art in the past – grace, youth, the elusive figures of reliefs and the graceful dancers in the paintings of Herculaneum – I now saw in real life. No predetermined plan could have led to an execution as skilful as the one taking place there by chance. Everything played its part, both time and place".[26]

The young woman observed by Tischbein, the daughter of Lord Cawdor, Duke of Argyll, was the mother of the two young ladies portrayed dancing by Lorenzo Bartolini around 1820 with the freshness and spontaneity derived from Canova's *Graces*. In this case, however, the Apollonian aura always instilled into marble by the modern Phidias gives way to a realistic perception in accordance with the principles of nascent Purism. The two figures appear to be linked by the trusting affection reflected in their reciprocal gaze. The contact of their bodies is endowed with fragrance by the rhythm of the garments, which suggest forms of already ripened beauty. The dance step (Bartolini describes them in a letter as dancing the waltz in Edinburgh) has the casual elegance of an opening chord on a domestic piano.[27] The organic union of form and nature developed by Neoclassicism has now given way to the impression of a quick, additional narrative. To return

Carlo Finelli, *The Three Graces*, c. 1820
Marble, 158 x 119 x 67 cm
Rome, Galleria Francesca Antonacci.

Luigi Bienaimé,
Dancing Bacchante, 1846
Carrara marble, 149 x 120 x 65 cm
Rome, Galleria Francesca Antonacci.

to the theatre, a clear-cut and recognizable architectural form in line with the convictions of the Neoclassical culture has been superseded by free, sentimental action capable of tormenting the heart with the unpredictability of forthcoming developments.

[1] See *Civiltà del '700 a Napoli 1734-1799*, exhibition catalogue, edited by N. Spinosa (Florence, 1980), vol. 2, p. 73; A. Ottani Cavina, "Il Settecento e l'antico", in *Storia dell'arte italiana* (Turin, 1982), vol. 6, II, pp. 642–46.

[2] J. J. Winckelmann, *Geschichte der Kunst des Alterthums*; reference is made here to the Italian translation by F. Pfister, *Il bello nell'arte. Scritti sull'arte antica* (Turin, 1973), pp. 68, 70.

[3] L. Capitani, "Canova e la danza", in *Polittico*, no. 2, 2002, p. 105.

[4] *Del Laocoonte o sia dei limiti della pittura e della poesia. Discorso di G. E. Lessing recato dal tedesco in italiano dal cavaliere C. G. Londonio* (Milan, 1833), pp. 143–45.

[5] For Foscolo's interest in figurative art, see R. P. Ciardi, "La cultura figurativa di Ugo Foscolo", in *Rivista di letteratura italiana*, II, nos. 2-3, 1985, pp. 291–325.

[6] Capitani 2002, pp. 99–100.

[7] Ibid., p. 100.

[8] Ibid., p. 102.

[9] Ibid., p. 99.

[10] E. Raimondi, "Il coreografo perduto", in *Le pietre del sogno. Il moderno dopo il sublime* (Bologna: il Mulino, 1985), p. 150.

[11] G. Pavanello, "Canova e la Danzatrice", in *Canova e la danza*, exhibition catalogue, edited by M. Guderzo (Crocetta di Montello: Terra Ferma Edizioni, 2012), pp. 25–26.

[12] E. Raimondi, "Un teatro terribile: Roma 1782", in *Le pietre del sogno* 1985, pp. 20–21, note 4.

[13] A. D'Este, *Memorie di Antonio Canova* (Florence, 1864; anastatic reprint edited and with an introduction by P. Mariuz, Bassano del Grappa, 1999), p. 48.

[14] Capitani 2002, p. 101.

[15] Ibid., pp. 111–12.

[16] Constant reference is made here to the in-depth examination of these subjects carried out by Lucia Capitani in connection with research for her doctorate thesis in the history of modern art, *La bellezza in movimento. La scultura di Canova tra mimica, danza e recitazione*, Pisa University, academic year 2003–04.

[17] Capitani 2002, p. 101.

[18] Raimondi, "Il coreografo perduto" 1985, pp. 123–57.

[19] Capitani 2002, p. 104.

[20] L. Capitani, "Arti dello spazio e arti del tempo: Canova e la danza", in *Canova e la danza* 2012, p. 33.

[21] See *Canova. L'ideale classico tra scultura e pittura*, exhibition catalogue, edited by S. Androsov, F. Mazzocca and A. Paolucci (Cinisello Balsamo, 2009), p. 266; description by A. Imbellone.

[22] I. Teotochi Albrizzi, *Opere di scultura e di plastica di Antonio Canova* (Pisa, 1823; anastatic reprint edited and with an introduction by M. Pastore Stocchi and G. Venturi, Bassano del Grappa, 2003), vol. 2, p. 43.

[23] M. Missirini, *Vita di Antonio Canova. Libri quattro* (Prato, 1824; anastatic reprint edited and with an introduction by F. Leone, Bassano del Grappa, 2004), pp. 193–94.

[24] See *Canova. L'ideale classico* 2009, p. 228; description by F. Mazzocca.

[25] Capitani, *La bellezza in movimento* 2003–04. For the poetics of drapery, see L. Capitani, "Il panneggio come 'luogo' delle passioni. La scultura di Canova tra mimica, danza e recitazione", in *Polittico*, no. 4, 2005, pp. 77–101.

[26] Capitani, *La bellezza in movimento* 2003–04, pp. 132–33.

[27] See *Lorenzo Bartolini. Mostra delle attività di tutela*, exhibition catalogue (Florence, 1978), p. 32.

Lorenzo Bartolini, *Emma and Julia Campbell (The Dancers)*, before 1821
Plaster, 167 x 95 x 74 cm
Florence, Galleria dell'Accademia.

VOLATILE BALANCE. DEGAS, RODIN, BOURDELLE AND THE DANCING REVOLUTION

FLAVIO ARENSI

PREAMBLE

To Ettore

Towards the end of the Second Empire, Parisians witnessed a series of artistic scandals that looked forward to the upheavals that were to alter the European aesthetic vision definitively. They did so also on walking past the building site of the new opera house, where the recently installed sculptures spoke even to hurried and perhaps preoccupied passers-by. The human memory is however capable of storing the most demanding messages, sometimes quite unconsciously, and waiting for them to ripen. What is certain is that among the four blocks of marble at the base of the façade's twin projections, Jean-Baptiste Carpeaux's *Dance* (1865–69) must have offended some. A bottle of ink was hurled at it by some irate, unknown hand shortly after its installation in 1869 and others showed equal alacrity and indignation in petitioning Napoleon III for its removal. The sculpture was placed in the heart of the city on one of the buildings emblematic not only of an era but also of the nation's ambitious drive for cultural hegemony. Unlike other works exhibited in a whole variety of events of limited duration, *Dance* was permanently on display, by no means unobtrusive, and impossible to conceal short of resolute elimination.

In giving vent to his deep loathing of mankind in the last scene of Claude Chabrol's film *Masks* (1987), Philippe Noiret defines obscenity as showing what should be hidden. What did the Ville Lumière not want to see? It was not just a question of covering up nudity but rather the truth of the new realistic approach that agitated

Antoine Bourdelle, *Dance, Bas-Relief for the Theatre of Champs-Élysées*, 1912
Bronze, proof no. 3 executed by Susse in 1977, 177 x 150 x 27 cm
Paris, Musée Bourdelle.

the closely-woven fabric of moralism by overriding the contemporary frameworks of the relationship between the viewer and the work of art, between the subject and the creative pretext. The most vivid and popular example of these dynamics today is unquestionably Édouard Manet's celebrated *Déjeuner sur l'herbe*, exhibited at the Salon des Refusés in 1863, due to the wanton presence of the naked woman seated between two clothed gentlemen with no need for a story, as though it were indeed something quite normal, and the lack of effort to integrate the four figures into the setting of a landscape summarily sketched with complete disregard of the basic rules of perspective or depth as taught in the academies.

At this point, faced with an icon of such established aura, it must be recalled that history sometimes reveals little, hidden, parallel pathways overlooked by the authors of books and the restricted attention span of the general public. There is in fact not one history of art but a variety of histories depending on the observer's vantage point. The great and widely known masterpieces serve as an initial point of access to the intricate tangle of art. It is, however, also a delight, once inside this great forest where the paths disappear from view and then return, to focus some attention also on what Martin Heidegger called *Holzwege*, interrupted paths that offer the pleasure of straying off the beaten track (assuming that there is one). This makes it possible to work back a little way in search of the context that may not have deliberately inspired Manet's *Déjeuner* but

must have prepared the social and intellectual terrain for the masterpiece capable of jointly communicating his teaching in all its weight.

Presented at the official Salon in 1847, where it caused a great outcry in the press ably orchestrated by the critic Théophile Gautier, Jean-Baptiste Auguste Clésinger's sculpture *Woman Bitten by a Snake* was a twofold challenge with implications for art criticism and society alike. It shows a reclining female nude writhing in agony with a snake coiled around her wrist. As attested by the cellulite visible on the upper part of the thighs and reproduced on the marble, the sculptor made use of a cast of the model, the courtesan Apollonie Sabatier, known as "La Présidente", who was also a muse of Baudelaire. In other words, the veil of abstraction was removed in pursuit of authenticity. The use of a cast taken from life for a sculpture was a procedure that still aroused the wrath of critics in the nineteenth century, as attested by the fact that Auguste Rodin (1840–1917) had to defend himself and be defended against this shameful accusation over his *Age of Bronze* nearly thirty years later. Clésinger plays astutely on two controversial themes. In the first

Auguste Clésinger, *Woman Bitten by a Snake*, 1847
Marble, 56.5 x 180 x 70 cm
Paris, Musée d'Orsay.

top
Auguste Rodin,
The Age of Bronze, 1877
Bronze, h 115 cm
Lyon, Musée des Beaux-Arts.

Jean-Baptiste Carpeaux,
Dance, 1863–68
Original plaster,
232 x 148 x 115 cm
Paris, Musée d'Orsay.

place, the imprint transposed into marble does nothing to conceal but rather proclaims the name of his ambiguous model, thus looking forward to the habit of frequenting – or claiming to frequent – the *demi-monde* in which some artists of later generations delighted. Examples include Picasso's *Célestine* and the brothel scenes of Toulouse-Lautrec. Though formulated by a sculptor who still pursued firm, classical features (maintained in the moulding of the head), this injection of realistic eclecticism shifted the point of observation from the sphere of ecstatic beauty to a more ordinary and less ideal level verging on plausibility.

Our interrupted paths take us back at this point to the scandal over Carpeaux's public work of sculpture. In the same year as Manet's *Déjeuner*, Charles Garnier, the architect who designed the new Théâtre de l'Opéra, commissioned four winners of the Prix de Rome to produce sculptural groups for the façade of the building. The excellent sculptor from Valenciennes was assigned the subject of dance, which he examined in depth in numerous studies, sketches and models for a good three years. At the end of long creative gestation, he produced an allegorical sculptural complex with women whirling in a ring around the genius of dance, vertical dynamics being employed for the male figure and horizontal for the female dancers. His primary concern was to suggest the idea of movement, immortalizing the festive group in an instant of realistic, playful dynamism. The outcry caused by the female nudity was, however, enough to distract attention from the innovative element of the sculpture, namely its rhythmic balance.

THE AGE OF BRONZE

The year 1875 marked the fourth centenary of the birth of Michelangelo, who had been the object of large-scale rediscovery for decades. It was his physicality that many artists harnessed to revitalize in their own work. Installed on the highly symbolic monument of the Arc de Triomphe halfway through the 1830s, François Rude's *Marseillaise* thus drew on the master for its vibrant and energetic manifestation of virility. Antoine Quatremère de Quincy's *Histoire de la vie et des ouvrages de Michel-Ange Bonarroti* (1835) enjoyed a vast readership. Eugène Delacroix used Vasari as a source for two essays on the Italian artist and painted *Michelangelo in his Studio*. Charles Calemard de Lafayette's *Dante, Michel-angel, Machiavel* was published shortly afterwards. The young Carpeaux copied plaster casts of *Day* and *Night* from the Medici Chapel for his master in 1850 and set off five years later for a long stay in Italy, returning with a series of formidable impressions that can be seen in his *Science* (1863–66) and the excellent *Ugolino* (1860). The Renaissance genius was esteemed not only as an artist but also as a poet due to the publication in Florence of his sonnets edited by Cesare Guasti, not to mention the admiration for him expressed by eminent intellectuals in numerous books on the Italian stage of the Grand Tour. Everyone was overawed by Michelangelo, but the most demanding artists were impressed not only by the exceptional sculptures and the paintings in the Sistine Chapel but also by his technical methodology, ideas and character. He had a particular influence on two young contemporary artists, namely Edgar Degas (1834–1917), who started out by copying the *Last Judgement* and the Louvre *Dying Slave*, and even worked on a reproduction of the *Battle of Cascina*, and Rodin, who celebrated the master in 1864 with a *Mask of a Man with a Broken Nose* inspired by the death mask of Michelangelo by Braghettone (Daniele Ricciarelli). It was in 1876 that Rodin made his first long trip to Italy, passing through Turin,

Genoa, Pisa, Siena, Florence, Rome and Naples and then returning to spend another day in the Tuscan capital before quickly visiting Padua and Venice. His study of the ancient masters focused above all on the sculptors Lorenzo Ghiberti, Donatello and of course Michelangelo, gaining insight into the master's philosophy and compositional intentions as well as the key significance of the *non-finito* work, which he developed in wholly modern terms to the point where it became a completely autonomous principle. Degas also studied Michelangelo's "unfinished" work but in terms of a rough draft and soon chose to escape from any subordinate position by offering an alternative to Michelangelo/Rodin as from 1880. Curiously enough, it was in the same year that Rodin exploded in a series of masterpieces that altered the very concept of sculpture. And I use the verb "explode" deliberately, as the French artist, learning from ancient art in the same way as contemporaries like Carpeaux and Medardo Rosso, introduced into sculpture a series of germs that opened up to the twentieth century and broke away definitively from the past. I refer to elements that all the great avant-garde masters took into consideration, as they regard not only the method of sculpture but also and above all its theoretical concept, introducing reflections that concern the materials of the profession and its practice. Rodin is an erupting volcano, and only the inhabitants of volcanic areas can fully understand what this means. Even though it is hard to conceive of any parallel with other artists, especially Medardo, who took a different approach, Rodin's revolution laid the foundations for a new aesthetic grammar. After his trip to Italy, the artist returned to Brussels, where he had already been working for numerous burgomasters for six years, and continued his work on the male nude of *The Age of Bronze*, begun in the October of the year before. The plaster was shown in Brussels at the Cercle Artistique in January 1877, but an article appeared in *L'Etoile belge* claiming that the artist had cast a living model. Rodin presented photographs and profile studies to refute the charge and distinguished figures like Dubois, Falguière, Carrier-Belleuse, Chapu, Chaplain, Thomas and Delaplanche testified to his innocence. Even though he succeeded in convincing critics of the quality of his work, Rodin was so annoyed that he stopped making life-sized sculptures while continuing to obey the principle that the artist must always take nature as his guide with the utmost sincerity, a fundamental tenet from which the master did not deviate even for the most daring works. He focused relentlessly on truth even in commissions for portraits, where he refused to comply with his clients' unrealistic wishes and sometimes magnified their flaws, choosing to show what they would have preferred to keep hidden. While *The Age of Bronze* has a solid structure, its centre of mass rests on an elegant classical balance, which is precisely what Rodin soon called into question in adopting an extreme interpretive stance drawn from his Italian experience.

SCIENTIFIC BALANCE, EMOTIVE IMBALANCE

The careers of Degas and Rodin ran parallel to one another. Curiously enough, while the former kept his sculpture private, confined exclusively to the studio, the latter concealed his paintings from the public. Apart from the anatomical exercises of the 1850s and the family portraits of the 1860s, the most interesting work dates back to the 1870s, when the sculptor captured the enchantment of the forest of Soignes just outside Brussels, which he portrayed as a magnificent structure comparable to a cathedral.[1] The series amounted to just over thirty paintings in oil on paper, glued onto cardboard after the artist's death, which he never sold or parted with. His painting is totally independent of his sculpture, serving neither as preparatory studies nor as compositional investigations into the rendering of volume and surface, and displays marked freedom of expression due to the lack of academic preparation, as is also the case with his sculpture.

While numerous visitors to Degas's studio mention his sculpture, the sole public showing was at the Impressionists' sixth group exhibition on 2 April 1881, which included his *Little Fourteen-Year-Old Dancer*, begun some years earlier. Though somewhat complex, the history of the painter's sculptures has been largely reconstructed. Suffice it to mention here that the originals, cast in bronze after his death, were a combination of wax and modelling clay on a metal frame with pieces of string, cork and sticks found in the studio. In a certain sense, the casting by the Italian Albino Palazzolo preserves their otherwise finite existence despite disobeying the express wishes of the artist, who envisioned them as works of ephemeral fragility. Moreover, as a result of questionable conduct on the part of the Hébrard foundry and Palazzolo, a considerable number of unauthorized bronzes came onto the market in the 1950s. (Sara Campbell's catalogue of 1955 gives an estimate of about 1,450 castings of 74 models.) Apart from all the attendant implications, the presentation of the artist's fourteen-year-old dancer is indicative of his aesthetic as a whole. When the wax figure in a tutu was shown, the public and critics were shocked by the highly realistic details and posture. Some were indeed reminded of Anna Maria "Marie" Tussaud, the sculptress from Alsace who left Paris early in the nineteenth century for London, where she founded the famous and still active waxworks museum. With a view to brutal emphasis, Degas placed the statue in a glass case (having presented an empty showcase at the Impressionists' fifth exhibition the year before), as he had seen done dozens of time with sculptures in the Louvre, especially an almost life-sized, polychromatic Etruscan sarcophagus studied in various pencil sketches. The painter would appear to have realized before Marcel Duchamp that the artwork involves not only representation but also presentation. At the same time, the expedient of the showcase makes it possible to accentuate the quasi-scientific intention to examine the dancer "as an ethnologist". Once again, the sheer realism disoriented critics and their attacks probably convinced Degas against presenting any more works of plastic art. In any case, it appears clear that he was intent on bringing everyday life into the heart of his three-dimensional work to perhaps a still greater extent than in his paintings. As from the end of the 1860s, Degas produced a constant stream of small figures that the Brazilian scholar Luiz Marques relates to Poussin, seeing in both cases "modelling as a prerequisite of painting, a moment of negation of the pictorial language – a negation that the painting incorporates in the progressive realization of its own truth".[2] In addition to the two primary subjects of horses and dancers, there were some portraits and a handful of other works, including the magnificent *Apple Pickers* relief of 1890, where the figures are barely outlined but endowed with great dynamism, and the woman bathing. While the reproductions executed for the first catalogue drawn up after the master's death still exist, it is a pity that we can only imagine the original models, kept standing by a series of minute and highly Spartan but fascinating engineering expedients which accentuate the impression of snapshots that all the images convey in the suspension of perfect balance wherein everything is held together. They are like stills of a film shown frame by frame, and it is no coincidence that the horses derive from the celebrated experiments of the English photographer Eadweard Muybridge, who used stop-motion photography to study the motion of animals and people. Degas developed in

Edgar Degas, *Little Fourteen-Year-Old Dancer*, 1921-1931 (original work 1865–81) Statue in bronze with patinas of various colours, tutu of tulle, pink satin hair ribbon, wooden base, 98 x 35.2 x 24.5 cm Paris, Musée d'Orsay.

sculpture a sort of investigation of motion and stability involving the harmony of form, which acts as an analytical probe.

Rodin also found the horse a formidable narrative stimulus. The equine form sometimes merges with the human in his drawings and the animal acts as a catalyst for speed of a swirling character. While an interest in the technique was transmitted to him by Pierre Leuset, "an old painter" of animals and friend of his father, the interest in animals as such, unconnected with other subjects, remained secondary, as attested by a few studies and a fairly negligible small oil painting of a motionless horse in the Saint-Marcel market near the Gobelin factory. A key part in the young Rodin's training was instead played by attendance at the École Impériale Spéciale de Dessin et de Mathématiques (Petite École) and acquaintance with the technique of Horace Lecoq de Boisbaudran, which was based on the practice of drawing from memory. This device enabled Rodin to develop his powers of observation and employ a sort of mnemonic paraphrase in order to reconstruct the original after reworking it mentally, as he did with the work of the great Italian sculptors, especially Michelangelo, on his return from Italy. The clearest evidence of this is provided by the long and intense work that led up to the creation of his *John the Baptist* (1880), in which "it is impossible not to see a representation of the artist, a man who walks and preaches in the wilderness".[3] The saint's pose is wholly unbalanced, the muscles of the arm pointing upwards are taut, and everything seems to be swallowed up in the emotive imbalance that was henceforth to characterize much of Rodin's work, where it is not perfection or fixity but the moment immediately before or after stasis that becomes a lyrical hallmark. While Degas saw photography as an allied instrument and practiced it personally as a way of capturing the fleeting moment and freezing movement, Rodin regarded it as a subordinate means serving for the two-dimensional correction of forms arising from sculptures, experimentation with settings to give them context and testing the effects of light and shadow.

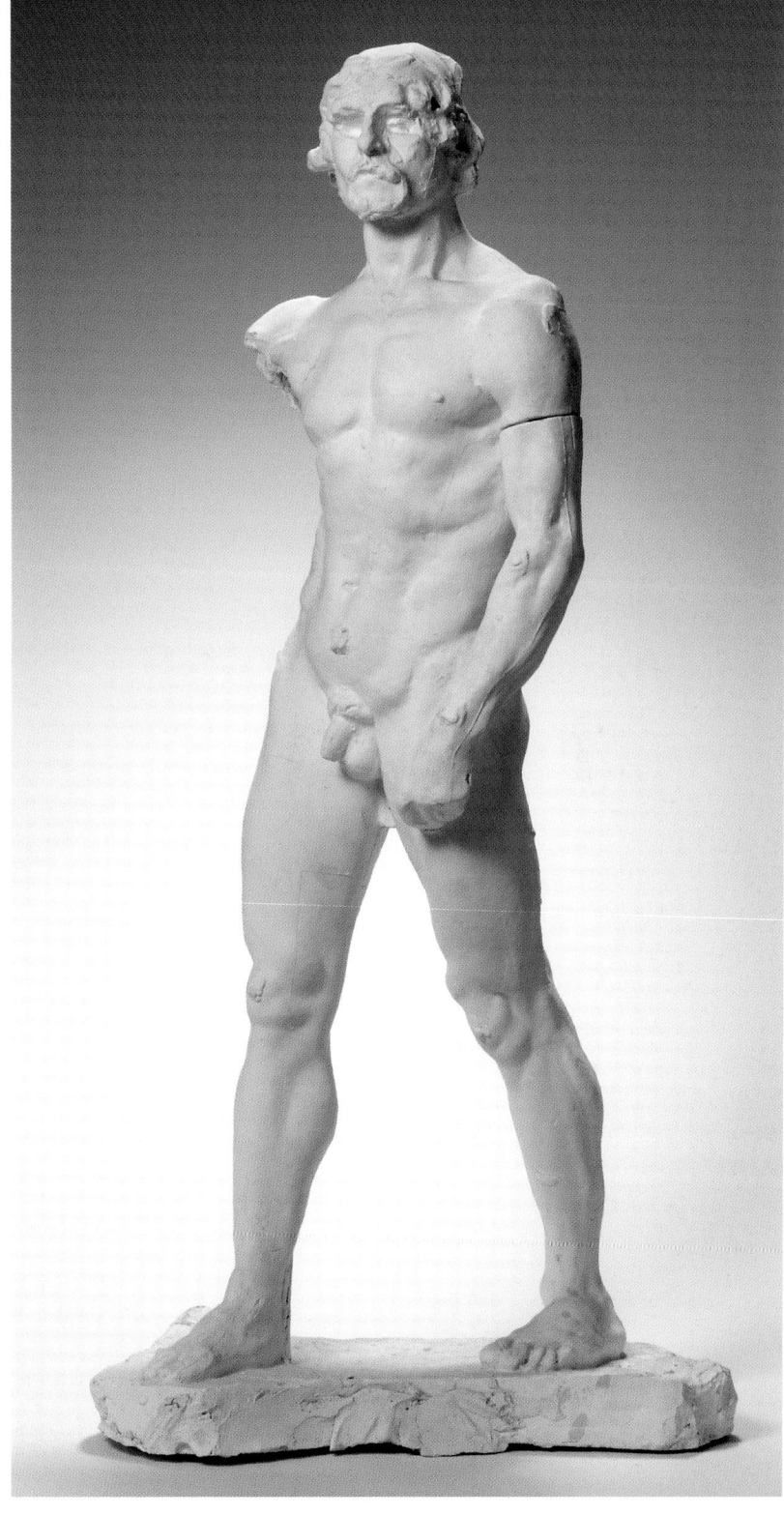

Auguste Rodin,
John the Baptist, 1878
Plaster, 97.3 x 45.5 x 24.7 cm
Paris, Musée Rodin.

Edgar Degas, *Dancer, Grande
Arabesque, Third Time*, 1921–31
(original work 1882–95)
Bronze (lost wax casting),
28.2 x 43 x 21 cm
Paris, Musée d'Orsay.

Edgar Degas, *Dancer, Arabesque over the Right Leg, Left Arm in Front, Second Study*, 1921–31
(original work 1882–95)
Bronze (lost wax casting),
29 x 39 x 14 cm
Paris, Musée d'Orsay.

He could not, however, be other than conceptually opposed to photography because he never sought the surgical coldness of fixity that eliminates naturalness. Rodin stated his views very clearly in a conversation with Paul Gsell on movement in sculpture: "The artist is honest and it is photography that deceives. Reality does not halt in time, and if the artist succeeds in producing the impression of a repeated act, his work is certainly far less conventional than a scientific image where time is halted abruptly. And this is also what condemns those modern painters who reproduce the poses supplied by photography in order to represent galloping horses".[4]

For the sculptor, movement is nothing other than "transition from one attitude to another".[5] If this attitude is now mature in the *John the Baptist*, it becomes more explicit in the poses of the *Burghers of Calais* (1889) and the extraordinary *Walking Man* (1907), often regarded as the symbol of pure creation finally freed from the weight of the subject, or more simply as the very image of motion. With the *Call to Arms* (1879) and *Gates of Hell* (1880–90) Rodin entered new territories where balance and imbalance are prior to the soul and the flesh. While remaining faithful to nature, he savoured and allowed others to savour pure existence, the absolute romantic impulse, Neoplatonic vitalism. Degas, born seven years earlier, was instead tied to reason. He led the public by the hand into the middle of scenes in which habit and the less noble actions prevail with an almost banal development that is, however, rigorously dissected by the eye. He mingled with the dancers and the women ironing, in some respects predating the normality of the Pop Art of Andy Warhol. As Baudelaire observed, he gave us "modern beauty and heroism in everyday life".[6] Degas therefore espouses the "abrupt" halting of time and examines it through the lens of the microscope. And it is everybody's time.

ARCHAIC DANCE

Degas's dance was the classical ballet taught in the old Parisian schools, with its self-imposed method of study to define the body and movement within an atmosphere of meticulous care that certainly does not conceal the athletic difficulties. He began to frequent the Opéra ballet school in 1872, but had already included some dancers in the background of a painting of some musician friends four years earlier. His objective was to examine the dynamics of a tiring, mechanical exercise through a scientist's eye. Rodin was instead interested in female Eastern dancers, like the Javanese performers portrayed during the Exposition of 1889 and the Cambodians to whom he devoted about 150 watercolours in the space of a few days in 1906. He also produced a series of nine small dance movements in 1911, going beyond the purely objective examination of Degas. As also happens with *Iris, Messenger of the Gods* (c. 1895), the bodies are opened out and contorted in extreme, frenzied poses derived from his acquaintance with the acrobatic Spanish dancer Alda Moreno of the Opéra Comique, the subject of over fifty drawings in 1910. These small sketches appear to have "happened" rather than been formed. The French art historian Aline Magnien traces Rodin's later strategy back to a childhood event recalled in a conversation with the American journalist Truman Howe Bartlett in 1887. Rodin's mother made him some oddly shaped pancakes when he was five years old and he was allowed to make little human figures of batter to fry. The extraordinary movements made by the pancake men were indelibly imprinted in his memory. Bartlett suggested that the shape and movement of the figure were fundamental elements in Rodin's art.[7]

Attracted by the new ballet of Loïe Fuller, Isadora Duncan and Vaslav Nijinsky, Rodin was in the audience at the Théâtre du Châtelet on 29 May 1912 when the Ballets Russes gave the first public performance of Debussy's *L'Après-midi d'un faune* with Nijinsky as the lead dancer and choreographer. The great étoile experimented with

Auguste Rodin, *Study for the Torso of The Walking Man*, 1878–79
Plaster, 52.2 x 25 x 17.8 cm
Paris, Musée Rodin.

a new scenic concept, reinterpreting the figures on an ancient Greek urn and requiring the *corps de ballet* to cross the stage in profile. The style of the ballet, in which a young faun meets, flirts with and pursues various nymphs, is deliberately archaic. The movements required by Nijinsky were, however, wholly extraneous to the dancers' repertoire and they found it hard to meet his demands. Ida Rubinstein gave up her leading part after the first session of rehearsals. The steps were complicated still further by the music, with which Nijinsky entered into an intricate relationship, using it solely as a sort of background to scenes where the dancers' movements became increasingly violent and discontinuous. It was only after numerous repetitions that the nymphs were able to dance fluidly with no awkwardness. The choreographer, who performed the leading role of the faun, instead confined himself to creating a dance with the feet set firmly on the ground, eliminating leaps and pirouettes. The break with the classical tradition and technical expertise to which the Russian professionals were accustomed was inevitable and evident. While critics and audiences once again failed to grasp the value of the new approach, Rodin championed the production and defended it in public. Nijinsky visited him on 31 May to express his gratitude but noted in his diary that Rodin had found fault with his body.[8] The sketches made during that meeting, which were long thought to have been lost, served as

Auguste Rodin, *Iris, Messenger of the Gods*, c. 1890–91
Plaster, 41.7 x 46 x 22 cm
Paris, Musée Rodin.

preparatory studies for a small sculpture made from memory of the young dancer leaping like a flame launched into space. Paradoxically enough, Nijinsky reminded sculptors of the aesthetic value and potential of the body understood as the construction of forms in which the *leçon de l'antique* (as he entitled his essay on classical art) takes on renewed relevance if developed within a rereading that subjects forms to a dynamics of abstraction. This was understood with great wisdom by Antoine Bourdelle (1861–1929), Rodin's favourite pupil, who developed the master's approach into monumental and architectural sculpture drawing upon two-dimensionality in order to regain the connection between form and the space occupied by mass.

Completed in 1913 to plans by the French architect Auguste Perret, the Théâtre des Champs-Élysées was intended by the journalist and impresario Gabriel Astruc to serve as a venue for contemporary music, dance and opera in sharp contrast with the more conservative institutions. It was indeed innovative even in the use of reinforced concrete, albeit with a marble facing. The new theatre hosted the Ballets Russes on 29 May 1913 with the world premiere of Stravinsky's highly controversial *Sacre du printemps*, choreographed by Nijinsky. The work features a pagan rite of sacrifice in ancient Russia at the beginning of spring, when an adolescent is chosen to dance herself to death in order to propitiate the gods and secure their benevolence for the new season. The two Russian artists created a situation that looked beyond modern civilization in search of its barbaric and primitive origins, where the order of traditional acts and forms gives way to the staging of an ancient ritual that ends in violence. Commissioned to create the metopes for the façade of the new theatre, Bourdelle immediately recalled the potential of the new artistic code as understood by him since the first performances of Isa-

dora Duncan, whom he saw in Gluck's *Iphigenia in Aulis* at the Châtelet in 1909. The resulting reliefs were conceived as running parallel to the wall and limited in their development so that "no gesture, plane, shadow or ledge should impair the smoothness of the wall",[9] clearly subordinating sculpture to architecture. The one entitled *Dance* (1912) suggests a strongly rhythmic choreography in which we can identify the artist's beloved Duncan (right) "with her delicate head tilted back … to dance within her pure emotion"[10] and Nijinsky (left) "in a wild impetus … imbued with the winged genius of birds".[11] Bourdelle's mature sculpture is endowed with rhythm enclosed in perfect geometry where the story develops with fluidity and concision. At this point

Auguste Rodin,
Dance Movement F, c. 1911
Plaster, 26.8 x 26.2 x 14.5 cm
Paris, Musée Rodin.

Auguste Rodin, *Nijinsky*, 1912
Plaster, 17.5 x 10 x 6 cm
Paris, Musée Rodin.

Ertruscan art, *Bronze of Hercules*,
fourth century BC
Cast bronze on wooden base,
24 x 12 x 12 cm
Soprintendenza per i Beni
Archeologici della Toscana –
Florence, Museo Archeologico
Nazionale.

it is no longer necessary to follow nature, albeit taken to extremes as in Rodin, but rather to take a further step that manifests itself in conceiving the model as governed by the architectural laws within whose boundaries the corporeal structures are oriented. As from the early years of the twentieth century, the sculptor sought a kind of organization that verges upon abstraction and turns to the past in order to identify systems capable of meeting the cogent requirements of the new times. Bourdelle mixes the hardened simplification of the hieratic iconography of the cathedral with the narrative fluidity of classical Greece, finding in archaism the premises for progress with respect to the trail blazed by Rodin, overcoming its limitations and at the same time adopting the framework of Greek culture as regards proportions and structural harmony (primary importance attaches in this connection to his relationship with his second wife, the model and sculptor Cléopâtre Sévastos).

It was in 1909 that the artist completed his *Head of Apollo*, begun in 1900, the key work of his career heralding the conquest of the new style. The head was conceived as a study when he was still a pupil and assistant of Rodin but already aware of the need to abandon the master's romanticism and follow a different path. This growth was certainly not painless and entailed a long period of reflection. On being shown the work, Rodin "was very shaken. He understood at once that I had broken away from him definitively and never forgave me".[12]

Perhaps shocked by Rodin's reaction, Bourdelle spoke of "drama" and the "isolation of sculptural thought", but could only accept the rift. Where Rodin accentuates shadows and projections, captures light and exaggerates muscles, he instead erects a whole of simplified form and line. The artist also attributed the head of Apollo an autobiographical dimension: "This sculpture is the drama of my life, one part completed and the other still in the phase of study. Prey to anxieties, austere, free from the past, whatever it may be, and from every contemporary contribution".[13]

Etruscan art, *Etruscan Red-Figure Stamnos*, fourth century BC
Ceramics, 30.7 x 30 cm
(Ø base 13.5 cm)
Soprintendenza per i Beni Archeologici della Toscana – Florence, Museo Archeologico Nazionale.

Antoine Bourdelle, *The Cello Player* (or *Music*), 1914
Bronze, proof no. 5 executed by Susse around 1990, 30 x 5.8 x 6 cm
Paris, Musée Bourdelle.

Antoine Bourdelle,
Dancing Isadora, n.d.
Pen and purple ink on tracing
paper, 25.8 x 20.1 cm
Paris, Musée Bourdelle.

Antoine Bourdelle, *Isadora*, [1909]
Pen and purple ink on tracing
paper, 22 x 14.2 cm
Paris, Musée Bourdelle.

L'HYMNE AU DELÀ DE LA VOIX
LES FORMES AU DELÀ DU CHANT

Antoine Bourdelle, *Isadora,*
Hymn beyond the Voice, Forms
beyond the Song, c. 1920
Pen, black ink and watercolour on
tracing paper, 27.4 x 20.8 cm
Paris, Musée Bourdelle.

Antoine Bourdelle, *Isadora,*
Movements Remaining
in My Memory, c. 1920
Pen, brown ink and watercolour on
tracing paper, 27.4 x 20.8 cm
Paris, Musée Bourdelle.

Antoine Bourdelle, *Nijinsky in the Role of Harlequin, the Carnival*, c. 1911
Pen and brown ink on tracing paper, 24.1 x 16.4 cm
Paris, Musée Bourdelle.

Antoine Bourdelle, *Nijinsky in the Role of Harlequin*, c. 1911
Pen and brown ink on tracing paper, 25 x 16.4 cm
Paris, Musée Bourdelle.

Roman art, *Bronze of Dancing Maenad*, fourth century BC
Cast bronze on wooden base, 18.5 x 6.5 x 6 cm
Soprintendenza per i Beni Archeologici della Toscana – Florence, Museo Archeologico Nazionale.

The work remained hidden for another decade before the author authorized reproductions despite the fact that it had marked a turning point for his art from the very outset: "one of my first works, one of those that to my eyes began to express what I really wanted to say".[14]

Realism was by now an outmoded issue and anti-naturalism a necessity. In 1909, the year of the Futurist Manifesto, Picasso produced the gold-painted plaster sculpture *Female Head (Fernande)*, which was then echoed a few years later in the portrait *Anti-Graceful* (1912–13) by Umberto Boccioni, painted while the Metaphysical Art of Giorgio de Chirico was establishing itself in Paris. Bourdelle hastened his detachment from the burdensome mastery of Rodin and marched towards new horizons through the same syntax sensed by Nijinsky. As he wrote, "Everything that is synthesis is archaism. The archaic is the opposite of the word copy and the sworn enemy of falsehood, of all this stupidly atrocious art of *trompe-l'œil* that turns marble into a corpse".[15]

THE END AND BEYOND

Bourdelle's work always expresses travail, a commitment also in conceptual terms that shines through in the teaching he began in 1900, when he founded a school with Rodin and Jules Desbois, and continued at the Académie de Grande Chaumière as from 1909. One of his pupils was Henri Matisse, introduced by his brilliant friend Eugène Carrière. Matisse immediately understood the master's theoretical precepts and clearly expressed them in his sculpture after an initial phase under the influence of Rodin. More than half of his work in this field took place between 1900 and 1909 with a particular focus on the human figure, as exemplified by *The Serf* (1900–04), where

the modelling recalls Rodin's *John the Baptist* while already displaying Bourdelle's influence in the drive for synthesis. Bourdelle influenced not only his sculpture, as the nude relief *The Back* (1916–17), but also his painting in the use of a stylized architectural or geometric setting. In this sense, the two versions of *Dance* (1909 and 1909–10) and *Music* (1910) opt for rhythm and balance, taking Bourdelle's archaism as far as possible. *Dance* in particular shows a joyful ring of dancers, like the one placed by Carpeaux on the façade of the Opéra, but devoid of any affectation. It is not known whether Matisse drew inspiration from the sculptural group, or indeed whether Picasso did for his *Three Dancers* (1925), where the central figure recalls the Opéra genius of dance and the bursting energy of the three figures recalls the festive abandon of Carpeaux's marble. What is certain is that the Spanish artist represents an amorous triangle in this work, and the links between the figures conceal the tragedy of a relation-

Henri Matisse, *Dance*, 1910
Oil on canvas, 260 x 391 cm
Saint Petersburg, The State
Hermitage Museum.

ship in which feelings, sex and death go hand in hand, transposing a playful iconography into a tragic metaphor. In an exceptional work of 1988 entitled *Musique et extase. L'audition mystique dans la tradition soufie*, the ethnomusicologist Jean During points out that what matters is not what you listen to but how you listen, who listens and with what intentions. This holds for all the arts and especially the visual ones, which occupy physical spaces from which they constantly address viewers, including the involuntary or distracted ones who fail to notice their presence. Nevertheless, the works continue to convey a message that penetrates the mind and lingers there. What counts is their silent communication, the constant effort to impart a message that crosses the barriers of ordinary language, as the great masters capable of reaching our innermost depths are well aware. The fact that between 1869 and 1913, a span of forty-four years, two Parisian theatres displayed as many sculptural elements, each of which regarding a precise way of seeing not only art but the very structure of society – harnessing the artistic vocabulary in order to translate existential and theoretical matters – provides curious proof that art history is not something confined to dusty tomes but a living experience. Separated by less than half a century, Carpeaux's *Dance* and Bourdelle's are the boundaries within which an upheaval took place in an aesthetic and collective canon and then continued in neighbouring but unforeseen directions, where synthesis became a story-telling element gradually converted into an exclusive narrative concept. On the eve of two wars, the Franco-Prussian War of 1870 and the First World War, the Ville Lumière underwent a change in awareness through a felicitous combination of the codes of music and the plastic arts to discover that balance, rhythm, imbalance, archaism and architecture were the terms of a new vocabulary of the soul and the mind. And, above all, the heart.

[1] B. Musetti, "Ceci n'est pas un peintre. Qualche riflessione intorno alla produzione pittorica di Auguste Rodin", in A. Magnien and F. Arensi, *Rodin. L'origine del genio*, exhibition catalogue (Turin: Allemandi, 2010), p. 70.

[2] "Mechanism and Classical Tradition in Degas' Bronze", in J. S. Czestochowski and A. Pingeot, *Degas sculptures: catalogue raisonné of the bronzes* (New York: Torch Press, 2002), p. 111.

[3] A. Magnien, "Le prugne e le frittelle. Ricordi di gioventù di Auguste Rodin", in Magnien and Arensi 2010, p. 22.

[4] *L'Art, entretiens réunis par Paul Gsell* (Paris: B. Grasset, 1911), p. 87.

[5] Ibid., p. 76.

[6] C. Baudelaire, *L'Art romantique*, 1869.

[7] T. H. Bartlett, *Auguste Rodin, Sculptor* [1889], quoted in Magnien 2010, p. 17.

[8] A. Le Normand-Romain, *Rodin et le bronze. Catalogue des œuvres conservées au Musée Rodin* (Paris: Éditions du Musée Rodin / Éditions de la Réunion des Musées nationaux, 2007), vol. 2, p. 551.

[9] V. Gautherin, "La Danse", in *Bourdelle*, edited by B. Manara (Rome: De Luca, 1994), p. 160.

[10] Ibid.

[11] Ibid.

[12] A. Bourdelle, *Apollon au combat*, typescript, Archives of the Musée Bourdelle, Paris.

[13] Ibid.

[14] Ibid.

[15] A. Bourdelle, *Ecrits sur l'art et sur la vie illustrés de dessins de l'auteur*, edited by G. Varenne (Paris: Editions d'Histoire et d'Art, Librairie Plon, 1955), p. 67.

DIVINE EQUILIBRIUMS. TERPSICHORE AND MERCURY

ELISA GUZZO VACCARINO

Hidden foot, visible foot; shod foot, bare foot; natural foot, artificial foot; poetic foot, prosaic foot. What seems to be born from the foot, which touches the chosen ground of the Court, or the sacred one of the Temple, to then trod upon the magical axes of the stage, is all the dance of the world. Hidden in the foot, in the relationship that is created when it touches the floor, from which it draws and with which it exchanges telluric energy, is the mystery of dance, in both its material and spiritual essence, the miracle of an expression of the self that is thought to be unnecessary to the survival of the bipedal species that colonized planet Earth. And yet there is no place or time without dance, from the caverns to the savannahs, from the forests to the cities. And always starting from the human foot. Concentrated therein is the certainty of the essential link with our planet's force of gravity; materialized therein is the desire to go beyond that gravity by taking aerial leaps and gathering strength from the marvellous spring that is the foot. To our Western eyes, the bare foot secretly vibrates with the subtle fragrance of scandal, as it reveals the person's intimacy. The foot's nakedness, on which the twentieth-century revolutionary modern dance focused with the overt intention of breaking away from the past, is usually dressed up on stage with various types of footwear-coverings, aesthetically coherent with the different dance genres, whether academic or contemporary, cultured or light, pop-folkloristic or ballroom.

As a countercheck, even in the Far East the bare foot is shod and decorated: in classical Indian dancing the contour of the foot is painted bright red with a red liquid dye known as Alta, so as to focus attention on the footsteps. The hands' fingertips are also painted for the purpose of emphasizing the gesture.

Rudolf Nureyev performs in George Balanchine's *Apollo* at the Uris Theatre in New York, 25 January 1975.

p. 144
Joan Miró, *Spanish Dancer*, c. 1960
Woollen tapestry, 200 x 152 cm
Albenga, Collection of Galleria d'Arte Moderna.

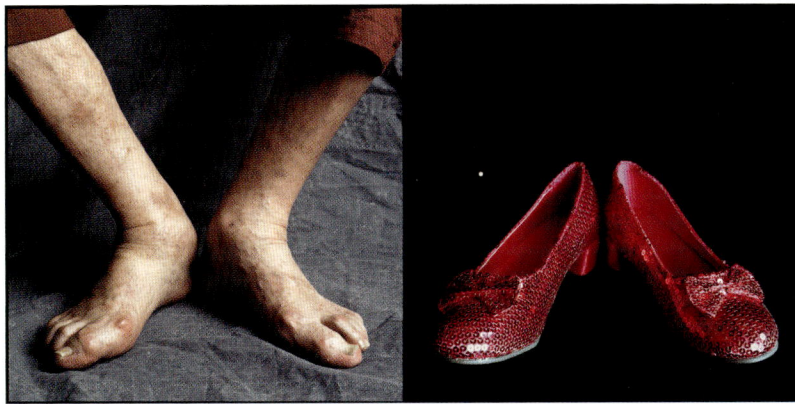

It has often been said that Pina Bausch,[1] a late-twentieth-century icon of expressive dance, in her spectacular *Stücke* lent a voice to the people and not to the characters. At the heart of her Tanztheater, she would have her women dancers – dressed in gorgeous evening outfits, yet worn creased and sideways – remove their stiletto heels, which sensuously alter the line of the leg, emphasizing femininity like a pedestal, as the object of male desire. Indeed, removing the heels denies a gender identity imposed by the eyes of someone else, and more or less willingly accepted and assimilated by seductive narcissism, whether biological or induced. Meanwhile, the Polish-German choreographer preferred to present men in everyday shirts and trousers, normal ones, but shoeless, unveiling in the direct contact with the ground their hidden fragility, without protection from the harshness of life and the public's insinuating and voyeuristic gazes.

In the foot, in its shape and its materialness lies a truth of the heart that cannot be hidden: this is the message that underlies so much of the stage dance that was invented in the past century.

After all, seeing the bare, worn and often painful or wounded foot of Rudolf Nureyev, "the flying Tatar", the brightest star of the period following the Second World War, or that of Antonio Gades, the late-twentieth-century genius of theatrical flamenco, or that of Mikhail Baryshnikov, who successfully crossed every possible expression of dance with unique lightness, or that of Sylvie Guillem, "the" ballerina of the passage to the new millennium, almost corresponds to a secret ceremony of the display of a "working tool" made of flesh and blood, soaked in love and hate – a ceremony to

which we have no right to be admitted. Only accredited photographers – beginning with the phenomenal French ballet dancer, who imposed a new physicality and a new image upon all the others of her generation and those to come – were able to snatch a few emotional details, laden with a private humanity to be concealed.

Indeed, an artist of the shoe like Salvatore Ferragamo, in his innate wisdom and in his mastery cultivated with determination, was fully aware of the fact that a foot can tell you a great deal about the personality, character, experiences – the soul, one might say – of each individual, in his or her uniqueness.

If all this shines through in the quality of a person's posture and gait, bearing in mind the morphology, the clothing, the natural and cultural motor habits, we need to emphasize that, in their particular case, the dancer and even more so the academic ballet dancer *en pointe*,[2] lifted up on the high arch of the foot that is indispensable to this art, have an especially intense relationship with their feet, a relationship that is analysed and perfected day after day, overcoming fatigue and torment, actually turning them into the pleasure of gloriously appearing in the limelight. It is the indomitable demon of dancing that a film like *Red Shoes* revealed even to the most unwitting audiences. And Dorothy's glittering, magical shoes in *The Wizard of Oz* are red as well: by wearing them and tapping her heels she can immediately take off

Roman art, *Bronze Foot*
Cast bronze, 11 x 21.5 x 13 cm (maximum amount of space)
Soprintendenza per i Beni Archeologici della Toscana – Florence, Museo Archeologico Nazionale.

Timothy Greenfield-Sanders,
Merce Cunningham Feet and Ruby Slippers, 2013
Diptych, digital print,
90 x 90 cm (each panel)
Courtesy Timothy Greenfield-Sanders, New York.

for wherever she wishes. It is within the shoe that dreams come true, as we know from many tales told about the seven-league boots. Indeed, wearing this same kind of tall and very soft boots Georgian male dancers hop on the tips of their toes: they don't seem to suffer at all as we admire their virile bravura. They probably honed this particular real-man skill because they had to learn how to prudently face the narrow and dangerous mountain trails of their land. The control over their feet is the source of their survival.

Another case in point, on the subject of the recondite meanings of the foot: Cinderella's stepsisters – indeed a universal tale[3] translated into a ballet to the music of Sergei Prokofiev in the twentieth century – would have made any sacrifice to be able to fit into the tiny glass slipper lost by the unknown beauty who had fled from the princely ball at the tolling of the midnight hour. The earliest version of the tale – which seems to come from China, where the feet of young girls (known

as Golden Lotuses or Golden Lilies because of their short, wavering footsteps) were bound to prevent further growth, cruelly folding the toes under the sole – envisioned the cutting off the big toe in order to fit into the tiny shoe, symbolizing the way to happiness, wealth, love, i.e. all that every beautiful, good girl aspired to by right (they weren't so good when they deceived a man and attacked their rival to get what they wanted).

And, on his part, a master of the classical milonguero-style tango (that is, with the traditional close embrace and the mindful maintaining of the line of dance), the Argentine Carlos Gavito, who passed away in 2005, never forgot to show the role of the foot in this dance of spiritual, sentimental, erotic communication, declaring that "what we dancers have the most in common, in the depth of the soul, is the elevation of the foot to an instrument of sublimation in the tango. The foot is the limbo between a detestable earth

The Wuppertal Tanztheater dancers in "Kontakthof" from the movie *Pina* directed by Wim Wenders, 2011.

Bernardo Buontalenti, *Delphic Couple* (models for the third Intermezzo of *La Pellegrina*), [1589], plates 13 and 17, 58.5 x 44.5 (including the passepartout) x 0.5 cm each Florence, Biblioteca Nazionale Centrale di Firenze.

– for the obtuse West – instead deserving of veneration in other cultures in the rest of the world, and an impalpable space, rich in energy. The foot, along with the pause, is the true principle of the tango; what happens between two steps is actually tango, what exists between two states, two decisions in choosing the 'thinking' position of the foot that is alive, that is no longer banished to secrecy as an academic and well-meaning dance would have it. Seeing how a foot moves everything around upsets us, because everything seems to follow it, the bodies as well as our freest internal motions".[4]

Pina Bausch, in turn, loved tango, music and dance, heels and splits, and had her dancer-actors study it with Tete Rusconi, another Buenos Aires-born high-class master. The axis and the energy that the couple share, as well as the balance that originates from the foot that takes and receives impulses from the floor, certainly lie at the heart of the tango.

The history of dance is also a wonderful path along which to examine the possible equilibriums, from verticality (toes and heels) to the barefoot use of the floor; from horizontality, when the whole body is parallel with the floor, to all the phases and all the intermediate stages in an exploration of the three-dimensional space, on oblique and broken lines at every level.

In the West, *basse danse*, or low dance, performed at the Court,[5] devoid of leaps upwards and theorized by the Italian writers of treatis-

es, postulates a perfect vertical control, while drawing in the ground the trajectories of the choreography, as if the feet were supposed to step precisely on the areas indicated by the floor prints, at times expressly projected in symbolic, allusive, celebratory forms.

Which shoes should be worn for this dance? The ones with the shining heel and buckle worn by Louis XIV the Sun King, the "Sovereign dancer", who created in 1661 in Paris the Académie Royale de Danse, from which academic or classical ballet would develop in all its expressions: romantic (*Giselle, Swan Lake*), "concertante" Neoclassical dance (founded by Russian-born George Balanchine in the United States), post-classical (founded by American-born William Forsythe in Germany), with its marked taste for the off balance, stretching the limbs as much as possible to exit and enter the equilibrium in axis.

When dance, which was no longer reserved for courtiers who were skilled at dancing just as they were at horseback riding and fencing according to the rules of etiquette, left the gala halls for the stage and became a profession for ballet dancers of humble origins, the heavy costumes, masks and high-heeled shoes were abandoned.

The first divas, Marie Camargo (1710–1770), the *technicienne*, capable of doing an *entrechat quatre* (leaping into the air and crossing her feet four times before landing) and Marie Sallé (1707–1756), the *tragedienne*, who, thanks to her remarkable acting skills, contributed to the birth of the so-called *ballet d'action* and who, among other things, was Terpsichore in Händel's *Il Pastor Fido*, wore crinolines and ballet slippers; Camargo would be the first, however, to use satin ballet shoes to leap to the sky.

Voltaire had this to say about the two dancers:

Carlo Blasis, *Manuel complet
de la danse* (Paris, 1830),
plate 37 ("Mercury")
and plate 38 ("Female Figure").

Giambologna, *Mercury*,
after 1580 (probably early
seventeenth century)
Disc-welded bronze, fastened to
simil-metal pedestal, h 58 cm
Florence, Museo Nazionale
del Bargello.

Ah, Camargo! How brilliant you are!
But Sallé, great gods, is ravishing!
How light your feet, how sweet hers!
She's inimitable and you are always new.
The Nymphs dance like you but the Graces dance like her![6]

Amid the outcry, in the days of the *coreodramma* of Salvatore Vi-ganò (1769–1821), clothing became lighter and shorter, and the legs were bared to show calves and ankles, while the feet would be visible in open footwear laced up in Graeco-Roman style, as can be seen in the portraits of Maria Medina, the choreographer's fiery Spanish com-panion. *The Creatures of Prometheus*, 1801, staged by the Neapolitan artist to the music of Beethoven, beguiled Stendhal, who, in *Rome, Na-ples and Florence* of 1817 wrote: "Not even the most beautiful tragedy by Shakespeare produces in me half the effect of a ballet by Viganò".

The nineteenth-century proto-romantic ballet (*La Sylphide*), later to become romantic, with the ethereal ballerina at the centre wearing her ever-shorter tutu, raised her up on pointe shoes, a seductive, ul-tra-sophisticated "utensil" that never stopped evolving technically and aesthetically from that time to our day and age.

During that phase, the codes for classical dance were established by observing classical sculpture. Carlo Blasis[7] (1797–1878), Vigano's excellent pupil and a dance instructor at La Scala from 1837 to 1850, in his *Trattato sull'arte della danza*, written between 1820 and 1830, accurately classified postures and figures, as well as the most effi-cient sequence of daily exercises, all of which still essential today. It was Giambologna's Mercury that inspired his *attitude*.

In the nineteenth century, the tension of the body's upward lines gradually took shape, both concretely and symbolically, thanks to the pointe shoes that allowed the dancer to perform on the tips of his or her toes, notwithstanding the tribulation and pain due to the material used at the time: light satin and only cotton and plaster for the padding.

La Sylphide and *Giselle*, ballets for supernatural women, were danced *en pointe*. The means suited to the ends developed at the same pace. To the extent that the modern *Giselle*, a 1982 remake of the Swedish Mats Ek, bares the "feet-emotions" of the brand-new lead dancer.

Today, two centuries after the flourishing of romantic ballet, pointe shoes vary not just based on technical-stylistic needs (Russian toes are distinguished by the tapered toe and low-cut toe box or vamp, while post-classical ballet in Forsythe style features the use of a square, wide toe), but also according to the individual morphology of the ballet dancer's foot.

Today's "pointes", to be used – based on science and the knowl-edge that comes from experience – only when the muscles of the foot and leg are "mature" and strong enough, are made up of several ele-

Johann Gottfried Schadow,
The Ballet Dancers Maria
and Salvatore Viganò, c. 1797.

pp. 150–51
Illustrations by Edgar Degas in Paul
Valéry, *Degas danse dessin* (Paris:
Vollard, 1936), 33.5 x 26.5 x 4 cm
Florence, Biblioteca Nazionale
Centrale di Firenze.

Degas

ments: the upper, which is usually made of satin, covering the outer part of the foot; the *empeigne*, the part closest to the toe, which can be of varying width and is adapted to the length of the toes; the *embout*, a protective encasing-padding made of layers of cotton, gauze, silicone; and the leather or polycarbonate sole, of varying stiffness, as well as the ribbons and/or elastic bands used to fasten the *chausson* to the ankle and throat of the foot, making sure to tuck the knot that keeps them in place under the ribbons – also because it might bring bad luck. Reinforcements made of neoprene with velvet insides also exist: anything that can relieve the pain caused by an artificial position and dynamic, which must always be exhibited to the public with "organic naturalness".

The dancer usually intervenes on his or her new ballet shoes – which are often changed – by embroidering the surface of the pointe so that it will adhere more closely to the floor; shoes are even broken in by striking them with a hammer to adapt them to the foot, which can be "Egyptian", that is, with the second toe longer than the others, or Greek, in which the big toe is the longest and the baby toe the shortest, or else square-shaped, and so on and so forth. In other words, each foot needs its own shoe and its own pointe.

Italian shoemakers, ingenious artisans and foot experts in terms of its overall complex shape, its specific expressiveness and the central role it plays with respect to the well-being of the whole body in movement (for instance, Salvatore Ferragamo was able to shod the un-

healthy foot with a perfect shoe, and the healthy foot with a shoe capable of generating the pain required to heal the other one), have played an important role in the modern pointe shoe and in the dance shoe in general.[8]

The psychophysical insight of Salvatore Ferragamo, who was strongly determined in his tireless search for forms and proportions in ideal balance, starting from the foot to reach the eu-functional totality of the interconnected parts of the human body, occurred at the same time as the remarkable interest in studies on the realignment of the body submitted to excessive stress in industrial-urban society. Indeed, numerous techniques aimed at pursuing a psychophysical balance were born in parallel with the studies and discoveries of the perspicacious Salvatore, ever the perfectionist down to the smallest detail.

Alexander, Pilates, Sutherland-Upledger, Feldenkrais, Mézières, names that indicate various accredited methods used to achieve mind/body harmony, reflected upon remedies for the malaise of modern life which was increasingly becoming "unnatural", and the harbinger of unhealthy and pathological tensions.

In the 1890s Australian-born Frederick Matthias Alexander (1869–1955), a Shakespearean actor who eventually lost his voice, studied curative re-training that started from self-observation and the discovery that his phonation problem stemmed from the contraction of his whole body. By working on the posture of the neck, on the axis of the body and on breathing, he found a solution for his problem that would also improve the health of others. He later published the results of his findings in physical education.[9]

Joseph Hubertus Pilates (1880–1967), from Germany, son of a gymnast and a naturopath, who suffered from asthma and rheumatic fever as a child, spent his entire life working on the use of a wide range

A photograph taken by Robbie Jack portraying Irina Golub and Mikhail Lobukhin in the ballet *In the Middle, Somewhat Elevated* by William Forsythe at Sadler's Wells Theatre in London, 13 October 2008.

Eadweard Muybridge, *Dancing (Fancy)*, 1887 Collotype print, 59.1 x 46 cm Kingston Museum and Heritage Service.

Plinio Nomellini, *Studies of a Dancer,
Isadora Duncan*, 1913
Pencil on paper, 37 x 26 cm each
(top left and bottom left and right),
37 x 24.5 cm (top right)
Florence, private collection.

Plinio Nomellini, *Studies of a Dancer, Isadora Duncan*, 1913
Pencil on paper,
37 x 26 cm each
Florence, private collection.

P. Nomellini

of disciplines to strengthen the body, from yoga to the martial arts such as Tai Chi Chuan, and dedicated himself to the circus, boxing and body-building. He too discovered the negative effects of bad posture and poor breathing, and thus conceived "contrology" to rectify their causes and outcomes, as discussed in his texts[10] on the relationship between the body and the mind seen as a journey through awareness.

William Garner Sutherland (1873–1954), an American phytotherapist and osteopath, is responsible for the craniosacral therapy developed in the 1970s by John Upledger, a treatment that aims at re-balancing the tension of the ligaments, realigning the body and correcting its altered coordination.

Moshe Feldenkrais (1904–1984),[11] judo master, is the inventor of a method of somatic education, initially developed as self-treatment after an accident, using biomechanics, the martial arts and psychology. His work focuses on the functional integration of every part of the body.

In 1947 French-born Françoise Mézières (1909–1991), masseuse and kinesitherapist, honed a technique to correct the deviations of the vertebral column by relaxing the tension and lengthening the muscles that were normally contracted using breathing exercises connected to specific movements involving the muscular chains. The concept and method of so-called anti-exercise is inspired by her work.

Clearly, then, the awareness of the close relationship between mind and body, and of the neuromuscular effort that involves both, advanced throughout the past century, harking back with renewed interest at the Far Eastern holistic disciplines and overcoming the philosophical-religious conception according to which the body, res extensa, is the servant of the mind, the much nobler res cogitans – the interpretation in a mechanically dualistic form of the eighteenth-century thinking of René Descartes, also known as Renatus Cartesius or Cartesio, the advocate of the rationalistic vision of knowledge by clear and distinct concepts, inspired by the accuracy and certainty of the mathematical sciences.

One cannot help but observe the points these different twentieth-century Western practices and theories have in common, all aimed at harmoniously reassembling the vision and the perception of the human being in his or her global essence, complete from every viewpoint, manifestly grafted on a common terrain of study.

If we instead examine the names that, during the same time period, strongly influenced the basics of the birth of modern dance,[12] we cannot overlook the importance of the Frenchman François Delsarte (1811–1871),[13] the inventor of a method used to teach actors and dancers, which was based on his own experience after he lost his voice, and focused on the emotion-gesture expressed in its vast array of gradations and intensities.

His system, inspired by the Trinity in a Christian-esoteric sense, is made up of static, dynamic and semiotic aspects; the body is observed from a physical, emotional and mental standpoint – all fields of investigation inherent to the legs, the trunk, the arms and the head. Movement, for Delsarte, could be eccentric, concentric or normal, as well as harmonious, of opposition or of succession. These principles – especially the one of opposition – were fruitfully applied by the very pioneers of modern dance in the United States.

Genevieve Stebbins (1857–1914?) from California was the one to spread Delsarte's thinking across the ocean. She was the student of one of the French master's protegés, the American actor Steele MacKaye, an advocate, in the United States, of Delsartian "harmonious exercise". In 1885 Stebbins published *The Delsarte System of Expression*, a very important book that was met by a huge success also and above all among the figures of a new, free and nobly expressive dance, such as Isadora Duncan (1877–1927), who was also from California. Ted Shawn was a Delsartian as well; together with Ruth St. Denis he was the soul of the so-called Denishawn school, the first of its kind dedicated to modern dance, founded in 1915 in Los Angeles, where famous movie stars such as Louise

Brooks studied, in a period when the centre of the American movie production was still on the West Coast.

Agnes DeMille, choreographer, a client – "with beautiful feet" – of Ferragamo[14] and the granddaugher of Cecil B. DeMille with whom Salvatore worked on the epic movie *The Ten Commandments*, is famous for saying: "Scratch a dancer and you find Denishawn!".

To Swiss-born Émile Jaques-Dalcroze (1865–1950) we instead owe eurhythmics, which involves teaching concepts of rhythm, structure and musical expression through movement.[15] His lessons, which were held since 1910 in the garden-city of Hellerau near Dresden, where the famous theatre designed by Adolphe Appia in the name of new compact multi-level structures and a new type of lighting (now the site of The Forsythe Company) was built, were attended by Marie Rambert, future assistant to Vaslav Nijinsky who helped him, when his dancers rebelled, to dominate the complex polyrhythms of Stravinsky's *Sacre du Printemps*, installed with scenes and costumes by the co-creator Nikolai Roerich.[16] A revolutionary and controversial choreography that was "too modern" for 1913 witnessed the hotly-debated and stormy Paris debut of this celebrated ballet with Sergei Diaghilev's Ballets Russes, unmissable for the *crème* of that day and age.

Rudolf Laban (1879–1975), a Hungarian dancer and choreographer, and an acute and prolific theoretician,[17] created his own system for movement analysis in order to describe, visualize and record (with the so-called "Labanotation") all the possible variables of human movement; the method is currently referred to as Laban Bartenieff (from Irmagard Bartenieff, physiotherapist, 1900–1981, who worked with those afflicted with polio, and was an expert in dance therapy[18]). Laban put together the components of a multidisciplinary approach, including anatomy, kinesiology and psychology to describe and classify effort and energetic dynamics, form, space, mobility/stability, interior/exterior, exercise/recovery, aimed at "total-body connectivity".

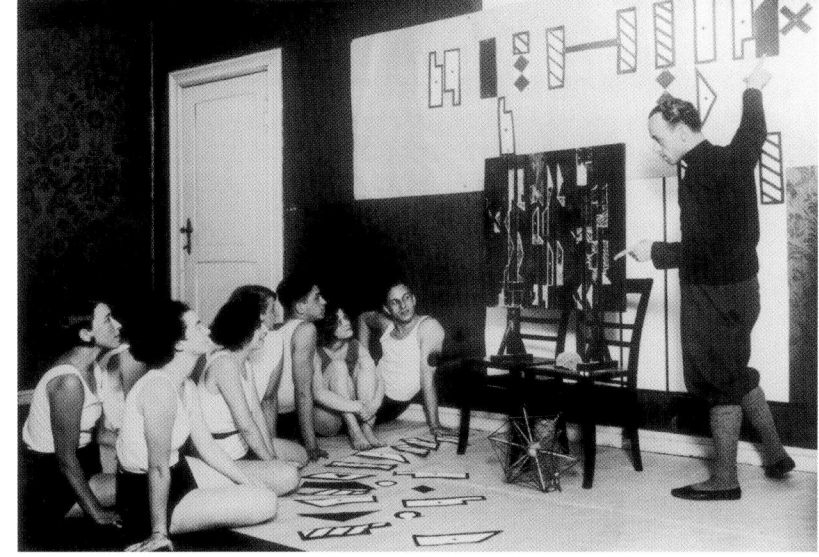

Worthy of note is the fact that in the early twentieth century Laban was at Monte Verità,[19] in neutral Switzerland, where naturists, anarchists, vegetarians, Dadaists, nudists, theosophists and anthroposophists would gather[20] to escape the devastation of the First World War in Europe. Even Salvatore Ferragamo may have come across Theosophy while in America (working for the movies, he met a number of actors who frequented Jiddu Krishnamurti, the "Maitreya Buddha") or, later, in Florence at the Accademia di Careggi, the cradle of hermeticists, Neo-Pythagoreans, and Neo-Platonists.[21]

Boris Kniaseff (1900–1975), an emigrated Russian ballet dancer, who was a member of the companies of Colonel de Basil, Bronislava Nijinska and the Ballets des Champs-Elysées, intentionally the continuers of the Ballets Russes,[22] which disbanded when their legendary impresario Diaghilev passed away in 1929, became a famous dance master in Paris. His students were the most enlightened talents of the Opéra and included Roland Petit, Zizi Jeanmaire, Yvette Chauviré

Rudolf Laban and
his students, c. 1920.

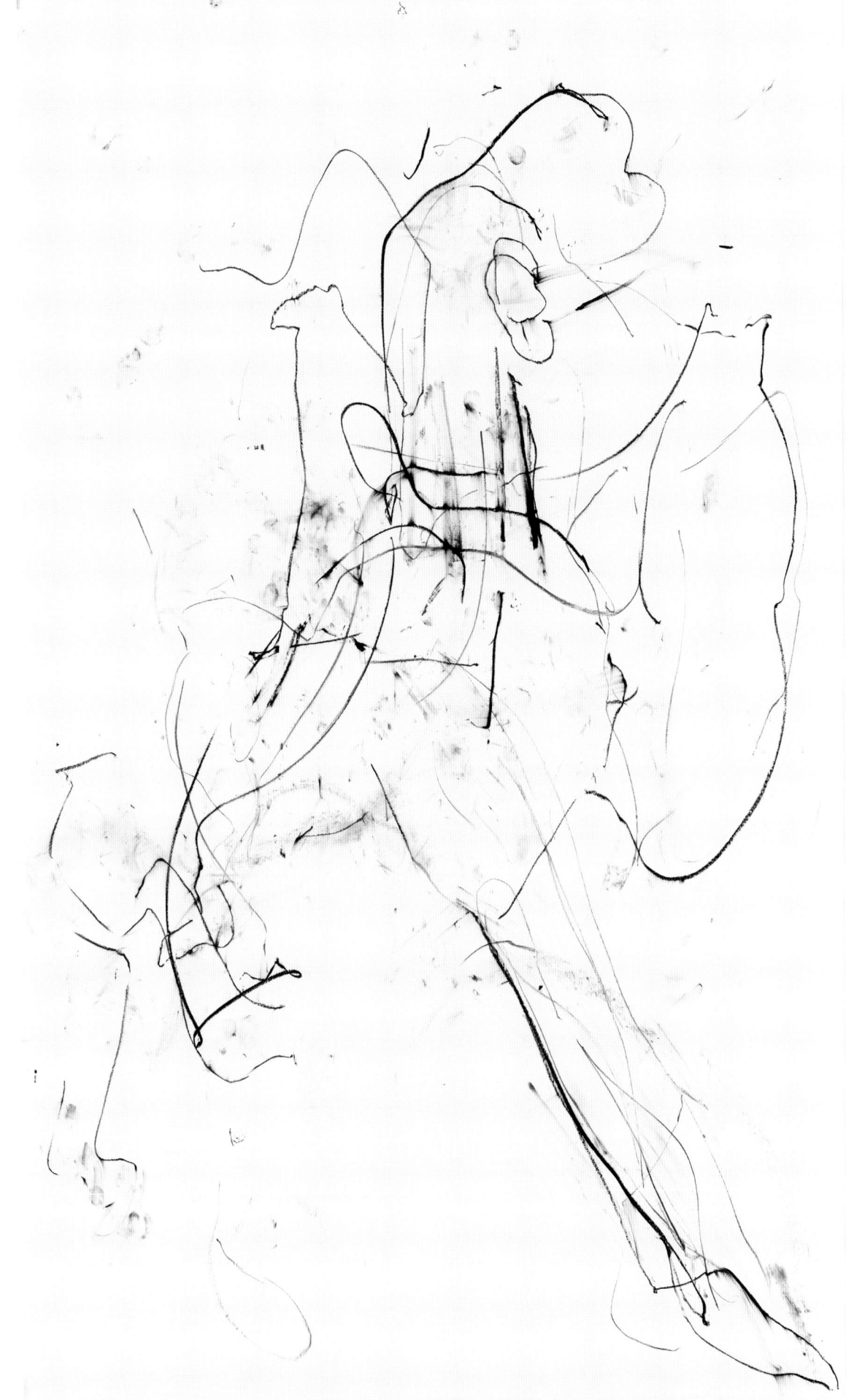

Trisha Brown, *Untitled
(Montpellier)*, 2002
Charcoal on paper, 329.6 x 271.1 cm
and 330.2 x 271.1 cm
Courtesy of the artist and Sikkema
Jenkins & Co., New York.

and even Brigitte Bardot. To avoid damage to the historical building in which he taught, Kniaseff invented his *barre à terre* class,[23] which proved to be of great use to working in detail on each limb without the gravitational load-bearing of vertical posture, thus developing proprioception as well as rehabilitating injured ballet dancers.

Meanwhile, the body in movement, also by virtue of the new technologies, such as photography, had since 1872 become the object of scientific investigation, and most importantly so in the work of the British Eadweard Muybridge, a precursor of cinema.[24] Using chronophotography, invented by Muybridge to show that during the gallop there is an instant when the horse doesn't touch the ground, human locomotion was broken down and analysed using 12 cameras, and later 24, arranged to take pictures in sequence. Boxing, weight-lifting, running and dance were thus reproduced and scanned.

Biomechanical theories and studies in the early twentieth century owe a great deal to this research – as well as to others that stemmed from it[25] – on the naked truth of movement: walking, running or leaping viewed as it begins, develops and ends. Biomechanics, from the analysis of the behaviour of physiological structures when they are submitted to static or dynamic stress, by way of posturology, kinesiology and bioengineering would be applied to the theatre thanks to the method developed starting from 1922 by Vsevolod Meyerhold. The Russian director and pedagogist wanted the actor's whole body to be the protagonist of the scene, as a means of artistic creation and an instrument of communication, with an eye to the rigorous accumulated codes in Oriental dance-theatre, Italian Commedia dell'Arte, classical ballet, the circus. In 150 physical exercises the would-be actor could scientifically learn the method required for the artistic theatrical process.

As for Russia, the long-repressed research that was carried out by the members of the art movement called Plast finally came to light. It all started with a remarkable project that began in Moscow within the Choreological Laboratory of the Russian Academy of Artistic Scienc-

es (RAChN, 1923–29), part of a broad utopian strand from that first revolutionary period, which preceded Stalin's closures, and was manifested in as many as four official and highly successful exhibitions devoted to the Art of Movement.[26]

Although the puritanical Soviet Union accused the fox-trot and the new American dances of being shocking, at first it didn't prohibit the dances for naked bodies of the choreographers who would later be labelled as being "decadent" – for instance, Kasyan Goleizovsky, whose experiments in Saint Petersburg inspired George Balanchine's new "Neoclassical" ballet forms, which he then developed further in the West, between Paris and New York. In Russia, instead, the innovative plastic forms of Goleizovsky, who would also devote himself to the musical, would serve to structure the regime's mass choreographies.

The collaboration of institutions such as the Russian Photographic Society (RFO), the Work Institute (CIT) and the Institute for the Study of the Brain in the scientific-practical research that peaked with the exhibitions on the Art of Movement mentioned before, and destined to be developed in the fields of biomechanics and the notation of movement, are proof of the originality of the path taken in Russia if compared with the so-called Free Dance of other countries.

In the period following the Second World War, it was the American post-modern[27] that, on the contrary, theorized – always starting from the dynamics of the body in performance – that everything is dance, every single everyday movement, even the least poetic, so that expressly theatrical places, choreography, sets, costumes, lighting and so on were no longer needed… at least in terms of the declared intentions, as it was always a question of performance.

This patrimony of culture that pervaded the twentieth century is unquestionably and intuitively included in the "spirit of the times" and the universe of science and the knowledge of the body. This is also true for the foot, to which a magnificent shoemaker like Salvatore Ferragamo instinctively reacted by acutely feeling the need to intelligently

rethink the support-cladding to be offered to the lower extremities – which are unique in every single person – careful to support the plantar arch in order to guarantee health and functionality and determine the best conditions for a new, advantageous and beneficial comfort.

The observations he made on the strong and arched feet of ballet dancers, repeatedly annotated in his adventurous autobiography, are enlightening and exact – he wrote about Alicia Markova, the famous French-born star Colette Marchant, the African-American Katherine Dunham, James Dean's teacher: all of whom had a magnificent arch and amazingly flexible ankles.

As for Italian shoemakers who were skilled at creating dance shoes, by chance (or perhaps by necessity) there have been many masters, excellent experts in the ballet foot.

In 1887 Salvatore Capezio, born in 1871 in Muro Lucano, opened a small shoe shop in a strategic part of Broadway, opposite the Metropolitan Opera House in New York. In addition to repairing ballet shoes, after marrying the Scala-trained ballet dancer Angelina Passone in 1902, he invented a type of pointe shoe that was revolutionary. This shoe associated the traditional shape, which was still inescapable, with the modern forms of advanced engineering, earning the confidence of some top-notch clients, the first among which being Anna Pavlova, the "dying swan"-cum-star of the Ballets Russes. And we mustn't overlook the fact that Capezio made the shoes for the Ziegfield Follies and for many Broadway shows; he also invented the split-sole, a device that, based on the instep of the dancer's foot, allowed for greater elasticity in the movements of the lower limbs. This

James Dean during his dance
lessons with Katherine Dunham,
New York, 1955.

pp. 162 63
Barbara Morgan immortalizes the
ballet dancer and choreographer
Martha Graham as she performs
in *Letter to the World* (*Swirl*; *Kick*),
New York, 1940. The work depicts
the life of the writer Emily Dickinson.
Getty Images.

Barbara Morgan, Martha Graham
in *Lamentation, Oblique*, 1935.
Getty Images.

was of fundamental importance to the jazz dance we are accustomed to seeing in the most popular musicals on stage and screen.

Also to be noted is the fact that even ice-skates today are made using sophisticated technologies that guarantee the maximum articulation possible for the foot, which is enclosed in a shell that has to hold up well during leaps and twirls, thus avoiding the risk of uncertain foothold before performing acrobatics on thin support blades, and even more so during the subsequent landing. Besides tap, ballroom and flamenco shoes, Capezio worked for hip hop dancers as well; these were forms and shapes taken from the street but carefully perfected for such planetary dancers as Michael Jackson. In New York there was the Peridance Capezio Center, a black box of 160 seats open to young people, workshops and the presentation of new works.

In 1903 Luigi Gamba, a waiter at the Savoy in London, which served important theatres like Covent Garden, became a shoemaker and affirmed himself in the wake of the Italian artisanal tradition and ancient school of ballet. Gamba could boast of clients like Vaslav

Nijinsky, the male supernova of the Ballets Russes who succeeded in undermining the central role of the "ballerina" in the early twentieth century. Nijinsky interpreted, among others, Shiva, *Le Dieu bleu*, the Indian god of dance, with very light slippers evoking the bare decorated foot of Bharata Natyam.

Rosa Repetto, a native of Nice, who lived a stone's throw away from the Opéra de Paris, created special dance shoes that fit like a glove for her son Roland Petit (a champion of the Neoclassical chic across the Alps) and his muse Zizi Jeanmaire, a spicy Carmen on toes, queen of the varieté with heels and thin stockings on fabulous legs. Beginning in 1947, Repetto ballet flats, called *sandales plateformes*, 135 grams of leather, just 7 millimetres of heel, became fashionable outside of theatres, worn by women who loved gracefulness. As for Italian manufacturers of dance shoes, Eugenio Porselli opened

left
Merce Cunningham in *Antic
Meet*, photographed by Robert
Rauschenberg in 1958.

Vasilij Kandinskij, *Field
Lines of Modern Dancer
Gret Palucca*, 1926
Pencil on paper, 16.5 x 16.5 cm
Dresden, Kupferstich-Kabinett,
Staatliche Kunstsammlungen.

Mikhail Baryshnikov practices
the movements created by
Merce Cunningham in *Signals*,
New York, 1994.

a dance shop in Milan close to the Teatro alla Scala, and since 1919 the "ballerine Porselli" have been the fashion for all women who like to wear colourful shoes.

All this was going on while modern ballet was beginning to blossom, pioneered by refined and very formal George Balanchine (1904–1983) in the United States and by the popular Maurice Béjart (1927–2007), the choreographer-star who left his native city of Marseilles to travel to every continent, beginning in the 1960s until the end of his days.

Today's post-classical ballet, in its multiple formats, when not wearing pointe shoes can toy with punk wedges such as those worn by Michael Clark, formerly the pearl of the Royal Ballet School in London, or be content to wear socks – which is what Forsythe does habitually now – often associated with minimal costumes, more similar to undergarments than to the couture that Diaghilev preferred. The Russian impresario had the friendship and support of a sponsor like Coco Chanel for *Le Train bleu*, 1924, an "athletic" ballet that is set at the seaside, signed by the choreographer Bronislava Nijinska, sister and accomplice to Vaslav. And pointe shoes, seen as an art object, also enchanted an angry anti-modernist like Jan Fabre, who interacted with design, painting, sculpture, performance, theatre, video and even ballet by decorating with a Bic pen the world's most fascinating "instruments of torture", with a special touch on the *Swan Lake* of the Royal Ballet of Flanders.

Modern dance[28] which, meanwhile, had undertaken a powerful dialogue with the floor, aimed – thanks to a different ethics of "real truth" – to openly show the bare foot's beauty and sincerity. The soft-landing ecstatic leaps of the sensuously spiritual Isadora Duncan, always portrayed with arms lifted skywards; Martha Graham's spiral of earthly energy, which began with breathing concentrated in the solar plexus constantly alternating contract and release, to use the whole body's movement to tell the tragedies of the human psyche; the tilts and twists of Merce Cunningham,[29] her rebellious student, the noble father

Fred Astaire practicing while rehearsing in the studio for a movie, 1941.

Gene Kelly dancing in the musical *Singing in the Rain*, 1951.

of post-modern dance, of the "dance that is sufficient unto itself", very pure, no longer dependent on music (his companion is John Cage, according to whom even mushrooms in the forest while silently growing make a sound) or on décor, the work of art in itself staged during dance, signed by Robert Rauschenberg, Jasper Johns, Andy Warhol; and the expression of modern European dance from Germany: everything hinges on the foot's grip on the ground capable of supporting the opposing movements of the various parts of the body at the same time, and of the various subjects in the group, down to the least in-axis, the most "dis-harmonious".

From the noble, free dance of the Californian Isadora to the "abstract" dance of Cunningham, nature becomes culture, starting – unsurprisingly – from the authentic beauty of the dancing foot. The long line of the forefoot: this is the key to jazz dance, the dance that is connected with the pleasures of entertainment in the American melting pot, with Fred Astaire's mirror shoes, the chicest tap dancer in the history of cinema who made the agility of African feet his own with upper class naturalness, in a deep relationship with Mother Earth. Talking feet, musical feet, which on the New Continent became an ingenious rhythm machine, bestowing shoes with metal blades or bottle caps, the way young kids do in the streets of New Orleans, and dancing wherever they want to, even under the rain, or with cartoon characters, as well as with the delightful French classical dancer Leslie Caron in *An American in Paris*, as did the athletic boy-next-door Gene Kelly, "America's fiancé".

There is no culture that hasn't set music to the feet with bells on the ankles as in Kathak, or clicking heels like in the flamenco, or clogs, as in Irish gigs, also found in comedy-ballet of country-like misunderstandings such as *La fille mal gardée* of 1789, the year of the French Revolution.

Rhythm is the secret driver of spectacular chorus lines, with legs, feet and shoes in a row, absolutely all the same, as we can clearly see in any of the images, for example, in the synchronized and impeccably English-language-speaking Tiller Girls.

In America, besides the refined custom-made footwear he produced for a very select clientele, Salvatore Ferragamo also created shoes for variety shows,[30] flat slippers and dance shoes with a whole array of toes, with either a strap or fasteners. It was the fruit of American glory, the right acknowledgement from a land conquered step by step with unfailing courage, destined to reverberate in Florence, which had been the cradle of the Renaissance, that was and would be the glory of Italy in the centuries and millennia to come. The acquisition of Palazzo Ferragamo, formerly Spini Feroni, etches in those bricks and stones, which are the *summa* of an outstanding, unique culture, the name of a figure who was just as unique, from every point of view.

How is such an epic to be told? How to synthesize the intelligent path, amid the incoercible drive to make things with his own hands, and the never-ending curiosity about the world of culture and the arts, between colour and form, epochs and styles and currents and battles between rival avant-garde movements in amazing the world, from exoticism to naturalism, from Futurism to Constructivism to the Bauhaus?

We can describe it in the simplest and most evident way, the way a master of impeccable style would have with the certainty of a taste that was always and in any case appropriate. From the bare or almost bare foot in Graeco-Roman footwear to the Baroque heel and the modern stiletto heel, to the pointe shoe to the hunting boot to the cork platform shoe, Terpsichore and Mercury – and, in name of and on behalf of this divinity, Salvatore Ferragamo – have glorified the extraordinary art of all ages and in all places that is born from the foot, in a kaleidoscopic journey between aesthetic and function, an endless hymn to the beauty and the intelligence of the body. Salvatore Ferragamo intoned the hymn that no-limits man dreams of with the humble doggedness of the labourer, and with the most contagious, creative joy of the artist in every instant of his outstanding life, magnificently stubborn and fanciful.

[1] L. Bentivoglio, *Tanztheater, dalla danza espressionista a Pina Bausch* (Rome: Di Giacomo, 1982); L. Bentivoglio, *Il teatro di Pina Bausch* (Milan: Ubulibri, 1985 and 1991); E. Vaccarino, *Altre scene, altre danze* (Turin: Einaudi, 1991); F. Quadri, *Sulle tracce di Pina Bausch* (Milan: Ubulibri, 2002); E. Vaccarino, *Teatro dell'esperienza, danza della vita* (Genoa: Costa & Nolan, 2005); P. G. Carizzoni and A. Ghilardotti, *Isadora Duncan, Pina Bausch* (Milan: Skira, 2006); L. Bentivoglio and F. Carbone, *Pina Bausch, vieni, balla con me* (Florence: Barbès, 2008); R. Giambrone, *Pina Bausch, le coreografie del viaggio* (Macerata: I libri dell'Icosaedro, Ephemeria, 2008).

[2] *Shoemaker of Dreams. The Autobiography of Salvatore Ferragamo* [1957] (Florence: Giunti, 1985), p. 84.

[3] *Cinderella* is the symbolic tale of an initiation to love and adult life, a paradigm of the journey in search of oneself, an itinerary of successful social redemption. According to Bruno Bettelheim, a great expert on the enchantment of fairy tales, the basic plot for Cinderella already existed in China in the ninth century BC; however, in Germany too we can find stories about men who work as fire servants and then become kings, without overlooking the Christian "ashes to ashes" that alludes to the cult of the deceased, in this case the main character's deceased mother. Thanks to its universal notoriety, in twentieth-century Russia Cinderella was transformed into a ballet set to music by Sergei Prokoviev, born from the need to build new repertory classics also in the centuries of modernity along the lines of the popular nineteenth-century masterpieces.

[4] Cf. E. Guzzo Vaccarino, *Il tango* (Palermo: L'Epos, 2010), p. 229; *Un tal Gavito*, 3 DVD, 2006.

[5] B. Castiglione, *Il libro del Cortegiano*, 1528 (Milan: Garzanti, 1981); D. da Piacenza, *De arte saltandi et choreas ducendi*, c. 1455; A. Cornazano, *Libro dell'arte del danzare*, 1455; G. Ebreo da Pesaro, *De praticha seu arte tripudii vulgare opusculum*, c. 1463; M. F. Caroso da Sermoneta, *Il ballarino* (Venice, 1581); C. Negri, *Le Gratie d'Amore* (Milan, 1602).

[6] "Ah! Camargo que vous êtes brillante!
Mais que Sallé, grands Dieux, est ravissante!
Que vos pas sont légers et que les siens sont doux!
Elle est inimitable, et vous toujours nouvelle;
Les Nymphes sautent comme vous, Et les Grâces dansent comme elle".

[7] C. Blasis, *Trattato dell'arte della danza*, edited by F. Pappacena (Rome: Gremese Editore, 2008).

[8] L. Scarlini, "Le geometrie dell'anima. I trionfi della moda e la percezione dell'altrove di Salvatore Ferragamo", in *Salvatore Ferragamo. Ispirazioni e Visioni* (Milan: Skira, 2011), p. 113.

[9] Cf. F. M. Alexander, *Man's Supreme Inheritance* (London: Methuen, 1910; revised editions New York, 1918, 1941, 1946, 1957 and London: Mouritz, 1996, reprinted 2002); *Constructive Conscious Control of the Individual* (USA: Centerline Press, 1923; revised edition London: Mouritz, 1946 and 2004); *The Use of the Self* (New York: E. P. Dutton, 1932; reprinted Orion Publishing, 2001); *The Universal Constant In Living* (New York: E. P. Dutton, 1941; reprinted London: Chaterson, 1942, 1943 and 1946; USA: Centerline Press, 1941 and 1986 and London: Mouritz, 2000).

[10] J. H. Pilates, *Your Health*, 1934; J. H. Pilates and W. J. Miller, *Return to Life through Contrology* [1945] (Bel Air: Christopher Publishing House, 1960).

[11] M. Feldenkrais, *Body and Mature Behaviour* (New York: International Universities Press, 1949; reprinted Berkeley: North Atlantic Books, 2005); *Awareness through Movement* [original in Hebrew 1967] (New York: HarperCollins, 2009); *Il Metodo Feldenkrais, conoscere se stessi attraverso il movimento* (Como: Ed. RED, 1991); *The Elusive Obvious* (Cupertino, California: Meta Publications, 1981); *The Potent Self* [1985] (San Francisco: Harper, 1992); *La saggezza del corpo* (Rome: Casa Editrice Astrolabio, 2011); M. Melucci, *Lezioni di Metodo Feldenkrais* (Milan: Biblioteca olistica, Xenia Edizioni, 2011).

[12] *La generazione danzante*, edited by S. Carandini and E. Vaccarino (Rome: Di Giacomo, 1997).

[13] http://www.lib.lsu.edu/special/findaid/r1301.inv.pdf; *François Delsarte, le leggi del teatro. Il pensiero scenico del precursore della danza moderna*, edited by E. Randi (Rome: Bulzoni, 1993); E. Randi, *Il magistero perduto di Delsarte. Dalla Parigi romantica alla modern dance* (Padua: Esedra, 1996).

[14] *Shoemaker of Dreams* 1985, p. 194.

[15] É. Jaques-Dalcroze, *Le rythme, la musique et l'éducation* (Paris, 1920).

[16] A. D'Adamo, *Danzare il rito, Le Sacre du Printemps attraverso il Novecento* (Rome: Bulzoni, 1999).

[17] R. Laban, *Modern Educational Dance* (Wokingham, Berkshire: Macdonald & Evans, 1948; reprinted 1963 and 1975, third edition edited by L. Ullman); *The Mastery of Movement on the Stage* (Wokingham, Berkshire: Macdonald & Evans, 1950; reprinted 1967, 1974, third enlarged edition edited by L. Ullman and 1980, fourth edition with introduction by R. Laban); *Principles of Dance and Movement Notation* [1956] (London: Macdonald and Evans, 1975); *Effort: Economy of Human Movement* (London: Macdonald and Evans, 1974), with F. C. Lawrence; DVD available at http://www.labanproject.com/

[18] I. Bartenieff, *Body Movement, Coping with the Environment* (USA and UK: Routledge, 1980).

[19] R. Landmann, *Ascona – Monte Verità* (Berlin: Ullstein, 1979); M. Green, *Mountain of Truth: The Counterculture Begins: Ascona, 1900–1920* (University Press of New England, 1986); *Highlights in the History of Monte Verità*, Museo Monte Verità, 2007; *Monte Verità. Le mammelle della verità*, edited by H. Szeemann (Locarno: Armando Dadò, 1978); K. Noschis, *Monte Verità: Ascona e il genio del luogo* (Bellinzona: Casagrande, 2013); *Corpo e potere. Körper und Macht*, exhibition catalogue, edited by G. A. Mina (Monte Verità, 2013); see also http://www.monteverita.org

[20] P. Giovetti, *Rudolf Steiner, la vita e l'opera del fondatore dell'antroposofia* (Rome: Edizioni Mediterranee, 1992); see also http://www.rudolfsteiner.it

[21] S. Risaliti, "Reminiscenze e ispirazioni di Salvatore Ferragamo", in *Salvatore Ferragamo. Ispirazioni e Visioni* 2011, p. 49.

[22] P. Veroli and G. Vinay, *I Ballets Russes di Diaghilev tra storia e mito* (Rome: Accademia Nazionale di Santa Cecilia, 2013).

[23] Y. Matsuyama and L. B. Ribeiro, *Danza classica e contemporanea con la didattica Kniaseff* (Milan: Gammalibri, 1984).

[24] H. C. Adam, *Eadweard Muybridge. The Human and Animal Locomotion Photographs* (Cologne: Taschen, 2010).

[25] G. Oliva, *Il laboratorio teatrale* (Milan: LED Edizioni Universitarie, 1999); E. Barba and N. Savarese, *L'Arte Segreta dell'Attore* (Milan: Ubulibri, 2005; reprinted and revised *Un Dizionario di Antropologia Teatrale*, Routledge: Centre for Performance Research and Bari: Edizionidipagina, 2011).

[26] N. Misler, *In principio era il corpo. L'arte del movimento a Mosca negli anni Venti*, exhibition catalogue (Milan: Electa, 1999); *V nachale bylo telo* [At the beginning was the body] (Moscow: Iskusstvo, 2010).

[27] S. Banes, *Tersicore in scarpe da tennis* (Macerata: Ephemeria, 1993).

[28] L. Bentivoglio, *La danza moderna* (Milan: Longanesi, 1977); *La danza contemporanea* (Milan: Longanesi, 1985).

[29] M. Cunningham, *Il danzatore e la danza* (Turin: EDT, 1990); C. Tomkins, *Vite d'avanguardia. John Cage, Leo Castelli, Christo, Merce Cunningham, Johnson Philip, Andy Warhol* (Genoa: Costa&Nolan, 1983); *Merce Cunningham*, edited by G. Celant (Milan: Charta, 2000); E. Guzzo Vaccarino, "John Cage, il libertador della danza", in *John Cage, Una rivoluzione lunga cent'anni*, edited by G. Fronzi (Milan and Udine: Mimesis, 2013).

[30] *Shoemaker of Dreams* 1985, p. 200.

CHAOS IN ORDER: CHAPLIN'S WALK

SANDRO BERNARDI

CHAPLIN AND PLATO, INEBRIATION AND DANCE

In the beginning, we might say, was dance. Actually, in the beginning were choruses and dance. According to the Greeks, from these two elements the arts were born: epic, drama, opera, and all the other arts. Plato believed that even law should be inspired by dance, as the quest for social harmony: dance, which comes from the movement of the stars, is an educational art, it teaches a person to distinguish between movements and between objects, whether beautiful or ugly. The contrast between beauty and ugliness, good and evil, pleasure and pain, is connected to the contrast between order and chaos. Beautiful are the harmonious movements that express virtue, ugly are the movements that express vice (*The Laws*, 654a). Plato distinguished between two types of dance: harmonious dance, inspired by the Muses and guided by Apollo, and Dionysian dance, in which restless and passionate moods are vented.

Indeed, we know how important wine was for Plato, as a source of enlightenment and spiritual elevation, but only when used "appropriately"; suffice it to recall the *Symposium*, which I will refer back to at the end of these thoughts. Hence, inebriation, music and education are three different yet closely interrelated things: "But the fact is that the right ordering of this could never be treated adequately and clearly in our discourse apart from rightness in music, nor could music, apart from education as a whole".[1]

"Our feet were soft
in flowers"
John Keats, *Endymion*
(in Charlie Chaplin,
Monsieur Verdoux)

The Circus, 1928
Little Tramp the acrobat.

When Nietzsche elaborated his theory of the birth of tragedy, he undoubtedly had in mind these and other passages written by Plato that dealt with Apollo and Dionysus. In his famous way of walking, Charlie Chaplin managed to visibly express this contrast between the Apollonian and the Dionysian worlds, between inebriation and dance.

AN ANCIENT MASK

Alcoholism, syphilis and madness are the dark shadows that accompanied the childhood of this great actor-director, and that throughout his life would be the best remedy against all vices (except, perhaps, against very, very young women…). He had been exposed to theatre as a child, as work and as entertainment, thanks to his mother who was an actress. To amuse him, she would tell him to sit by the window and look at the passers-by, which she would then imitate and parody. We might say that for Chaplin imitation and parody have always been a symbolic channel that enabled him to stay in touch with his mother's image, which he carried inside, and from which he probably drew his energy. If we consider this early moment in the story of his life to be essential – and it actually is, as he himself chose to recall it in his autobiography – then it is no accident that Chaplin was born into the world of entertainment by specializing in walking like a drunk. Pantomime, his starting point, was a mute recital, but at the same time a dance, a game between showing and hiding, saying and not saying, a representation that brought together many distant threads: from shadow play to the Commedia dell'Arte. Already as a member of Fred Karno's company of mimes, both in Great Britain and during the long tours in America, the drunk was always his main act. And he would continue to perfect it, without ever abandoning it, not even when walking slowly away from the camera and towards the guillotine in the part of Monsieur Verdoux. As Chaplin himself told us, it was precisely the pantomime drunk that allowed him to enter movies: "Mr. Charles Kessel, one of the owners of the Keystone Comedy Film Company, said that Mr. Mack Sennett [the director, *Editor's note*] had seen me playing the drunk in the American Music Hall on Forty-second Street and if I were the same man he would like to engage me to take the place of Mr. Ford Sterling [who was leaving Keystone, *Editor's note*].[2]

However, like every great author, Chaplin gathered and transformed many aspects of the past theatrical tradition, from the Italian Commedia dell'Arte to the English harlequinades, which became especially popular thanks to the London Christmas Pantomimes in vogue from the late eighteenth through the nineteenth centuries. Harlequin's acrobatic style is clearly recognizable, with his leaps and above all the typical knees apart pose of Tristano Martinelli (the inventor of Harlequin), as well as the backward kick, the brash stamping of the legs on the ground to make himself noticed by others. Proof of the legacy of Harlequin, the eternally enamoured acrobat, can be found in a number of Anglo-American studies. According to Bryony Dixon, Chaplin's clothing is a fusion between two masks, those of Harlequin and Clown: the former is in love with Columbine, the latter is an out-and-out thief, often hired to watch over Columbine, but in truth Harlequin's accomplice. (Clown was originally the name of an English mask, to later become the common name for a certain type of circus character.) Dixon says that some of Chaplin's traits, including the famous cane he used to walk as well as to beat with, stem from the tradition of the harlequinades, among which those of a famous eighteenth-century Clown named Joseph Grimaldi (1778–1837): "The physical attributes and the comic business of the harlequinade can be traced quite easily in early film comedy: clown costumes, exaggerated makeup, comedy policemen, the hat, the cane, the magic bat or slapstick, outsize or undersize clothing, drag … Various transforming devices were used to comic and dramatic effect such as Harlequin's bat or slapstick, magic lamps or buttons, magic wands, potions, *intoxications (a favorite of Chaplin's)*".[3]

So, a long line of mimes and popular comic characters – who entertained the public for a few pennies in plays out in the open, in the squares, at the fairs – links these comic actors and acrobats to Chaplin, their great heir. Wittkower described this very long tradition, actually impossible to reconstruct, as a full-fledged migration of symbolic forms, which crossed the centuries to join not only the visual arts but also the most popular and humblest comic actors to the great master of cinema. For this reason as well, Chaplin's walk, his "regal posture" or his "indefinable something", as it was labelled as early as 1915, were from the outset recognized by the masses of spectators of every class, especially the lower classes since they had belonged to them for centuries.

If the game of leaps and jests came from the Commedia dell'Arte, the Edwardian costume, complete with top hat and frock-coat, walking stick and spats belonged to the British heritage: it was the symbol of the teddy boy, the petty criminal who dressed the way the wealthy did, but as an unwitting parody, because his clothes were torn rags. To combine the drunk with the teddy boy, Chaplin invented the entrance with his back to the audience: "I entered with my back to the audience – an idea of my own. From the back I looked immaculate, dressed in a frock-coat, top hat, cane and spats – a typical Edwardian villain. Then I turned, showing my red nose. There was a laugh. That ingratiated me with the audience".[4]

DRUNK OR DANCER?

In the movies, the difference between Chaplin and the other actors emerged from the very start, from his first year as Little Tramp (1914), his mask, which turns a hundred this year; a difference that had already appeared in the conflict between Chaplin's and his employer Mack Sennett's understanding of cinema. The dominant trait of Sennett's comedy was the famous chase: thieves, victims of theft, policemen, con-artists and their victims, husbands, wives and lovers, policemen and criminals would, at a certain point, be overcome by a dizzying pursuit scene that was the high point of the performance and the climax to every film: everyone would suddenly start chasing everyone else. But Chaplin wasn't fond of the chase, he saw it as being a clumsy movement devoid of rhythm, far-removed from the theatrical pantomime of which he was already a master at the age of twenty-four, when he first went into the world of movies. This was indeed the difference: Chaplin wanted to develop his own version of the chase, turn it into a ballet, add leaps, acrobatics and grimaces, he wanted to add ornament to his actions in a musical sense. Mack Sennett couldn't control him anymore, and was on the verge of firing him, when he suddenly started receiving tons of telegrams from New York asking him for more of Chaplin's gags. From the year the persona was created, in 1914, Little Tramp had a number of imitators and many attempted to parody him. His first imitator was his own companion, Mabel Normand. In one of the Keystone stunts, *Mabel's Married Life*, alone in her apartment Mabel despairs of her ineffectual spouse (played by Chaplin), and addressing the camera directly, she points to his shoes, puts her hand on her head and mocks his way of walking. Dan Kamin, a Chaplin expert and himself a mime, describes Mabel's parody as being "a cross between a hop and a waddle". While the imitation reminds us of Chaplin's way of walking, it is really just a faint resemblance: "The fact is that anyone could do a recognizable approximation, but actually moving like Chaplin was another matter. His walk is far more complex and subtle than the stiff side-to-side duck waddle assumed by Mabel and most of his imitators. Given the peculiarity of his posture and costume, he walks surprisingly lightly, with a pleasing economy of movement. If he wants to tilt from side to side, he moves from the hip joint, not the waist, as Mabel does; it is much more elegant and structurally sound to move from the hip joint".[5]

The rhythm of Chaplin's way of walking could be felt, but it couldn't be broken down or even less so repeated: hops, backward kicks, skidding to a stop with one leg hopping and grimaces were repeated at a calculated distance. From the beginning his style wavered between uncertainty, the weakness of the vagabond drinker, and the arrogance of the sophisticated gentleman who need not demonstrate his scorn for others because he truly scorns them. But Chaplin's differentiated walk, which continually ranged from fear to malice to arrogance, was never quite understood by Sennett, who was used to a much more banal and stereotyped style. And this is why it didn't take long for Chaplin to understand that to avoid the savage cuts to his pantomime, he needed to begin directing his own films as soon as possible, even before leaving Keystone. At first Sennett had Mabel Normand direct him, but then he gave up and let him direct himself.

IN THE BEGINNING
IT SEEMED LIKE CHAOS

Or, we might say, "in the beginning it was chaos". But it's really the same thing, because in Little Tramp's way of walking like a drunk we find both: the order and the disorder of life. Chaplin's drunk is prey to an intoxication of the Platonic sort, noble and confused at the same time; it is an act during which aggressive drives are released, but a great deal of self-control persists, quickly turning every aggressive gesture into a game and a joke. With his falsely innocent smile, the young Chaplin is like a cat, simulating indifference before and after each attack. His Little Tramp character – nicknamed "Charlot" by French poets and painters, such as Léger, who adored him and even devoted a film to him, *Charlot cubiste* (1924) – is entirely a game of chaotic elegance, a contradiction without end, a toing and froing without an end-point, without a direction, a game that breaks down

the space and muddles the paths; it is an acrobatics in which each false step is transformed into a new equilibrium – one, in turn, precarious and uncertain. In other words, it is rhythm without rhythm, order tuned to disorder, an oxymoron in which all the negative and destabilizing aspects offer an opposite sensation in the end, the sensation of a great self-assuredness and overall stability. If Paul Valéry described the dance as being "an ornament of space", in correspondence to music that is the "ornament of time", I think it is safe to say that Chaplin's gait is a dance without dance. To paraphrase Valéry, we could call it the "ornament of chaos",[6] or else a parody of dance itself, which is always outlined or disavowed in a game of contrasts in which we recognize the unstable side of life, where chaos and harmony are continually at odds or, better yet, united and comfortable together. Little Tramp was even capable of murdering someone while dancing, while reciting poetry. And this indeed is what he would do in *Monsieur Verdoux* (1947), reciting Keats while dancing his way into the bedroom of one of his wives, Lydia, to murder her: "Our feet were soft in flowers".[7] More than once, after all, he has been compared to a legendary creature, a hero, at times even to a god, to the god Pan, if not to Dionysus himself.[8]

But does Chaplin's way of walking evolve? I would say it does, and I think its evolution is long and complex. In this essay I will try to distinguish at least four phases, each of which could naturally be broken down into some shorter sub-phases, seeing that each film is part of a different phase. But for the sake of simplicity, let's call them:

a) The Sennett period (1914), characterized by the legacy of the music halls and the combination of contradictory forms and styles, with no harmony, a mass of dances, kicks, grimaces, endless chases and escapes – a hybrid yet anarchical style at the same time, surreal and, therefore, also very free and fascinating.

b) The first period during which Chaplin finally directed his own films (1915–23), and gradually abandoned Sennett's magmatic way to seek

Kid Auto Races at Venice, California, 1914. Little Tramp's first appearance as a tramp who keeps getting in the way of the movie camera.

more harmonious and musical forms to join different registers, the comic and the tragic; a style that I would refer to as "visual counterpoint".

c) The achievement of complete harmony between comic and tragic styles (1925–40), which leads to the creation of great and sublime images.

d) The final works (1947–52) in which chaos and fear of life return, but subsumed within the perfection of art. This harmony of disharmony, expressed in his walk, has by now become a peculiar approach to the things of the world.

a) ANARCHICAL LITTLE TRAMP. The first time the Little Tramp persona appeared wearing his complete outfit and walking in his own special way was, curiously, in a movie about cinema: *Kid Auto Races at Venice, California* (1914). A film crew is filming the junior version of an automobile racing event, but, as we are told by the first caption, a curious fellow tries to make himself seen in every way possible, always getting into the shot and hampering the filming. He stands in front, to the side, he shakes all over so that he'll be noticed as he walks in the middle of the race track, he pulls all sorts of faces, looks into the camera, waves. The director or the operators throw him out each time, but he invariably returns in with his back towards us, as though he were there by chance. The contradictory aspects are all there, albeit still nebulous and imprecise. The future "gentleman ruffian" isn't really much of a gentleman. He makes his way among the people by force, he vainly sashays like a woman, he leans on his cane with an absent-minded look on his face, he refuses to give up, he ignores the people's threats and insults, he spreads his knees and kicks making sure everyone can see his long clown shoes.

At the Keystone Film Company, in less than a year (1914), Chaplin made 33 short comedies and a full-length film, *Tillie's Punctured Romance*, in which he co-starred with Marie Dressler. It's the story of a poor country girl whom he, dressed as a teddy boy, seduces and

Tillie's Punctured Romance,
1914. A drunk's tango.

Mabel's Strange Predicament, 1914.
A drunk's walk.

then abandons. There's a little bit of everything in the movie, even an actual orgy among millionaires where Chas gets drunk and dances a mad tango, which includes acrobatics, leaps and goofy antics; he falls and gets back up without touching the floor. He is a fireworks display in which you can't make out the different dance steps, the kicks and the punches. In *Tango Tangles* (also made in 1914) Little Tramp enters the dance hall in a drunken state and causes the same pandemonium. Dancing, falling, punching are all closely intertwined, and without being able to say which of them prevails over the others, so harmoniously are they carried out in a single, multiple gesture. It's not hard to see the dance steps he attempts and then immediately gives up even as he's walking. Sometimes it looks like he's doing a *grand battement*, stretching his leg – almost always the left one – out in front of himself, without ever completely getting his knee straight. The same can be said about his foot, which is often flexed, so that it's vertical instead of with the toe pointed forward as it should be in the *battement*. He often places his right foot *en dehors*, with the toe turned outward, as is typically done in classical ballet, but he does so while standing with his back to the movie camera, the opposite of what the dance step normally requires.

His partners are tough and stiff. With Fatty (Roscoe Arbuckle) Chaplin often creates duets; in *The Rounders* the two drunks swagger from one bar to the next, but while Fatty, who's big and fat (as the nickname says) drags himself from one place to another in a stupor, Chaplin has a rhythm that is clearly visible, constant, he plays and dances with everything. *Twenty Minutes of Love* is one big brawl, but it's also visual music. Even when the punches and kicks he receives make him fall to the ground, with a somersault he's back on his feet again, like at the circus, which was another one of his sources along with the figure of Clown. His big over-long shoes contribute to his shaky equilibrium, and they remind us of the acrobatic shoes worn by clowns, thanks to which he can lean as far forward as he wants

Caught in a Cabaret, 1914.
His first jumps with one leg hopping.

without ever falling. In one of the first gags, *Mabel's Strange Predicament*, where he doesn't play the leading role but is the third wheel between two couples, he wanders around a hotel drunk, he bothers the ladies, he can't even stay seated, he slips to the ground from all the armchairs, he kicks, slides in everywhere, always on the brink of falling: yet he never does. Nonetheless, what also transpires from his state of human dereliction is the motif of fallen nobility, one that perhaps was never born: a sovereign disdain for those who pass him by, a harmonious gesture that allows him to offer a lovely lady a flower he just stole right before her eyes. *The Star Boarder* and *His Favorite Pastime* involve acrobatic games so that he can stay by the side of a woman, and fights with husbands and boyfriends twice his size. *The Fatal Mallet* (1914) is a war in which rubber bricks are hurled and a rubber mallet is used, in which he too becomes a rubber marionette.

Little Tramp often wears a disguise: the tramp becomes a gentleman, or else an elegant lady, but his walk is always the same, and it unequivocally tells us that every disguise is an illusion and can fool no one: you can never not be yourself. In *Face on the Barroom Floor*, Charlie is a drunken tramp who aspires to be a great artist, but the acrobatic movements he makes to stay standing fail desperately, and he ends up with his face on the floor, revealing the artist inside the tramp, and the potential tramp inside the artist. In *Caught in a Cabaret* Little Tramp is the waiter in a seedy cabaret who dons a gentleman's clothing. But once again it is his gait that betrays him: if it is true that beneath the waiter there is always a gentleman slumbering there, when the gentleman finally does appear we discover that there's the soul of a waiter inside.

Chaplin plays a woman for the first time in *The Masquerader*. To be able to work in the movies Little Tramp dressed up as a woman, a very beautiful one, filled with charm and sweet nothings. Everyone courts her, but she walks as though she were dancing, the way Little Tramp does, and just like him she can't help kicking. Here is another side to Chaplin's way of walking: his hidden femininity. He sways his hips and sashays with his shoulders with a hint of vanity; he is feminine without the woman. He'll do the same thing in *The Perfect Lady* (1915), in which he sashays around in a woman's clothing with a prissy look on his face so that he can be with his girlfriend – he even ends up also seducing her father. Elsewhere we find the military figure without the soldier, such as the hussar in *A Film Johnnie*, the parody of a soldier in dance form (which would be perfected in *Shoulder Arms*, 1918). But now, while he is still at the beginning, he is only a fake hussar who from the front smiles the way a gentleman would to a lady, while behind he kicks like a horse. Another classical ballet position he seems to parody is the *attitude*, which he uses his left leg to execute. The *attitude* involves raising the leg at a 90-degree angle to the body while keeping the foot *en dehors*, that is, open outwards, and extended and raised as high as the other knee. These rhythmical interruptions also often appear in Chaplin's gait, especially in the way he moves his shoulders. The ambiguity here is total. Indeed, we'll never be able to say whether he escapes pretending to dance or whether he dances pretending to escape.

While working with Sennett, Chaplin learned how to use very different techniques all at the same time: the pantomime of the drunk, the ballet in Karno's variety shows, the chase, the escape, the fight, a hybridization that was often confusing and anarchical of very different movements and registers that stay together miraculously, or by chance, combining the high with the low, the loafer with the hero.

b) COUNTERPOINT AND MUSIC FOR THE EYES. After he left Keystone, in the three years that followed, Chaplin worked hard to eliminate from his pantomime the earlier, grotesque aspects that audiences enjoyed so much, but that were also rather stereotyped. The Apollonian spirit (dance) and the Dionysian spirit (inebriation) were already there and very close to each other. It was a question of joining them, interweaving them,

Caught in the Rain, 1914.
Jumping with a backward kick
in the manner of Harlequin.

The Knockout, 1914. Little Tramp is the referee at a boxing match, stopping two boxers with a pose that is a parody of the ballet's *grand battement*.

trying to find a new form of acting in which the contradictory aspects could finally be integrated. Gradually, Chaplin's very original and surreal style surfaced: a counterpoint of different rhythms. After this three-year period, when he began to put together some of his greatest works, such as *Shoulder Arms* in 1918, or *The Kid* in 1921, Little Tramp's character was by then far-removed from the hybrid mask it had been at Keystone. Although he dressed the same way, in apparent continuity, it was an entirely new figure, whose walking was just enough to create his very personal counterpoint: a dissonant harmony.

For the start of the movie *The Tramp* (Essanay Film Manufacturing Company, 1915), Chaplin found the model that he would so often go back to: a scene out in the open and in depth of field in which we see the tiny man walking towards us, from far away to a close-up shot, looking right into the camera. And when the little man, instead of joining us, walks away form the camera and into the distance, he will create the greatest farewell scenes of all his masterpieces, in which his incompatibility with the world finds its visual synthesis. These are the great finales of *The Pilgrim*, *The Circus*, *Modern Times*, *Monsieur Verdoux*. However, this model had already appeared back in 1917, in *The Count*, where he runs off, hopping and kicking backwards, until he becomes very small, invisible. But before moving away from his viewers, for many years Chaplin would have to come towards them, to conquer their liking and admiration. Nothing is taken for granted, nothing is born suddenly: grace must be conquered a few steps at a time. Since 1915, the fact that he was directing his own films allowed him to organize ballet and the parody of the ballet. The years he had worked for Essanay (1915–16) and then for Mutual (1916–17) were years of freedom from the dry residues of the variety show, years of constant elaboration for Chaplin, who was already working every single day of the year, the way he always would, and filming and printing all the rehearsals he made, at great expense, so that he could see himself on the screen, as he didn't

even trust himself. The relationship between film used and film made was already close to a ratio of 1:100, something never to be equalled by anyone else. Only Eisenstein, Kubrick and Fellini would come close, and only in a few of their movies.

The drunk was hard to get rid of. Actually, he never quite disappeared. We find him in *A Night in the Show* (1915) where he keeps walking by in the theatre, burping and making all the people seated stand up each time, or in *A Night Out*, also made in 1915, in which he loafs around all night long, lurching from one café to another, bumping into everything: fountains, clothes hangers, people; now he does so with a more musical rhythm, more dancingly – but it's still ambiguous. When he cocks his leg and hops is he dancing or imitating a dog peeing? Hard to say, actually impossible to say. In the parody of *Carmen* (1915), in which Chaplin plays the part of Don José, he is a duck-like soldier who walks with his knees spread apart and feet flat: he's lucky to stay standing, but never falls.

Little Tramp's solos, the repetitions, the rhythms continued to grow. The period peaked with *One A.M.* (1916), his first totally solo performance. In the movie he gets home drunk and endlessly repeats his attempts to climb the stairs: he tries in every possible and imaginable way, before giving up. A puppet full of grace, he seems to be held up by invisible strings. *The Cure* (1917) is the apotheosis of the drunk, which had made him so successful. By making an entire hotel get drunk he transforms the lobby into a group party. If dance and partying are transformed into battles, his battles, his boxing has a great deal in common with dancing: *The Knockout*, where he plays the referee but gets punched much more than the contenders, as well as *The Champion*, and *City Lights*.

The musical motion of his walk becomes clearer with every movie he makes. Chaplin has the grace that Kleist talks about in his famous essay on puppets: that unwitting natural elegance that can only be found in the absence of human conscience, or that can only be

The Pilgrim, 1923. Hopping along the border between the United States and Mexico.

achieved at the end of a very long working pathway, in the quest for simplicity.[9] Kleist tells us that nothing is harder to find than naturalness, that delicate elegance that we have at times had for an instant, without knowing it. But Chaplin the puppet-cum-man does find it. As of 1917 he was no longer just a mime, he was also a comic and tragic poet who gave shape to his own world, to all that surrounded him, both people and landscapes. His way of walking would arouse the comments and endless reflections of so many theoreticians and philosophers. Béla Balázs remarked as follows: "Because in such films the rhythm of the walk of the wayfarer who looks at the landscape plays a very important part and may determine the entire atmosphere of the scene ... The hero's walk will reveal what happened; his walk will be a confession and a soliloquy expressing his reaction to the scene just experienced, more completely and more sincerely than if he had been shown on the spot itself. ... Can anyone who has seen it ever forget the last shot of Charlie Chaplin's *Circus*, when Charlie goes away? Once more he has been left alone, deceived by life; he has lost everything and has started out again all by himself. But the whole wide, free world lies before him, and anything might still happen, and the sun shines and the distant dreams lure him on. Charlie waddling away into the infinite was a beautiful, optimistic visual poem".[10]

Or Pasolini: "Let us recall the boxing scene in *City Lights*, in which Little Tramp is up against a champion who is as always much stronger than him: the marvellous drollery of Charlie's dance, his tiny steps a bit here and there, symmetrical, useless, heart-breaking, and irresistibly ridiculous".[11]

Chaplin's way of walking was gradually honed, refined, perfected. From a simple contrast between dance and drunkenness it became the quest for order in chaos, or chaos in order, and finally it was transformed into an admirable synthesis of opposites: joy into pain, happiness into despair. His walk was becoming a visual monologue on life's contradictions.

Modern Times, 1936. The second great farewell finale.

c) **TOWARDS THE SUBLIME.** One of Chaplin's first medium-length films, *The Pilgrim* (1923), ends with Little Tramp's cheerful and desperate escape as he hops along the border between the United States and Mexico, a foot on either side. On one side, the American police are after him, on the other, bandits are shooting at everyone and everything, hidden behind the brush of the Mexican desert. Forced to dance on the metaphorical rope of the border, he offers us one of his richest figurations, symbolizing the man who "walks on a tight-rope" (Nietzsche, *Thus Spake Zarathustra*). The border is a metaphor: on one side there is order, the law, but also prison, on the other side there is anarchy, freedom, but also fear. It's impossible to choose, you need to walk with one foot on either side.

Later, in the movie *The Circus* of 1928 (which it took him five years to make), after having gone through unspeakable acrobatic feats to defend his love, when he realizes that she, Merna, loves the real acrobat, the young and strong one, he goes off on his own, the same way he had arrived, and we see him from behind, as he walks off into the endless plains of the world. But pain is just a game, a pantomime: his hips sway so joyfully they are perhaps telling us that he has in actual fact given up nothing and everything starts over again from scratch. Even work can make a person drunk: in *Modern Times* (1936) Chaplin leaves the factory hopping and buttoning the skirts of the women he meets along his way. The ambiguity between inebriation and dance that he had learned when he was working for Sennett now becomes a symbolic image that represents the inability to adapt and the impossibility of not doing so. How can one live like this? But how can one live otherwise? *Modern Times* also ends with Chaplin walking away with his back to us, until he disappears into the distance, this time in the company of a beautiful street urchin (Paulette Goddard): another great farewell to the world that could also, however, be an entrance into it.

The Circus, 1928. The first great
farewell finale, amidst joy and pain.

Everything remains, everything comes back. The very famous dance with the globe in *The Great Dictator* (1940), where the tyrant in monkey-like fashion climbs up the curtains, hops onto the desk and with his bottom strikes the world turned into a balloon, is a perfected version, a development of the original comic scenes: the mischievous kicks he would deliver his rivals have become delicate prods he gives to the whole world to adjust it to his liking. In the same movie, as the Jewish barber (the dictator Adenoid Hynkel's good double) escapes along with the Nazi colonel Schultz – due to that inconceivable misunderstanding that will lead him to command the Nazi troops – he never forgets to interrupt the escape along the roofs, hopping and raising his left leg, in the position of the *grande attitude*. But once again we don't really know whether it's a dance step or his imitation of a dog peeing.

d) A SYMBOLIC FORM. We are close to the end, and *Monsieur Verdoux* turns all of Little Tramp's comedy into tragedy. The poor bank clerk who is fired and, to be able to support his family, becomes the murderer of rich widows, manages to win over the sympathy of his spectators thanks to the many slapstick scenes added to this tragic tale, inspired among other things by a real person, the famous French serial killer Henri Landru. Chaplin forgets nothing of his old experiences, and from Karno to Mack Sennett he preserves all of the old along with the new. When Verdoux walks away from the camera towards the dark and gloomy guillotine in the background (the story is set in France), as usual we see him walking away with his back to us. But this time he's not in the countryside, he's inside, walking along death's corridor, between two gendarmes. Verdoux's walk is uncertain, wobbly, it reminds us of the old Charlie, except that now we see the white hair of an aged tramp, which reveals all the hidden ferociousness – but not too much so – in the little man of the past. We wish to shout out along with André Bazin: "Verdoux was Charlie! They're going to guil-

Monsieur Verdoux, 1947. The third great farewell finale: society has condemned Little Tramp to death!

Limelight, 1952. The fantasy: in poor Calvero's tormented dreams, Chaplin recalls his own debut with Fred Karno's Pantomime Troupe in London in 1914, with the old pantomimes of the gentleman tramp.

lotine Charlie! ... This same road to nowhere, always taken from film to film by the little fellow with the cane, which some see as the road of the wandering Jew while others prefer to identify it with the road of hope – now we know where it ends".[12]

Chaplin's message is always ambiguous, just like his way of walking. A few years later, in 1952,[13] *Limelight* would not just be a tribute to the old vaudeville theatre; it would above all be a subverting of the despair and solitude of Monsieur Verdoux, an exhortation to be brave, to be strong, to struggle and to laugh, actually to smile, sovereign and victorious, even in the face of death. In this last testamentary film, unsurprisingly set in London in 1914, as we read in the initial caption (and we know that 1914 was the year Chaplin left London, the year he bid farewell to theatre, the year of Little Tramp's birth), Chaplin went back to the old pantomimes, with the two scenes of the flea trainers and the gentleman tramp, whom old Calvero dreams of during his long nights, and especially with the great finale in which Calvero,

along with Buster Keaton, translates into visual form the whole desperate search for harmony and art against the chaos of the world. The two players can't play and end up smashing their instruments – piano and violin – into pieces. These are the two great symbols of the struggle against time and the storm of life, an image worthy of being placed alongside the myth of Ocnus, condemned to spend eternity weaving a rope of straw, which the donkey standing behind him eats as fast as it is made, in the bas-relief at Villa Pamphilj in Rome.[14] Each of the two players bears within the conflict between art and the world, order and disorder. It is again the drunk's way of walking, with which Calvero enters the start of the movie, and it is the memory of old times. But it doesn't seem like nostalgia. Rather, it seems to be saying that only in street theatre, in the music of street musicians, in popular entertainment, in celebrations, in taverns, in the songs of old drinkers can art find the strength to start over again. Not by chance the three travelling musicians in the movie, very similar to the three Parches, are always performing in front of a "Rum House".

So let's try to conclude the way we started, with Plato, a lover of wine, equilibrium, dance. Let's read his words in the *Symposium*: "Socrates was discoursing to them. Aristodemus was only half awake, and he did not hear the beginning of the discourse; the chief thing which he remembered was Socrates compelling the other two to acknowledge that the genius of comedy was the same with that of tragedy, and that the true artist in tragedy was an artist in comedy also. To this they were constrained to assent, being drowsy, and not quite following the argument. And first of all Aristophanes dropped off, then, when the day was already dawning, Agathon. Socrates having laid them to sleep, rose to depart".[15]

The common themes between these final page of the *Symposium* and Chaplin's work are many: in the first place, the drinking, which is the moment of risk from which wisdom and inspiration are born; then the deep-seated analogy between tragedy and comedy: if one is a poet of the former, he is also a poet of the latter; and, lastly, the farewell, which is of the same type: also in many of Chaplin's movies, as we have seen, the main character walks away, alone or in company. Furthermore, Verdoux, before heading towards the guillotine, says his last line, which seems to be filled with irony if we think of Little Tramp the drunk: "Just a moment. I never tasted rum".[16]

Chaplin's way of walking as he moved away from the screen over the years gradually came to symbolize the inability to live life. But perhaps what Socrates and Chaplin along with him wanted to say was exactly the opposite: we mustn't take either comedy or tragedy too seriously, as they are just two literary genres. And with this I can end my discourse. After all, I like wine too! Appropriately.

[1] Plato, *The Laws* (London: Penguin, 1970).
[2] C. Chaplin, *My Autobiography* (London: Penguin, 1964), p. 138.
[3] B. Dixon, "Harlequin in the New World", in *Chaplin's Limelight and the Music Hall Tradition*, edited by F. Scheide and H. Mehran (Jefferson, North Carolina and London: Mc Farland & Co., 2006), pp. 148 ff.
[4] Chaplin 1964, p. 100.
[5] D. Kamin, *The Comedy of Charlie Chaplin* (Lanham, Maryland, Toronto, Plymouth: The Scarecrow Press, 2008), p. 24.
[6] P. Valéry, *Degas danse dessin* [1936], now in *Oeuvres*, vol. 2 (Paris: Gallimard, 1960), p. 1172.
[7] L. Nepi, "Un Chaplin moderno: Monsieur Verdoux tra Keats Schopenhauer e il fermo-immagine", in *Annali del dipartimento di storia delle arti e dello spettacolo*, XI, 2010, pp. 253–64.
[8] R. Payne, *The Great God Pan. A Biography of the Tramp Played by Charles Chaplin* (New York: Hermitage House, 1952).
[9] H. von Kleist, "On a Theatre of Marionettes", 1811.
[10] B. Balázs, *Early Film Theory: "Visible Man" and the Spirit of Film* [© 1952] (Oxford and New York: Berghahn Books, 2010).
[11] P. P. Pasolini, "Il cinema di poesia" [1965], now in P. P. Pasolini, *Empirismo eretico* (Milan: Garzanti, 1972), p. 185.
[12] A. Bazin, *What Is Cinema?* [1958] (Berkeley: University of California Press, 2004), p. 109.
[13] Although the opening credits state that the movie was made in 1951, it was not approved by the Board of Censorship until the following year.
[14] For the myth of Ocnus, see J. J. Bachofen, *Il simbolismo funerario degli antichi* [1854], edited by M. Pezzella (Naples: Guida Editori, 1989), pp. 509 ff. In the Etruscan funerary relief at Villa Pamphilj a man is seated weaving a rope of straw while the donkey next to him eats it as fast as it is made. In this version of the myth of the Three Parches, Bachofen sees a symbolic representation of the conflict between the principle that weaves and the one that devours, an image of human labour against time, and nature that destroys all of man's work.
[15] Plato, *Symposium* (Oxford: Oxford University Press, 2008).
[16] After kindly saying no to the guards who try to help him find the courage he needs to face the guillotine by offering him a glass of rum, he changes his mind and accepts.

Modern Times, 1936. The second great farewell finale.

BALANCES: LEARNING TO LEARN. FELDENKRAIS' GREAT LESSON

EMANUELE ENRIA

TOUCHING AS A SOCIAL ACT: FEAR

Touching. Being touched. There is no inhabitant of a large city who is unfamiliar with the varied, pleasant or oppressive, curious or violent, friendly or diffident feeling of being touched, brushed against, caressed, pushed, embraced, jerked, approached or kept at a distance in the many different moments of everyday life. It happens when we find ourselves on the metro at rush hour, face to face with a stranger only a few centimetres away from us. It happens in the street, in shops, at the stadium, in museums. At school, in church, in parks, on trains. At home. It happens from the time we are newborn babies; and the way in which we are touched, even verbally, and rocked, cuddled, scolded and rewarded influences our development – as revealed by Freud.

We are constantly touched or brushed against by others, and our body – along with our nervous system (its radar, we might call it, the code of access for movement, whose main function is to inform the body as to what is going on around it) – conforms to this in its actions, attitudes and postures, including unconscious ones. Thus like a rising citadel, we come to stand on our two feet after having experienced the relation between our body and space. The relation between the "inside" of the body, which governs its structure and connections (skeleton, musculature, brain), and the "outside" that surrounds it, was called "self-image" by Moshe Feldenkrais, who argued that this is what determines our every behaviour.

August Sander, *Kinetic Researcher from Vienna (Max Thun-Hohenstein)*, 1930.

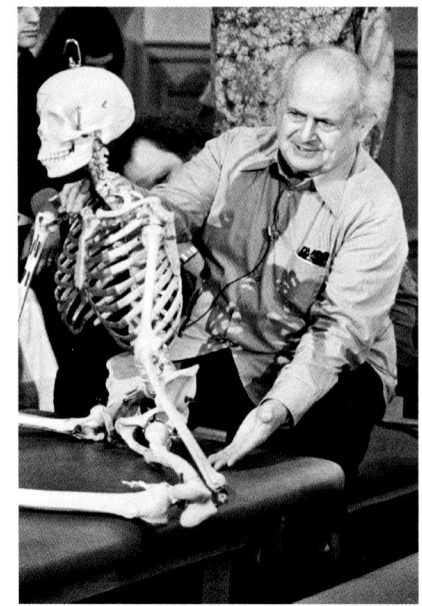

But what did he mean by such word? "I would argue that it is a body image; namely, it is the shape and relationship of the bodily parts, which means the spatial and temporal relationships, as well as the kinesthetic feelings. Included with this are feelings and emotions and one's thoughts. All of these form an integrated whole".[1]

According to Feldenkrais, our usual neuromotor patterns merge with our emotional and mental habits: "Bodily attitude and movement, emotions and thoughts, are only different aspects and functions of the same reality: through movement, the nervous system merely implements those distinctions which lead to the preference or choice of given actions or forms of behaviour".[2]

The industrial revolution has shaped urban fluxes and urban dwelling once and for all through factories and machines, including the two most amazing extensions of the human eye: the photo camera and the video camera, mobile eyes that extend or recoil according to one's wishes. It is impossible to understand or even broach the theme of balance, of walking and dancing, without taking account of this new environment, which has come to influence people's body, postures and attitudes – and indeed clothes and footwear, this second skin as thin as a glove or as thick as armour, a tangible sign of our relation with touch and contact, and with our own self-image. Every inhabitant must deal with these variables in order to find his or her balance – before losing it again and again.

Instability is an inescapable law of walking and balancing. An erect posture, of the sort developed by mankind, is the most precarious system of locomotion. We only need to consider the skeleton supporting us: it shows that through the act of standing on two feet instead of four like animals – an achievement of our species, ensured by the vertical superimposition of pelvis, trunk and head

– the centre of gravity is brought as far up as possible, thereby reducing the moment of inertia required to move it. While many land animals are much faster than man, only an erect posture enables one to move across space in any direction. Yet because our centre of gravity is as high up as it could be given our skeleton – we would be far more stable if we reverted to sitting with our arms resting on the ground, thus lowering the centre of gravity – our balance will always be precarious when we stand. For this reason, walking invariably means "walking and falling" at every step, to quote a song by Laurie Anderson. Besides, science has shown that the body is always in motion, even when it appears to be still, through what is known as "postural oscillation".[3] We constantly seek to adapt our point of balance: a climber's point of support will probably fall just in front of his or her feet, that of a dancer will shift again and again, and that of a person wearing heels will be set considerably forward: the brain must constantly appraise how the muscular framework of the body as a whole is rebalancing its posture.

Moshe Feldenkrais (1904–1984) is among the most versatile figures of the twentieth century. Having set off from Poland at the young age of fourteen to reach Palestine on foot, he discovered one of the most effective ways of putting the body, and hence the mind, in a condition to continue learning. Feldenkrais developed a capacity to analyse the human condition, starting from the very way in which we move, in certain respects carrying on – a century later and through his own discoveries, inventions and his very personal synthesis – the unfinished work begun in 1833 by Honoré de Balzac with *Théorie de la démarche*. With a striking degree of versatility, Feldenkrais combined his work as an engineer and physicist, martial arts instructor (the first to be authorized by Kano, the founder of judo, to

Moshe Feldenkrais in
San Francisco, Training
Courtesy of the International
Feldenkrais Federation Archive.

teach the sport in France) and researcher in Paris under the future Nobel laureate Frédéric Joliot-Curie with the theories of Freud and Pavlov, Frederick Matthias Alexander's technique, biology, ethology, Darwin's notion of heredity, and contemporary neuroscience.

Touch, however, also played a crucial part in Feldenkrais' work. In fact, he envisaged it as the code of access to the nervous system – that "system more ancient than ourselves", as he used to call it, which played as fundamental a role for him as the unconscious did for psychology. But in this case, there was no need for words. Already Elias Canetti, in his extraordinary book on the formation of groups and crowds and their relationship with power, had stressed that "Humans fear nothing more than being touched by the un-known".[4] This fear constitutes the foundation on which man builds his own habits, home and office, or fills his city with lights, in the constant fear of being touched by what he does not know. Not only that, but his fear of being touched, of finding himself too close or too far, of being invaded or overrun, is the constitutive mechanism at work between crowds and power.

Why, is this not how political leaders, Hollywood stars, celebrities or the pope construct their identity today, through physical distance or proximity, also in terms of touch, of whether people can touch them? Thus sitting, standing, lying, kneeling, standing on a podium (like an orchestra director, who will stand on a podium alone, with his back to the public; and at a given moment in his career, the great Jazz trumpet-player Miles Davis also started playing with his back to the public, as though to claim the same authority and attention due to a classical musician – how revealing postures can be...), or even the marching of an army, which in the mind of a German may evoke "a walking forest",[5] all reveal a kind of group identity, understood as a single social body. These variables make up the personality of individuals, which manifests itself through their posture (Feldenkrais was to show in more detail how conflicts are always rooted in the body). There is one variable that Canetti overlooked, however, aside from the fear of being touched: the fear of collapsing. This second variable too deeply influences whole social systems as well as individuals. Do governments, myths and generations not fall? Are not symbolic monuments toppled over, as in the case of statues of leaders during revolutions? Every society and social system will behave and organize itself in such a way as to resist falling. Sociologist Pierre Bourdieu also realized this, noting that bodily posture (*héxis*) conforms to social and political requirements.[6] Individuals face this fear right from their childhood, starting from the moment in which they are picked up and held in someone's arms with the utmost attention, lest they should fall.

Feldenkrais, then, spent his whole life investigating the acquisition of the motor patterns through which we move and conduct

Feldenkrais, Judo Rolls
Courtesy of the International
Feldenkrais Federation Archive.

ourselves. He noted the striking similarity between the reactions of newborn babies afraid of having their support removed from under them – namely, a violent contraction of all flexor muscles, with a sudden stop in breathing – and adults' reactions of fright or fear. Feldenkrais even hypothesized that this kind of reaction to falling might be present from birth, and hence that it might be inborn and independent of individual experiences. The eighth cranial nerve is the seat of both the cochlear nerve of the ear, which is related to hearing, and the vestibular nerve, which is related to balance. When this nerve receives a strong stimulus connected to this fear, the impulse further expands, reaching the tenth cranial nerve, the upper olive, which controls breathing. The early experience of anxiety is therefore related to the stimulation of the vestibular branch, and hence to balance and the fear of falling. Even from an evolutionary perspective, this factor might be interpreted as a kind of transmitted survival instinct: for how could the newborns of an arboreal primate have survived the constant risk of falling from trees, if not by making their rib cage flexible by suddenly contracting their abdominal muscles, holding their breath, and bending their head forward? This ensured a kind of protection of "vulnerable" areas such as the head by making the body only hit the ground with "a strongly arched spine somewhere in the region of the lower thoracic vertebrae or lower, nearer the centre of gravity. The shock will therefore be transformed into a tangential push along the spinal structure, on either side of the point of contact, and absorbed in the bones, ligaments, and muscles, instead of being transmitted directly to the internal organs, and so injuring the body fatally".[7] Interestingly, the position that is taught in judo in order to reduce the impact of falls is exactly the same as the one evoked in newborns by the feeling of falling. This shows how Feldenkrais succeeded in wonderfully bringing together a range of different elements, combining them in such a way as to form what we would call an "organic whole".

Freud identified anxiety as the central problem behind all forms of neurosis. And let us also consider what Paul Schilder had argued, as quoted by Feldenkrais himself: "Dysfunction of the vestibular apparatus is very often the expression of two conflicting psychic tendencies; dizziness occurs, therefore, in almost every neurosis. The neurosis may produce organic changes in the vestibular sphere. Dizziness is a danger signal in the sphere of the ego, and occurs when the ego cannot exercise its synthetic function in the senses, but it also occurs when conflicting motor and attitudinal impulses in connection with desires and strivings can no longer be united. Dizziness is as important from the psychoanalytic point of view as anxiety. The vestibular apparatus is an organ, the function of which is directed against the isolation of the diverse functions of the body".[8]

To return now to the relation between crowd-formation and power, as described by Canetti, or the formation of an individual's posture and his or her personality, was it still possible to overlook the psychological as well as physical fear of falling? Was not the same relation that defines each sphere of life – the private, the sentimental and the social – at play here, namely the relation between stability and instability? Feldenkrais calls these the "biological aspects of posture". Stability, which for a physicist/engineer like him meant the amount of resistance required in order to move or be moved. But from a psychological perspective, all this easily translates into a stability that brings safety, a feeling of protection – and therefore dependence upon such state. And instability, which entails that a thing can easily move, physically shift. Yet it also implies a risk in psychological terms. Were not both necessary? Did systems not fall ill when they no longer had both options and found themselves confined to only of the two? Was this not the origin of totalitarian states? This is where anxiety as the "fear of falling" emerged. Both aspects are biologically important. Yet it was necessary to define to what extent the most complex system of all, that of the human being, consisting of

Andrea Vesalio, *De humani corporis fabrica libri septem* (Venice: Francesco de Franceschi & Johann Criegher, 1568), 32.3 x 23.2 x 5.5 cm Florence, Biblioteca Nazionale Centrale di Firenze.

a brain and nervous system, is willing to forego the two parameters in order to gain acknowledgement, or even a minimum of acceptance, from its own cultural group, starting from the family. This was revealed to Feldenkrais by none other than his greatest teachers: children – by the way in which a newborn baby develops his motor skills to the point of being able to stand and walk.

NOTES ON THE NERVOUS SYSTEM

In order to fully understand the way in which Feldenkrais developed what was destined to become his method, it is necessary to sum up – also in the light of all the discoveries made in the twentieth century – what happens when we perform an action, i.e. when we move. This is a matter that had already been explored by the great Aristotle. How is the intention of movement transmitted to the muscles that will move the skeleton?

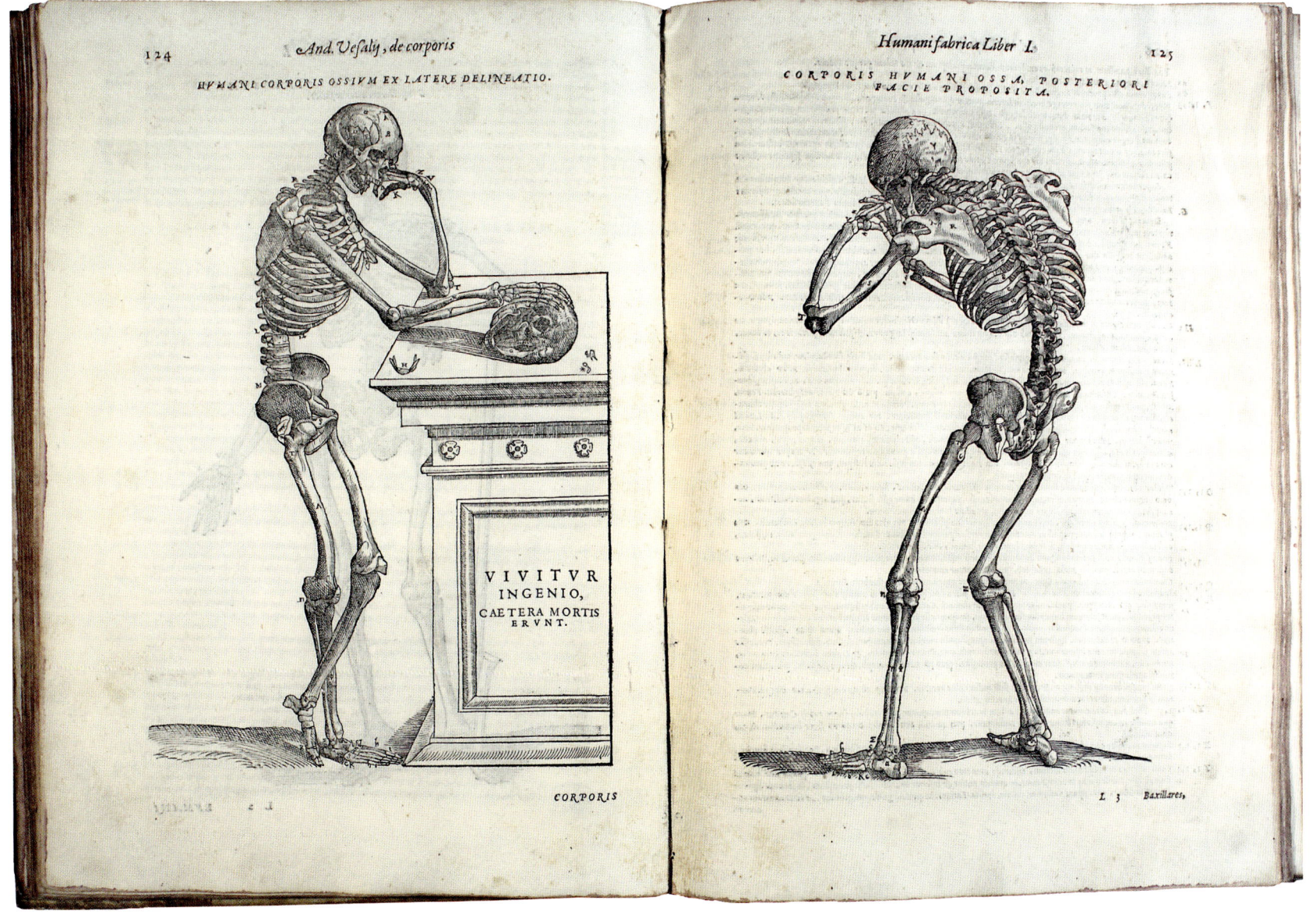

"Our muscles are controlled by pyramidal neurons (neurons shaped like a pyramid), which are located on the motor cortex, or Ronaldo's cortex – after the Italian anatomist who was the first to describe it. From this area of the cortex orders are issued to individual muscles. The orders are channelled through the pyramidal neural pathways that reach the spinal marrow, enter in contact with the motor neurons located there and then continue on to the muscle fibres. When we bend a finger, the order comes from the motor cortex and reaches the flexor muscles in a fraction of a second … the harmonious arrangement of movements is ensured by the ganglion cells of the base and cerebellum: these neural structures record the memories of those motor organs (the triggering of a sequence of muscles) that will enable us to strike a moving ball, execute a vault on the parallel bars, or perform a crawl-stroke or butterfly-stroke … located in front of the motor cortex are the premotor cortex and the supplementary motor cortex, areas which select what movements are to be executed by the motor cortex. … Before movement occurs, the moment in which we are thinking of performing it but have yet to do so, the premotor cortex – located in front of the motor cortex in the frontal lobe – becomes active. This shows that there are areas of the brain that prepare movement and areas that implement it: the premotor cortex prepares the motor action without this being executed, without the motor cortex being activated. In other words, thinking a movement means half implementing it".[9]

THINKING, ACTING, TOUCHING

What we find, then, is that thinking a movement already entails a form of muscular behaviour. What's more, thought implies a physical function supporting the mental process. The very speed of thought is closely related to the speed of the functions of the motor cortex.[10]

When we count in silence, for instance, we realize that the passage from 0 to 10 is in any case faster than that between two-digit numbers, since our thought already brings the pertaining musculature into play. Cognitive functions are always integrated with motor functions. And if we wished to change the way we act, or act in a different way, for instance by changing the way we walk, would this not mean automatically changing the very thought of walking, of the various body parts involved, such as our feet, and hence operating on the self-image discussed by Feldenkrais?

So how can a body be prevented from reproducing many of the habitual patterns it has acquired? Once again – and this was Feldenkrais' brilliant intuition – it is a matter of balance and gravity: a position must be found in which the body will not activate the anti-gravity mechanisms that trigger many of these habitual patterns. This is explicitly revealed by the posture the body keeps in the specific situation in question. For if we make a person lie down on the stomach or back, on a horizontal floor or on a stable bed, what supports his or her body is the floor. After a few moments the nervous system will recognize that there is something stable supporting it and that therefore it has no need to activate all its anti-gravity mechanisms. In such a way, the usual pressure exercised on the sole of the feet is removed (i.e. the way in which we usually keep our feet to support us when standing), along with their influence on the muscle joints of the body.

Feldenkrais noticed, however, that even in this position the body will preserve many of its habitual motor patterns. He did so by experimenting with the body, by touching it and making it move. For example: he would ask people lying on their stomach to bend the legs in such a way as to make the soles of their feet face the ceiling; in most cases, their feet would not be parallel to the ceiling. Feldenkrais thus noted that even when people's musculature was not supporting any weight – as in this case their feet – it still did not adopt a neutral body schema in relation to the force of gravity. The reason for this is

LIST OF WORKS EXHIBITED

PATENTS, SHOES AND WORK TOOLS

From the Salvatore Ferragamo Archive. The original patents are kept in Rome, Archivio Centrale dello Stato

Patent: 1,399,606, granted in the United States on 6 December 1921
Application: 427740, 2 December 1920
Surgical instrument.

Patent: 1,479,536, granted in the United States on 1 January 1924
Application: 512612, 3 November 1921
Orthopaedic device.

Patent: 281241, granted on 7 January 1931
Application: 6243, 9 July 1929
System to reinforce the arch of the sole of a shoe, known as the waist.

Patent: 354889, granted on 13 December 1937
Application: 7496, 17 September 1937
Shoe in the form of a wedge so that the sole and the heel are not separated.
The patent is for one of Salvatore Ferragamo's most famous inventions, the wedge heel.

Ghillie, 1937
Brown suede upper. Vamp divided in the middle by a calfskin strip. Cork wedge heel covered in brown calfskin, 26 x 4.5 cm.

Patent: 15953, granted on 10 March 1939
Application: 671, 6 July 1938
Sole for sandals or the like made of several layers of cork covered in fabric.

Sandal, 1938
Black satin sandal with wedge heel made of several layers of shaped cork, covered in alternating gold and silver calfskin, 25 x 6 cm.

Patent: 17366, granted on 21 February 1940
Application: 667, 16 June 1939
Woman's shoe with wedge heel support, rounded in the back.

Sandal, 1939
Red suede upper. Border padded with gold kid around the V-shaped collar. Squared cork insole, referred to as "wedged", lined with red suede and trimmed with gold kid, 20 x 6.5 cm.

Sandal, 1939
Black suede upper. Border padded with gold kid around the V-shaped collar. Squared cork insole, referred to as "wedged", lined with black suede and trimmed with gold kid, 20 x 6.5 cm.

Patent: 18366, granted on 31 October 1940
Application: 559, 14 May 1939
Woman's summer shoe, woven raffia with cuff and triple wedge heel.

Patent: 19736, granted on 23 July 1941
Application: 161, 11 February 1941
Woman's sandal, with layered raised sole and connection between heel and said sole.

Sandal, 1940
Black velvet and kid upper. Wooden and cork insole and heel covered in gold and silver kid, 24 x 13 cm.

Patent: 390769, granted on 19 December 1941
Application: 2830, 18 April 1941
Direct stitching of the upper to the sole. So as to avoid contact between the stitches and the surface on which the foot rests.

Sandal, 1942–44
Sandal with upper comprised of four crossed burgundy calfskin strips, two of which used as fasteners. Three-layered cork wedge insole covered in burgundy calfskin. White felt sole sewn and tacked, 22 x 6.5 cm.

Patent: 22930, granted on 15 May 1943
Application: 289, 18 March 1943
Mule with upper made of ribbons that serve as holes through which to pass the lace that holds the foot still.

Sandal, 1943
Upper made of strips of gold kid and an orange silk cord fastened around the ankle. Wooden wedge insole carved with geometric motifs and painted in various colours. The wedge is divided between the sole and the arch of the foot to make movement easier, 22.5 x 8 cm.

Patent: 25505, granted on 16 November 1946
Application: 524, 12 March 1946
Insole for children's shoes with areas that are lighter and help grip the ground on which the wearer walks. The sole has several variously shaped holes that ease support and act as suction cups.

Primi passi (First Steps), 1946
Light brown calfskin derby, 12.5 cm.
The shoes belonged to Ferruccio Ferragamo.

Patent: 26673, granted on 10 May 1947
Application: 158, 16 January 1947
Support shape for woman's shoe sole, with central spur and support for the heel.

Patent: 426001, granted on 17 October 1947
Application: 944, 28 January 1947
Procedure to make uppers using a continuous length of thread, string, ribbon or the like.

Invisibile, 1947
Gold kid and nylon sandal, "F" wedge insole covered with kid, 19 x 9.5 cm.
Thanks to this model, Ferragamo was awarded the Neiman Marcus Award for fashion.

Invisibile, 1947
Black suede and nylon sandal. Wooden heel covered in suede, 21.5 x 6 cm.
Thanks to this model, Ferragamo was awarded the Neiman Marcus Award for fashion.

Invisibile, 1947
Red calfskin and fishing line sandal. Wooden wedge heel covered in calfskin, 23 x 6 cm.
Thanks to this model, Ferragamo was awarded the Neiman Marcus Award for fashion.

Invisibile, 1947
Pink kid and nylon strand sandal. Wooden heel covered in pink kid, 20 x 8.5 cm

Patent: 426545, granted on 29 October 1947
Application: 4962, 27 May 1943
Procedure to fit the upper in women's wooden mules so that they look like leather shoes with conventional stitching.

Patent: 455690, granted on 9 March 1950
Application: 7004, 1 August 1949
Soles for segmented shoes with elevated thickness and flexibility.

Podca, 1949
Laced shoe with three-piece brown suede upper. Vamp divided in the middle by a calfskin strip. Wooden heel and insole covered in brown calfskin. The insole is divided into three parts, 22 x 3.5 cm.

Patent: 41888, granted on 7 July 1952
Application: 1252, 4 April 1952
Shoe with half sole, waist lined with the same leather as the upper.
The model is sturdy but also as flexible as a glove, and known by the name "gloved arch".

Pump, 1952
Rust calfskin upper with wooden heel and waist covered in calfskin. Top piece and forepart sole in leather, 22 x 8 cm.

Patent: 482872, granted on 13 July 1953
Application: 14390, 24 December 1951
Metal cage heel for women's shoes.

Sandal, 1955
Black satin upper, stiletto heel made from a brass cage-like structure. Insole and waist lined with black satin, 22.5 x 8.5 cm.

Patent: 485546, granted on 16 October 1953
Application: 4173, 3 April 1952
Woman's shoe with sole limited to the front part of the toe spring, and procedures to make such shoes.

Patent: 546657, granted on 28 July 1956
Application: 111, 4 January 1956
System for the underneath of a shoe, with metal heel and waist, and connected anterior sole.

Sandal, 1956
18-k gold upper with vamp made of two intertwined cords, fastened with a Gladiator-style ankle strap. Laminated gold waist and chiselled, carved and embossed high heel, 24 x 9.5 cm.

Patent: 547227, granted on 20 August 1956
Application: 191, 7 January 1956
Heel for women's shoes including an outer metal structure made of thread-like elements.

Patent: 556791, granted on 8 February 1957
Application: 11145, 23 July 1956
System to interchange heels for woman's shoes: one portion of the heel is fixed, another can be removed and replaced.

Patent: 569431, granted on 20 November 1957
Application: 4613, 27 March 1957
Heel for woman's shoe comprised of two pieces, one of which moveable and replaceable.

Patent: 67890, granted on 22 April 1958
Application: 4620, 4 October 1957
Dancer's pointe shoe with sole featuring a tapered border glued to the sides of the upper and forming a reinforcement at the toe.

Flat, 1955–57
Beige kid upper with string around the collar threaded through the trimming to form a bow at the centre of the collar. Leather shell sole, 22 cm.

Ballet shoe, 1957
Black satin upper with bow. String around the collar threaded inside the grosgrain trimming to form a bow at the centre of the collar. Leather reinforcement and sole. Red satin lining and insole, 23 cm.

Patent: 578173, granted on 20 June 1958
Application: 14373, 31 October 1957
Shoemaking system that guarantees a match between the sole glued to the upper.

Patent: 71816, granted on 21 March 1959
Application: 2726, 27 May 1958
Multiple "X" reinforcement in the structure used to stiffen the shank of the shoe so as to support the weight of the body and contain the metatarsal joint.

Patent: 71817, granted on 21 March 1959
Application: 2727, 27 May 1958
Multiple "Y" reinforcement in the structure used to stiffen the shank of the shoe so as to support the weight of the body and contain the metatarsal joint.

Patent: 78902, granted on 9 June 1960
Application: 4600, 31 August 1959
Woman's shoe in which the structure of the sole extends along the instep of the foot in the area corresponding to the arch of the waist so as to increase the stiffening of the shoe and the support of the foot.

Pump, 1959
Black suede upper with wooden heel covered with genuine cowhide. Genuine cowhide shell sole that extends along the instep of the foot in the area corresponding to the waist, dividing the upper into two parts, 22.5 x 6 cm.

One of the original pendulums used by Salvatore Ferragamo in his studies, 1950–60.

1930s scale from the Pharmacy on Via Porta Rossa in Florence, used by Salvatore Ferragamo to test the lightness of his shoes.

100 wooden lasts of Ferragamo clients' feet, from the 1920s onwards, including:
Lauren Bacall, Abhisheck Bacchan, Drew Barrymore, Ingrid Bergman, John Bingham, Buitoni, Priyanka Chopra, Bette Davis, Marlene Dietrich, Duchess of Aosta, Duchess of Windsor, Greta Garbo, Ava Gardner, Prince George Alexander Louis of Cambridge, Rita Hayworth, Audrey Hepburn, Katharine Hepburn, Mrs. and Mr. Horenstein, Patty Hou, Angelina Jolie, Michael

Jordan, Kang Dong-won, Sonam Kapoor, Carina Lau Kar-ling, Ayako Kawahara, Nicole Kidman, Mrs. and Mr. Kierman, Kim Hye-soo, Kim So Yeon, Kim Yunjin, Ang Lee, Lee Mi Yeon, Tony Leung, Chi Ling Lin, Sophia Loren, Madonna, Anna Magnani, Riho Makise, Carmen Miranda, Karen Mok, Marilyn Monroe, Margherita Pasquini, Dev Patel, Claretta Petacci, Mary Pickford, Freida Pinto, Princess Soraya of Persia, Gloria Swanson, Katsunori Takahashi, Lana Turner, Ken Watanake, Wong Kar-Wai, Yoo Ji-tae, Zhang Ziyi.

DEVELOPMENT OF FIT

Development of a woman's shoe fit in five sizes, four for men.

Development of a woman's model from the last to the final phase with eleven different heel heights.

SHOES

Pump, 1930–35
Dyed dentex skin upper and heel, 22 x 8 cm.

Sandal, 1936
Kid and canvas upper embroidered with a patchwork effect, octagonal cork heel covered in blue kid, 23 x 4 cm.

Laced shoe, 1936–38
Hemp upper. Blue kid laces, trimming around the collar and insole. Three-layered medium wooden heel covered in red, green and blue suede, 24 x 5 cm.

Sandal, 1936–38
Crimson satin and gold kid upper. Cork wedge insole in padded layers covered in satin and kid, 22 x 7 cm.

Sandal, 1936–38
Upper made of gold kid stripes and circles with two black satin strips crossing it. Cork wedge three-layered insole covered in gold kid and black satin, 23 x 7 cm.

Ankle-strap sandal, 1937
Gold kid and black satin upper, high ankle-strap and peep-toe. Gladiator-style black satin lacing. Layered cork wedge covered in gold kid and black satin, 22.5 x 7.5 cm.

Sandal, 1937–38
Red raffia upper with multicoloured crocheted raffia decorative motifs around the collar. Red raffia laces. Wedge insole covered in rope, 23 x 3 cm.

Sandal, 1938
Gold kid upper. Wedge insole covered in black satin with hand-painted floral motifs, 22.5 x 8 cm.

Sandal, 1938
Upper made of two gold and silver kid bands. Gladiator-style ankle-strap. Cork wedge insole covered in red velvet with hand-embossed brass and rhinestone framework, 22 x 8 cm. Created for the Maharani of Cooch Behar.

Ninfea, 1938
Ankle boot with black suede upper divided into six bands to form a corolla. Crimson satin cuffs around the collar. Cork wedge insole covered in black suede, 22 x 8 cm.

Sandal, 1938
Gold kid and red suede upper. Gold kid asymmetrical strap. Cork wedge heel in three layers covered in red suede and gold kid, 21 x 6.5 cm.

Rainbow, 1938–39
Gold kid sandal with cork platform and layered heel covered in multicoloured suede, 22 x 13 cm. The model was created for Judy Garland.

Ankle boot, 1938–39
Black suede upper with satin cuffs around the collar. Wooden heel and insole lined with suede, 22 x 10.5 cm.

Two pieces, 1938–39
Black kid upper, platform insole and cork wedge heel covered in kid, 23.5 x 8 cm.

Sandal, 1939
Black satin and gold kid upper. Insole and cork heel covered in small pieces of golden mirror, 22 x 10 cm. The model was created for Carmen Miranda.

Fiamma, 1939
Pump with black suede upper perforated on the vamp and counter by brown kid strips. Wooden heel covered in suede, 22 x 8 cm.

Pump, 1939
Patchwork crimson kid upper topped with suede rectangles in the same colour, 22 x 8.5 cm.

Sandal, 1940
Black velvet and kid upper. Wooden and cork insole and heel covered in gold and silver kid, 24 x 13 cm.

Laced shoe, 1940
Black suede upper with three asymmetrical loops on the vamp. Tubular red kid strip. Cork wedge insole covered in red kid, 23 x 3 cm.

Wodka, 1940
Closed shoe with black suede upper and red kid cuffs around the collar. Cork wedge insole covered in red kid, 23 x 3 cm.

Laced shoe, 1940
Black suede upper divided by a red kid band that, starting from the vamp, extends to the wedge. Red kid strap. Cork wedge insole covered in red suede and kid, 22 x 7.5 cm.

Sandal, 1940
Black satin upper. Cork wedge insole with vertical grooves covered in gold kid, 20 x 6 cm.

Sandal, 1940
Upper made of alternating bands of silk crepe de chine and black satin. Pentagonal heel and cork platform covered in black silk crepe de chine, 23 x 9.5 cm.

Sandal, 1940
Upper made of white antelope bands and black silk grosgrain ribbons arranged in stripes. Cork wedge insole covered in black grosgrain and white antelope, 22.5 x 7 cm.

Attica, 1941
Closed shoe with blue calfskin upper in two pieces stitched in the middle of the upper and fastened to one side. Cork wedge insole covered in red kid, 22 x 4.5 cm.

Sandal, 1942
Red suede upper. Insole in padded cork layers covered in suede, 21.5 x 8 cm. The model was created for Carmen Miranda.

Sandal, 1942–44
Crocheted black cellophane upper. Three-layered cork wedge insole covered in reptile skin, 22 x 7 cm.

Sandal, 1943
Blue suede upper, carved and painted wooden wedge platform, 22.5 x 7.5 cm.

Sandal, 1943
Red suede upper, carved and painted wooden wedge platform, 21 x 7 cm.

Thong sandals, 1944
Sandal made of four green suede strips and a doubled orange kid strip, all twisted into a fastener. Cork wedge insole covered in green suede. White felt sole, 22 x 2 cm.

Sandal, 1944
Closed Oriental-style toe in blue and red kid. Raw hemp vamp with red and blue decorative motifs woven into the collar. Cork wedge insole in three padded layers covered in red, white and blue kid, 22 x 6 cm.

Arabesca, 1944–45
Sandal with red suede upper. Black and gold kid Oriental-style toe. F-shaped wooden wedge insole covered in red suede, 22.5 x 6.5 cm.

Closed shoe, 1945–47
Black suede upper with points on the instep of the foot. Patent leather strip crossing the vamp. Wooden insole and heel covered in patent leather, 22 x 7.5 cm.

Sandal, 1945–47
Upper made of white and black patent leather kid bands. Vamp around the ankle. Three-layered cork wedge insole covered in black and white kid, 23 x 7 cm.

Sandal, 1945–47
Upper formed by seven black suede strips sewn and braided together at the centre to form a band. Platform insole covered in black suede, 22 x 10.5 cm.

Guelfa modificata, 1945–47
Closed shoe with black suede upper in two parts placed over the vamp to form a side wing. Wooden heel covered in suede, 22.5 x 7 cm.

Medicea, 1946
Loafer with whole brown suede upper forming two folds to the sides of the vamp, high around the ankle. Rectangular copper buckle. Heel covered in suede, 22 x 3 cm.

Sandal, 1946
Black suede upper made of bands edged with ornamental stitching arranged transversally over the instep of the foot and joined with triangular stitches tone upon tone. F-shaped wooden wedge insole in two pieces covered in suede, 21.5 x 6.5 cm.

Sandal, 1947
Upper formed by alternating rectangles of blue, red, yellow and green suede sewn in a patchwork with a white kid strip. F-shaped wooden wedge insole covered in blue and red suede, 22 x 6 cm.

Sandal, 1947
Gold kid upper. Wooden heel and cork platform insole covered in gold kid, 24 x 11.5 cm.

Sandal, 1947
Silk velvet and black satin upper. Wooden heel and platform insole covered in black satin, 24 x 11 cm. The model was created for Ava Gardner.

Sandal, 1950
Upper made from nylon mesh and gold kid strip. Platform sole and heel covered in kid, 24 x 10 cm.

Onda, 1950–52
Pump with black suede upper. Scalloped vamp. Tapered wooden heel, called "Louis XV", covered in black suede, 21.5 x 8.5 cm.

Elsie, 1950–55
Closed shoe with burgundy suede upper, separated at the waist. Shawl vamp with central seam. Upturned toe sewn to the vamp. Wooden heel covered in suede, 22 x 6 cm.

Fiamma, 1950–55
Pump with black suede upper perforated on the vamp and counter by white kid strips. Wooden heel covered in suede, 22.5 x 9 cm.

Artia, 1952
Closed shoe with red suede upper and cuff around the collar. Oval heel and waist covered in red suede, 22.5 x 6 cm. The model was created for Audrey Hepburn.

Sinfony, 1952–54
Pump in black suede with a cross at the centre of the vamp formed by two bands of black kid trimmed with green kid. Wooden heel covered in black kid, 22 x 8.5 cm.

Pump, 1952–54
Black suede whole upper. Low-cut vamp. Wooden heel covered in black suede, 21.5 x 8 cm.

Duria, 1953
Brown antelope laced shoe. Vamp trimmed with beige kid. Beige calfskin lace and round heel. Waist covered in brown antelope, 24.5 x 4.5 cm.

Prua, 1953–54
Pump with black suede upper. Collar edged in white calfskin with a bow at the centre. Tapered heel called "Louis XV" covered in white calfskin, 23 x 8.5 cm.

Pump, 1954
Black suede upper. Vamp made of small black suede squares joined by Tavarnelle lace, 22.5 x 8.5 cm.

Frania, 1955
Pump with white suede upper. Fringed collar crossed by brown calfskin strip. Waist covered in suede. Heel covered in brown calfskin, 24.5 x 3 cm. The model was created for Anna Magnani.

Claudia, 1955
Pump with peep toe. White calfskin upper. Vamp gathered at the centre and garnished with a knot. Wooden heel covered in calfskin, 22 x 8 cm.

Pump, 1955
Black suede upper with peep toe. Gold metal wiring pom-pom on the vamp, 22 x 10 cm.

Pump, 1955–56
Black suede upper. Heel covered in red mica, 22 x 9 cm.

Abbe, 1955–60
Beige calfskin pump with pearly effect. Brown kid toe and strip crossing the vamp horizontally with mother-of-pearl effect. At the centre of the vamp a ruffle of the material used to make the upper. Wooden stiletto heel covered in brown calfskin, 24 x 9.5 cm.

Laced shoe, 1956
Red sea leopard upper with tongue made of the same material and colour, 24 x 3 cm.

Viatica, 1957
Stiletto heel pump with grey suede and calfskin upper, 21.5 x 10 cm. The model was created for Marilyn Monroe.

Viatica, 1957
Pump with grey suede upper, brown crocodile toe and counter, 24.5 x 10.5 cm. The model was created for Marilyn Monroe in *Some Like It Hot* by Billy Wilder.

Chianti, 1957
Pump with red suede upper and collar edged with yellow, pink and blue stitch. Stiletto heel covered in red suede, 22 x 11 cm.

Pump, 1957
Yellow suede upper decorated with circular perforations trimmed with multicoloured kid strips. Stiletto heel covered in yellow suede, 23.5 x 10.3 cm.

Pump, 1957
Black calfskin upper with white calfskin band around the collar sewn with kid strips in alternating colours (black, red, black). Stiletto heel covered in white calfskin on which the same decorative motif is repeated, 22 x 11 cm.

Pump, 1958–59
Black suede upper with small calfskin appliqués and stiletto heels, 24.5 x 11 cm. The model was created for Marilyn Monroe.

Pump, 1958–59
Pump with brown crocodile upper. Stiletto heel covered in crocodile, 24.5 x 11 cm. The model was created for Marilyn Monroe.

Pump, 1960
White calfskin upper. Stiletto heels, 24.5 x 11 cm. The model was created for Marilyn Monroe in *The Misfits* by John Huston.

Unica, 2011 (1939)
Copy of black suede ankle boot, shaped wedge platform and insole covered in white suede, 23 x 9 cm.

Viatica, 2012–13
Red patent leather pump, 22 x 11 cm. New version of the original model created by Salvatore Ferragamo for Marilyn Monroe.

Equilibrium, 2014
Facsimile of an original model by Salvatore Ferragamo with a multicoloured half-stitch upper with patchwork effect, known of from a 1930 photograph.

PLASTERS, SOLES, STUDIES AND A DEVICE TO MEASURE LASTS FROM THE 1950S
From the Jerry Ferragamo Archive.

WORKS EXHIBITED

Marina Abramović, *Shoes for Departure*, 1991
Amethyst, 26 x 50 x 20 cm
Paris, Collection Enrico Navarra.

Marina Abramović and Ulay, *The Lovers, The Great Wall Walk*, 1988/2010
Two-channel video, colour, 16:45 mins
Based on the 1988 performance *90 Days, the Great Wall of China*
Courtesy of the Marina Abramović Archives and Murray Grigor.

Dante Alighieri, *La Divina Commedia*. Commentary by Cristoforo Landino, illustrations attributed to Sandro Botticelli (Florence: Nicolò di Lorenzo della Magna, 1481), 40 x 28 x 8.7 cm
Florence, Biblioteca Nazionale Centrale di Firenze.

Dante Alighieri, *La Commedia*, membranaceous manuscript, fifteenth century, 25.8 x 17 x 5.5 cm
Florence, Biblioteca Nazionale Centrale di Firenze.

Dante Alighieri, *La Divina Commedia*. Commentary by Francesco da Buti, membranaceous manuscript, fifteenth century, 38 x 28 x 13 cm
Florence, Biblioteca Nazionale Centrale di Firenze.

Alessandro Allori, *Animated Skeleton*, c. 1564-1565
Black pencil on white paper, 42 x 28 cm
Florence, Gabinetto Disegni e Stampe degli Uffizi, inv. 6700 F.

Alessandro Allori, *Animated Skeleton*, c. 1564-1565
Black pencil on white paper, 43 x 28.7 cm
Florence, Gabinetto Disegni e Stampe degli Uffizi, inv. 6709 F.

Alessandro Allori, *Animated Skeleton*, c. 1564-1565
Black pencil on white paper, 41.5 x 27.8 cm
Florence, Gabinetto Disegni e Stampe degli Uffizi, inv. 6710 F.

Australopithecines on the move. Model based on a series of footprints found at Laetoli in Tanzania (Africa) on a volcanic layer deposited 3.6 million years ago. Model by Lorenzo Possenti.

Cecil Balmond, *Equilibrium*, installation and mixed media preparatory drawings. Curved stainless steel structure, stainless steel cables and central knot plus two wooden plinths with acrylic mirror sides, 113.8 x 220 x 88.5 cm
Created for the *Equilibrium* exhibition, 2014.

Cecil Balmond, *H-edge*, 2011
Steel and aluminium installation exterior to the exhibition.

Roberto Barni, *Continuous*, 2001
Patinated bronze, 19 x 39 x 63 cm
from 1 to 7 + 1 artist's proof signed A.P.
Florence, collection of the artist.

Roberto Barni, *Enterprise*, 2010
Patinated red bronze, 56 x 22 x 12.5 cm
Florence, collection of the artist.

Roberto Barni, *Joke*, 2013
Patinated bronze, 145 x 23 x 23 cm
Florence, collection of the artist.

Roberto Barni, *Razor 2*, 2003
Bronze, 71 x 18 x 50 cm
from 1 to 6 + 1 artist's proof signed A. P.
Florence, collection of the artist.

Pietro Berrettini (called Pietro da Cortona), *Tabulae anatomicae a celeberrimo picture Petro Berretino* (Rome: A. de Rubeis, 1741), 44.4 x 32.5 x 3.4 cm
Florence, Biblioteca Nazionale Centrale di Firenze.

Govard Bidloo, *Anatomia humani corporis…* (Amsterdam: the Widow of J. Van Someren, the Heirs of J. Van Dyk, H. Boom and the Widow of T. Boom, 1685), 53 x 38 x 5.7 cm
Florence, Biblioteca Nazionale Centrale di Firenze.

Luigi Bienaimé, *Dancing Bacchante*, 1846
Carrara marble, 149 x 120 x 65 cm
Rome, Galleria Francesca Antonacci.

Antoine Bourdelle, *Dance, Bas-Relief for the Theatre of Champs-Élysées*, 1912
Bronze, proof no. 3 executed by Susse in 1977, 177 x 150 x 27 cm
Paris, Musée Bourdelle, inv. MB BR. 778.

Antoine Bourdelle, *The Cello Player* (or *Music*), 1914
Bronze, proof no. 5 executed by Susse around 1990, 30 x 5.8 x 6 cm
Paris, Musée Bourdelle, inv. MB BR. 2201.

Antoine Bourdelle, *Dancing Isadora*, n.d.
Pen and purple ink on tracing paper, 25.8 x 20.1 cm
Paris, Musée Bourdelle, inv. MB D 6484.

Antoine Bourdelle, *Isadora*, [1909]
Pen and purple ink on tracing paper, 22 x 14.2 cm
Paris, Musée Bourdelle, inv. MB D 7100.

Antoine Bourdelle, *Isadora, Hymn beyond the Voice, Forms beyond the Song*, c. 1920
Pen, black ink and watercolour on tracing paper, 27.4 x 20.8 cm
Paris, Musée Bourdelle, inv. MB D 2963.

Antoine Bourdelle, *Isadora, Movements Remaining in My Memory*, c. 1920
Pen, brown ink and watercolour on tracing paper, 27.4 x 20.8 cm
Paris, Musée Bourdelle, inv. MB D 2967.

Antoine Bourdelle, *Nijinsky in the Role of Harlequin, the Carnival*, c. 1911
Pen and brown ink on tracing paper, 24.1 x 16.4 cm
Paris, Musée Bourdelle, inv. MB D 4757.

Antoine Bourdelle, *Nijinsky in the Role of Harlequin*, c. 1911
Pen and brown ink on tracing paper, 25 x 16.4 cm
Paris, Musée Bourdelle, inv. MB D 4758.

Trisha Brown, *Untitled (Montpellier)*, 2002
Charcoal on paper, 330.2 x 271.1 cm
Courtesy of the artist and Sikkema Jenkins & Co., New York, inv. TB 11411.

Trisha Brown, *Untitled (Montpellier)*, 2002
Charcoal on paper, 329.6 x 271.1 cm
Courtesy of the artist and Sikkema Jenkins & Co., New York, inv. TB 11403.

Trisha Brown, *It's a Draw*, 2008
Video performance
Minneapolis, Walker Art Center

Bernardo Buontalenti, *Delphic Couple* (models for the third Intermezzo of *La Pellegrina*), [1589], plates 13 and 17, 58.5 x 44.5 (including the passepartout) x 0.5 cm each
Florence, Biblioteca Nazionale Centrale di Firenze.

Alexander Calder, *Gouaches et Totems* (Paris: Maeght, 1966), 38.5 x 29 x 2 cm
Florence, Biblioteca Nazionale Centrale di Firenze.

Alexander Calder, *Stabile-Mobile*, 1973
Painted metal, 80 x 80 x 80 cm
Florence, private collection.

Antonio Canova, *Five Dancers Holding Hands*, 1799
Tempera, 35 x 80 cm
Possagno, Museo e Gipsoteca Antonio Canova, inv. 128.

Antonio Canova, *Dancer Holding Veil*, 1799
Tempera, 29 x 25 cm
Possagno, Museo e Gipsoteca Antonio Canova, inv. 128.

Antonio Canova, *Dancer with Her Arms about Her Head*, 1799
Tempera, 29 x 35 cm
Possagno, Museo e Gipsoteca Antonio Canova, inv. 128.

Antonio Canova, *Dancer with Cymbals*, 1799
Tempera, 29 x 25 cm
Possagno, Museo e Gipsoteca Antonio Canova, inv. 128.

Adriano Cecioni, *First Steps*, c. 1869
Plaster, 71 x 30 x 27 cm
Florence, Galleria d'Arte Moderna di Palazzo Pitti, inv. Giornale 5612.

Mario Ceroli, *Untitled*, 2002
Russian pinewood, 144 x 75 x 25 cm
Private collection, courtesy of Tornabuoni Arte, Florence.

Edgar Degas, *Dancer, Grande Arabesque, Third Time*, 1921–31 (original work 1882–95)
Bronze (lost wax casting), 28.2 x 43 x 21 cm
Paris, Musée d'Orsay, inv. RF 2072.

Edgar Degas, *Dancer, Arabesque over the Right Leg, Left Arm in Front, Second Study*, 1921–31 (original work 1882–95)
Bronze (lost wax casting), 29 x 39 x 14 cm
Paris, Musée d'Orsay, inv. RF 2066.

Albrecht Dürer, *The Great Fortune*, c. 1501–02
Burin, 30.5 x 23.2 cm
Pavia, Musei Civici, inv. St. Mal. 284.

Albrecht Dürer, *The Small Fortune*, c. 1495–96
Burin, 10.6 x 5.7 cm
Pavia, Musei Civici, inv. St. Mal. 285.

Ertruscan art, *Bronze of Hercules*, fourth century BC
Cast bronze on wooden base, 24 x 12 x 12 cm
Soprintendenza per i Beni Archeologici della Toscana – Florence, Museo Archeologico Nazionale, inv. 19.

Etruscan art, *Statuette of Laran (Etruscan Mars)*, fourth century BC
Bronze on wooden base, 29 x 8.5 x 7 cm
Soprintendenza per i Beni Archeologici della Toscana – Florence, Museo Archeologico Nazionale, inv. 350.

Ertruscan art, *Etruscan Red-Figure Stamnos*, fourth century BC
Ceramics, 30.5 x 31 cm (Ø base 13.3 cm)
Soprintendenza per i Beni Archeologici della Toscana – Florence, Museo Archeologico Nazionale, inv. 4102.

Etruscan art, *Etruscan Red-Figure Stamnos*, fourth century BC
Ceramics, 30.7 x 30 cm (Ø base 13.5 cm)
Soprintendenza per i Beni Archeologici della Toscana – Florence, Museo Archeologico Nazionale, inv. 4103.

Pericle Fazzini, *The Dancer*, 1936–37
Wood, 164 x 95 x 34.5 cm
Siena, Collection of Banca Monte dei Paschi di Siena, inv. 5048489.

Carlo Finelli, *The Three Graces*, c. 1820
Marble, 158 x 119 x 67 cm
Rome, Galleria Francesca Antonacci.

Alberto Giacometti, *Tight-rope Walker*, 1943 Pencil and charcoal on paper, 36.8 x 28.6 cm New York, Yoshii Gallery.

Julio González, *Dishevelled Dancer*, 1935 Forged and welded iron, 53.5 x 37 x 20 cm Nantes, Musée des Beaux-Arts, inv. 958.7.2.S.

Julio González, *Dancer with Daisy*, c. 1937 Bronze, 46 x 30.5 x 9.5 cm Madrid, Museo Nacional Centro de Arte Reina Sofía, inv. AS03118.

Antony Gormley, *Domain LXVIII*, 2009 Welded stainless steel rods, 188 x 64 x 29.5 cm Florence, private collection.

Greek art, *Relief with Dancing Maenads*, first century AD Marble, 57 x 69 x 2.5 cm Rome, Museo di Scultura Antica Giovanni Barracco.

Greek art, *Satyr with Maenads Playing and Dancing*, fragment of a relief on the side of a sarcophagus Plaster, 70 x 98 x 15 cm Rome, Museo della Civiltà Romana.

Wassily Kandinsky, *Circle of Friends of the Bauhaus*, 1932 Drypoint, 20 x 24 cm Albenga, Collection of Galleria d'Arte Moderna.

Paul Klee, *Tight-rope Walker*, 1923 Lithograph on paper, 52.7 x 37.5 cm Düsseldorf, Beck & Eggeling International Fine Art.

Daniel Leclerc and Jean-Jacques Manget, *Bibliotheca anatomica sive recens in anatomia inventorum thesaurus…* (Geneva: J. A. Chouët, 1699), vol. 2, 37 x 24 x 8.5 cm Florence, Biblioteca Nazionale Centrale di Firenze.

Le Corbusier, *Le Poème de l'angle droit; lithographies originales* (Paris: Tériade, 1955), 44 x 34.2 x 5.2 cm Florence, Biblioteca Nazionale Centrale di Firenze.

Fernand Léger, *Cirque: lithographies originales* (Paris: Tériade, 1950), 44 x 34 x 4.2 cm Florence, Biblioteca Nazionale Centrale di Firenze.

Jacques Lipchitz, *Mother and Child*, 1912 Pencil on paper, 32 x 22.6 cm Prato, Museo di Palazzo Pretorio, inv. 22720.

Jean-Jacques Manget, *Theatrum anatomicum…* (Geneva, 1716–17), 2 vols., 44 x 29 x 8.3 cm Florence, Biblioteca Nazionale Centrale di Firenze.

Marino Marini, *Dancer*, 1953 Polychrome plaster, 171.5 x 44.5 x 30 cm Florence, Museo Marino Marini.

Marino Marini, *Dancer*, 1953 Polychrome plaster, 149 x 65.5 x 36 cm Florence, Museo Marino Marini.

Marino Marini, *Juggler*, 1939 Polychrome bronze, 161.5 x 63.4 x 13.6 cm Florence, Museo Marino Marini.

Henri Matisse, *Study of a Foot*, c. 1909 Bronze, h 30 cm Saint Petersburg, The State Hermitage Museum, inv. H.sk. 2392.

Fausto Melotti, *Equilibriums*, 1971 Gold and enamel on Plexiglas base, 49 x 43 x 15 cm Courtesy Galleria Christian Stein, Milan.

Fausto Melotti, *The Invisible Acrobat*, 1980 Brass, 66 x 28 x 22 cm Florence, private collection.

Joan Miró, *Spanish Dancer*, c. 1960 Woollen tapestry, 200 x 152 cm Albenga, Collection of Galleria d'Arte Moderna.

Barbara Morgan immortalizes the ballet dancer and choreographer Martha Graham as she performs in *Letter to the World (Swirl)*, New York, 1940. The work depicts the life of the writer Emily Dickinson. Getty Images.

Barbara Morgan, Martha Graham in *Lamentation, Oblique*, 1935. Getty Images.

Barbara Morgan immortalizes the ballet dancer and choreographer Martha Graham as she performs in *Letter to the World (Kick)*, New York, 1940. The work depicts the life of the writer Emily Dickinson. Getty Images.

Ugo Mulas, *Calder Circus*, 1963–64 Fotografie Ugo Mulas © Eredi Ugo Mulas. Tutti i diritti riservati Courtesy Archivio Ugo Mulas, Milan – Galleria Lia Rumma, Milan/Naples.

Eadweard Muybridge, *Animal Locomotion* (plate 747): sequence with a baboon walking, 1887 Florence, Raccolte Museali Fratelli Alinari (RMFA) – Palazzoli Collection.

Eadweard Muybridge, *Animal Locomotion* (plate 540): sequence with a naked boy walking, 1887 Florence, Raccolte Museali Fratelli Alinari (RMFA) – Palazzoli Collection.

Eadweard Muybridge, *Animal Locomotion* (plate 355): sequence with a man walking and carrying a rifle, 1887 Florence, Raccolte Museali Fratelli Alinari (RMFA) – Palazzoli Collection.

Eadweard Muybridge, *Animal Locomotion* (plate 559): sequence with a naked man walking, 1887 Florence, Raccolte Museali Fratelli Alinari (RMFA) – Palazzoli Collection.

Eadweard Muybridge, *Animal Locomotion* (plate 369): sequence with a naked man moving his right leg, 1887 Florence, Raccolte Museali Fratelli Alinari (RMFA) – Palazzoli Collection.

Bruce Nauman, *Walking in an Exaggerated Manner around the Perimeter of a Square*, 1967–68 Black and white 16mm video, silent, 10 mins. Courtesy of Electronic Arts Intermix (EAI), New York.

Bruce Nauman, *Slow Angle Walk (Beckett Walk)*, 1968 Black and white video, sound, 60 mins. Courtesy of Electronic Arts Intermix (EAI), New York.

Plinio Nomellini, *The Strike*, 1889 Pen drawing on paper, 9 x 13.3 cm Florence, private collection.

Plinio Nomellini, *Study of a Dancer, Isadora Duncan*, 1913 Pencil on paper, 37 x 26 cm Florence, private collection.

Plinio Nomellini, *Study of a Dancer, Isadora Duncan*, 1913 Pencil on paper, 37 x 26 cm Florence, private collection.

Plinio Nomellini, *Study of a Dancer, Isadora Duncan*, 1913 Pencil on paper, 37 x 24.5 cm Florence, private collection.

Plinio Nomellini, *Study of a Dancer, Isadora Duncan*, 1913 Pencil on paper, 37 x 26 cm Florence, private collection.

Plinio Nomellini, *Study of a Dancer, Isadora Duncan*, 1913 Pencil on paper, 37 x 26 cm Florence, private collection.

Plinio Nomellini, *Study of a Dancer, Isadora Duncan*, 1913 Pencil on paper, 37 x 26 cm Florence, private collection.

Giulio Paolini, *Carte Noire*, 1999–2000 Silk-screen, coloured pencils and collage on black paper, 163 x 223 cm (nine framed elements 53 x 73 cm each) Turin, collection of the artist.

Pablo Picasso, *The Acrobats*, 1905 Drypoint, 22.8 x 32.6 cm Albenga, Collection of Galleria d'Arte Moderna.

Pablo Picasso, *Salome*, 1905 Drypoint, 40 x 34.8 cm Albenga, Collection of Galleria d'Arte Moderna.

Raphael (school), *Dancer*, fifteenth century Pen on burnished white paper, 23 x 13.9 cm Florence, Gabinetto Disegni e Stampe degli Uffizi, inv. 1251 E.

Auguste Rodin, *Dance Movement F*, c. 1911 Plaster, 26.8 x 26.2 x 14.5 cm Paris, Musée Rodin, inv. S 1053.

Auguste Rodin, *Nijinsky*, 1912 Plaster, 17.5 x 10 x 6 cm Paris, Musée Rodin, inv. S 890.

Auguste Rodin, *Iris, Messenger of the Gods*, c. 1890–91 Plaster, 41.7 x 46 x 22 cm Paris, Musée Rodin, inv. S 835.

Auguste Rodin, *Study for the Torso of The Walking Man*, 1878–79 Plaster, 52.2 x 25 x 17.8 cm Paris, Musée Rodin, inv. S 3203.

Auguste Rodin, *Study for Saint John the Baptist*, 1878 Plaster, 97.3 x 45.5 x 24.7 cm Paris, Musée Rodin, inv. S 2205.

Roman art, *Bronze of Dancing Maenad*, fourth century BC Cast bronze on wooden base, 18.5 x 6.5 x 6 cm Soprintendenza per i Beni Archeologici della Toscana – Florence, Museo Archeologico Nazionale, inv. 78350.

Roman art, Aretine production, *Cup Matrix in Terra Sigillata from the Arezzo Area*, first century BC – first century AD Ceramics (positive in wax), h 6.7 cm (Ø 10 cm) Soprintendenza per i Beni Archeologici della Toscana – Florence, Museo Archeologico Nazionale, inv. 72697.

Roman art, *Bronze Foot* Cast bronze, 16.5 x 32 x 12.5 cm (maximum amount of space) Soprintendenza per i Beni Archeologici della Toscana – Florence, Museo Archeologico Nazionale, inv. 1982.

Roman art, *Bronze Foot* Cast bronze, 11 x 21.5 x 13 cm (maximum amount of space) Soprintendenza per i Beni Archeologici della Toscana – Florence, Museo Archeologico Nazionale, inv. 1985.

Roman art, *Bronze Foot* Cast bronze, 8 x 7 x 18 cm Soprintendenza per i Beni Archeologici della Toscana – Florence, Museo Archeologico Nazionale, inv. 1983.

Roman art, *Right Foot in Gilded Bronze Belonging to a Nike*, from the Forum of Augustus, inaugurated in year 2 BC Gilded bronze (cast) on bronze solid tenon joint and iron hinge, 76 x 40 x 49 cm Sovrintendenza Capitolina ai Beni Culturali Rome, Museo dei Fori Imperiali – Trajan's Markets.

Georges Rouault, *Cirque de l'étoile filante; eaux-fortes originales set dessins gravés sur bois* (Paris: A. Vollard, 1938), 45.5 x 34.5 x 5.2 cm Florence, Biblioteca Nazionale Centrale di Firenze.

George Segal, *Red Woman Acrobat Hanging from a Rope*, 1996 Bronze with red patina, 223.5 x 104.1 x 49 cm New York, The George and Helen Segal Foundation.

Gino Severini, *The Acrobat* (or *Masks and Ruins*), 1928 Oil on canvas, 160 x 145.5 cm Siena, Collection of Banca Monte dei Paschi di Siena, inv. 5048307.

Paul Valéry, *Degas: danse dessin*, illustrations by Edgar Degas (Paris: Vollard, 1936), 33.5 x 26.5 x 4 cm Florence, Biblioteca Nazionale Centrale di Firenze.

Juan Valverde, *Anatomia del corpo umano* (Rome: A. Salamanca and A. Lafrery, 1560), 31 x 22 x 3.7 cm Florence, Biblioteca Nazionale Centrale di Firenze.

Andrea Vesalio, *De humani corporis fabrica libri septem* (Venice: Francesco de Franceschi & Johann Criegher, 1568), 32.3 x 23.2 x 5.5 cm Florence, Biblioteca Nazionale Centrale di Firenze.

Andrea Vesalio, *Suorum de humani corporis fabrica libro rum epitome* (Basileae: Ioannis Oporini, 1543), 50 x 35.7 x 3.4 cm Florence, Biblioteca Nazionale Centrale di Firenze.

Bill Viola, *Inner Passage*, 2013 (tribute to Richard Long) Colour HD video on plasma display mounted vertically on wall, stereo sound, 155.5 x 92.5 x 12.7 cm, 17:12 mins. Performer: Blake Viola Photo: Kira Perov.

FILMS

Equilibri reversibili (Reversible Equilibriums), film directed by Francesco Fei. Dancer Paul Lee. Teatro Cango – Cantieri Goldonetta, Florence, directed by Virgilio Sieni. This is a film about the Korean dancer Paul Lee, whose steps and movements are inspired by Moshe Feldenkrais's method. Directing, photography and editing: Francesco Fei. Produced by Apnea Film. HD 1920 x 1080, 3 video channels.

Equlibrium?, video interviews directed by Francesco Fei. Interviews by Emanuele Enria on the subject of equilibrium: Wanda Miletti Ferragamo, Eleonora Abbagnato, Cecil Balmond, Reinhold Messner, Philippe Petit, Will Self. Directing and photography: Francesco Fei. Editing: Claudio Bonafede. Music: Massimiliano Fraticelli. Colourist: Luca Parma. Produced by Apnea Film. HD 1920 x 1080.

Video interviews with Jerry and James Ferragamo, directed by Francesco Fei.

The videos made with archive footage have been edited by Daniele Tommaso.

The video **Salvatore Ferragamo e il comfort** (Salvatore Ferragamo and Comfort) was mounted with images excerpted from the report "Scarpe di lusso nascono a Firenze", included in the *Settimana Incom* newsreel (no. 00555, 15 July 1951) produced by Istituto Luce, as well as from the report "Cenerentola ha fatto scuola" from a 1950s SEDI newsreel.
The audio contains parts of interviews with Salvatore Ferragamo on fit and comfort, conducted by various Australian radio stations in 1958, during his promotional trip there.
The video **Salvatore Ferragamo con le dive** (Salvatore Ferragamo with the Divas) was made from archive photographs showing Salvatore at Palazzo Feroni with such famous clients as Gloria Swanson, Audrey Hepburn, Sophia Loren, Anna Magnani, Ira Fürstenberg, Valentina Cortese.

One Small Step. This is the video of Mission Apollo 11 (16–24 July 1969), with sound and filming of the first steps on the Moon by American astronauts Neil Armstrong and Buzz Aldrin, beginning at 8:20 p.m. on 20 July 1969. Also audible is the voice of the third astronaut, Michael Collins, who had stayed on the lunar module Eagle, as well as that of space engineer Bruce McCandless from Houston. In 1984 the latter participated in the Shuttle Challenger STS–41–B mission, becoming the first astronaut to make an untethered free flight using the MMU (Mannned Maneuvering Unit).

The Ways of Walking of Famous Twentieth-century Figures including: Fidel Castro, Winston Churchill, Elizabeth II Queen of England, Mohandas Gandhi, Adolf Hitler, John Paul II, John F. Kennedy, Martin Luther King, Nelson Mandela, Benito Mussolini, Iosif Stalin, Mother Teresa of Calcutta, Margaret Thatcher, Mao Tse Tung

In Honour of Charlie Chaplin, Celebrating Little Tramp's Centenary: *Kid Auto Races at Venice, California*, 1914, directed by Henry Lehrman; *Caught in the Rain*, 1914, directed by Charlie Chaplin; *Caught in a Cabaret*, 1914, directed by Mabel Normand; *The Rounders*, 1914, directed by Charlie Chaplin; *Face on the Barroom Floor*, 1914, directed by Charlie Chaplin; *Tango Tangles*, 1914, directed by Mack Sennett; *Tillie's Punctured Romance*, 1914, directed by Mack Sennett; *The Knockout*, 1914, directed by Charles Avery; *The Pilgrim*, 1923, directed by Charlie Chaplin; *The Gold Rush*, 1925, directed by Charlie Chaplin; *The Circus*, 1928, directed by Charlie Chaplin; *Modern Times*, 1936, directed by Charlie Chaplin.

BIBLIOGRAPHY

E. Abbagnato, *Un angelo sulle punte* (Milan: Rizzoli, 2009).

H. C. Adam, *Eadweard Muybridge. The Human and Animal Locomotion Photographs* (Cologne: Taschen, 2010).

G. Agamben, *Ninfe* (Turin: Bollati Boringhieri, 2007).

F. M. Alexander, *Man's Supreme Inheritance* (London: Methuen, 1910; revised editions New York, 1918, 1941, 1946, 1957 and London: Mouritz, 1996, reprinted 2002).

F. M. Alexander, *Constructive Conscious Control of the Individual* (USA: Centerline Press, 1923; revised edition London: Mouritz, 1946 and 2004).

F. M. Alexander, *The Use of the Self* (New York: E. P. Dutton, 1932; reprinted Orion Publishing, 2001).

F. M. Alexander, *The Universal Constant In Living* (New York: E. P. Dutton, 1941; reprinted London: Chaterson, 1942, 1943 and 1946; USA: Centerline Press, 1941 and 1986 and London: Mouritz, 2000).

S. Almécija et al., "The femur of Orrorin tugenensis exhibits morphometric affinities with both Miocene apes and later hominins", in *Nature Communications*, no. 4, art. 2888, December 2013.

L. Anisimova and P. Khoroshilov, *Nudo per Stalin* (Rome: Cangemi Editore, 2009).

L'Art, entretiens réunis par Paul Gsell (Paris: B. Grasset, 1911).

N. Ashton et al., "Hominin Footprints from Early Pleistocene Deposits at Happisburgh, UK", in *PLOS One*, February 2014.

J. J. Bachofen, *Il simbolismo funerario degli antichi* [1854], edited by M. Pezzella (Naples: Guida Editori, 1989).

B. Balázs, *Early Film Theory: "Visible Man" and the Spirit of Film* [© 1952] (Oxford and New York: Berghahn Books, 2010).

C. Balmond, *Informal* (Munich, Berlin, London and New York: Prestel, 2002).

C. Balmond, *Element* (Munich, Berlin, London and New York: Prestel, 2007).

S. Banes, *Tersicore in scarpe da tennis* (Macerata: Ephemeria, 1993).

E. Barba and N. Savarese, *L'Arte Segreta dell'Attore* (Milan: Ubulibri, 2005; reprinted and revised, *Un Dizionario di Antropologia Teatrale*, Routledge: Centre for Performance Research and Bari: Edizionidipagina, 2011).

R. Barni, *Gambe in spalla* (Prato: Gli Ori, 2008).

R. Barni and C. Mazzonis, *Mappe del cielo tatuato* (Prato: Gli Ori, 2004).

I. Bartenieff, *Body Movement, Coping with the Environment* (USA and UK: Routledge, 1980).

T. H. Bartlett, *Auguste Rodin, Sculptor* [1889], quoted in A. Magnien and F. Arensi, *L'origine del genio*, exhibition catalogue (Turin: Allemandi, 2010).

C. Baudelaire, *L'Art romantique*, 1869 (as vol. 3 of the first edition of Baudelaire's collected works).

A. Bazin, *What is Cinema?* [1958] (Berkeley: University of California Press, 2004).

G. Belli and E. Guzzo Vaccarino, *La danza delle Avanguardie. Dipinti, scene e costumi, da Degas a Picasso, da Matisse a Keith Haring*, exhibition catalogue (Milan: Skira, 2005).

L. Bellosi, *La sede storica della Banca Monte dei Paschi di Siena. L'architettura e la collezione delle opere d'arte* (Siena: Banca Monte dei Paschi di Siena, 2002).

M. Belpoiti, *Camera straniera. Alberto Giacometti e lo spazio* (Milan: Johan & Levi Editore, 2012).

W. Benjamin, "Il narratore", in *Angelus Novus, saggi e frammenti* (Turin: Einaudi, 2006).

W. Benjamin in G. Didi-Huberman, *Ninfa moderna. Saggio sul panneggio caduto* [2004] (Milan: Abscondita, 2013).

L. Bentivoglio, *La danza moderna* (Milan: Longanesi, 1977).

L. Bentivoglio, *Tanztheater, dalla danza espressionista a Pina Bausch* (Rome: Di Giacomo, 1982).

L. Bentivoglio, *La danza contemporanea* (Milan: Longanesi, 1985).

L. Bentivoglio, *Il teatro di Pina Bausch* (Milan: Ubulibri, 1985 and 1991).

L. Bentivoglio and F. Carbone, *Pina Bausch, vieni, balla con me* (Florence: Barbes, 2008).

C. Blasis, *Trattato dell'arte della danza*, edited by F. Pappacena (Rome: Gremese Editore, 2008).

I Borghese e l'Antico, exhibition catalogue, edited by C. Brook and V. Curzi (Milan: Skira, 2011).

A. Bourdelle, *Apollon au combat*, typescript, Archives of the Musée Bourdelle, Paris.

A. Bourdelle, *Ecrits sur l'art et sur la vie illustrés de dessins de l'auteur*, edited by G. Varenne (Paris: Editions d'Histoire et d'Art, Librairie Plon, 1955).

D. M. Bramble and D. E. Lieberman, "Endurance running and the evolution of Homo", in *Nature*, no. 432 (7015), 2004.

R. Calasso, *La follia che viene dalle ninfe* (Milan: Adelphi, 2005).

E. Canetti, *Massa e potere* (Milan: Rizzoli, 1972).

Canova. L'ideale classico tra scultura e pittura, exhibition catalogue, edited by S. Androsov, F. Mazzocca and A. Paolucci (Cinisello Balsamo: Silvana Editoriale, 2009).

Canova e la danza, exhibition catalogue, edited by M. Guderzo (Crocetta di Montello: Terra Ferma Edizioni, 2012).

L. Capitani, "Canova e la danza", in *Polittico*, no. 2, 2002.

L. Capitani, *La bellezza in movimento. La scultura di Canova tra mimica, danza e recitazione*, doctorate thesis, Pisa Univeristy, academic year 2003–04.

L. Capitani, "Il panneggio come 'luogo' delle passioni. La scultura di Canova tra mimica, danza e recitazione", in *Polittico*, no. 4, 2005.

L. Capitani, "Arti dello spazio e arti del tempo: Canova e la danza", in *Canova e la danza*, exhibition catalogue, edited by M. Guderzo (Crocetta di Montello: Terra Ferma Edizioni, 2012).

D. Capresi, B. Cinelli and D. Sogliani, *Dipinti, sculture e disegni del Novecento. Esperienze di collezionismo nelle Raccolte della Banca Monte dei Paschi di Siena e della Fondazione Banca Agricola Mantovana*, exhibition catalogue (Milan: Skira, 2012).

F. Careri, *Walkscapes. Camminare come pratica estetica* (Turin: Einaudi, 2006).

P. G. Carizzoni and A. Ghilardotti, *Isadora Duncan, Pina Bausch* (Milan: Skira, 2006).

M. F. Caroso da Sermoneta, *Il ballarino* (Venice, 1581).

B. Castiglione, *Il libro del Cortegiano* [1528] (Milan: Garzanti, 1981).

L. L. Cavalli Sforza and T. Pievani, *Homo sapiens. La grande storia della diversità umana* (Turin: Codice Edizioni, 2011).

G. Celant, *Melotti. Catalogo Generale*, 2 vols. (Milan: Electa, 1994).

A. Cecioni, *Scritti e Ricordi con lettere di Giosuè Carducci, Ferdinando Martini ecc.*, foreword and notes by G. Uzielli (Florence: Tipografia Domenicana, 1905).

C. Chaplin, *My Autobiography* (London: Penguin, 1964).

R. P. Ciardi, "La cultura figurativa di Ugo Foscolo", in *Rivista di letteratura italiana*, II, nos. 2–3, 1985.

Civiltà del '700 a Napoli 1734–1799, exhibition catalogue, edited by N. Spinosa (Florence, 1980).

J. Clair, "Parade et palingénésie. Du Cirque chez Picasso et quelques autres", in *La Grande Parade*, exhibition catalogue, edited by J. Clair (Paris: Gallimard, 2004).

G. Comis and B. Della Casa, *Klee–Melotti*, exhibition catalogue (Berlin: Kehrer, 2013).

"Complete Fourth Metatarsal and Arches in the Foot of Australopithecus afarensis", in *Science*, 331 (6018), 2011.

"Conversazione con Drassaï. La mia ultima visita a Giacometti" and "Conversazione con Pierre Schneider. Al Louvre con Giacometti", in *Alberto Giacometti. Il mio lungo cammino*, edited by E. Grazioli, introduction by M. Pozzati (Cernusco Lombardo: Hestia Edizioni, 1998).

A. Cornazano, *Libro dell'arte del danzare*, 1455.

Corpo e potere. Körper und Macht, exhibition catalogue, edited by G. A. Mina (Monte Verità, 2013).

V. Crapanzano, *Orizzonti dell'immaginario* (Turin: Bollati Boringhieri, 2007).

M. Cunningham, *Il danzatore e la danza* (Turin: EDT, 1990).

J. S. Czestochowski and A. Pingeot, *Degas sculptures: catalogue raisonné of the bronzes* (New York: Torch Press, 2002).

A. D'Adamo, *Danzare il rito. Le Sacre du Printemps attraverso il Novecento* (Rome: Bulzoni, 1999).

A. D'Este, *Memorie di Antonio Canova* (Florence, 1864; anastatic reprint edited and with an introduction by P. Mariuz, Bassano del Grappa, 1999).

D. da Piacenza, *De arte saltandi et choreas ducendi*, c. 1455.

Da Vela a Medardo Rosso. I grandi scultori italiani dell'Ottocento, exhibition catalogue, edited by R. Bossaglia (Milan: Skira, 1998).

H. de Balzac, "Théorie de la démarche" [1833], in *Oeuvres diverses de Honoré de Balzac*, vol. 2 (Paris: Louis Condard, 1938).

G. Deleuze, *Francis Bacon. The Logic of Sensation* (London and New York: Continuum, 2003).

E. De Luca in *Le parole che sono importanti* (Milan: Feltrinelli, 2014).

Degas. La vita e l'arte. I capolavori, series I classici dell'arte, edited by V. Gavioli (Milan: Rizzoli-Skira 2003).

Del Laocoonte o sia dei limiti della pittura e della poesia. Discorso di G. E. Lessing recato dal tedesco in italiano dal cavaliere C. G. Londonio (Milan, 1833).

D. Demetrio, *Filosofia del camminare. Esercizi di meditazione mediterranea* (Milan: Raffaello Cortina Editore, 2005).

D. Demetrio, *I sensi del silenzio. Quando la scrittura si fa dimora* (Milan: Mimesis, 2012).

J. Denoyar and R. Kendall, *Degas and the ballet. Picturing Movement*, exhibition catalogue (London: Royal Academy of Arts, 2011).

V. Di Bernardi, *Virgilio Sieni* (Palermo: L'Epos, 2011).

G. Didi-Huberman, *Ninfa moderna. Saggio sul panneggio caduto* (Milan: Il Saggiatore, 2004).

G. Didi-Huberman, *L'immagine insepolta. Aby Warburg, la memoria dei fantasmi e la storia dell'arte* (Turin: Bollati Boringhieri, 2006).

E. Fisset, *L'Ivresse de la marche: Petit Manifeste en faveur du voyage à pied* (Paris: Éditeur Transboréal, 2008).

B. Dixon, "Harlequin in the New World", in *Chaplin's Limelight and the Music Hall Tradition*, edited by F. Scheide and H. Mehran (Jefferson, North Carolina and London: Mc Farland & Co., 2006).

I. Duncan, *Pina Bausch. Danza dell'anima, liberazione del corpo* (Milan: Skira, 2006).

D. Dupuy, *La Sagesse du Danseur* (Paris: Jean-Claude Béhar éditions, 2011).

G. Ebreo da Pesaro, *De praticha seu arte tripudii vulgare opusculum*, c. 1463.

T. Espedal, *Camminare. Dappertutto (anche in città)* (Milan: Ponte alle Grazie, 2009).

M. Feldenkrais, *Body and Mature Behaviour* (New York: International Universities Press,1949; reprinted Berkeley: North Atlantic Books, 2005).

M. Feldenkrais, *Awareness through Movement* [original in Hebrew 1967] (New York: HarperCollins, 2009).

M. Feldenkrais, *The Elusive Obvious* (Cupertino, California: Meta Publications, 1981).

M. Feldenkrais, *The Potent Self* [1985] (San Francisco: Harper, 1992).

M. Feldenkrais, *Il Metodo Feldenkrais, conoscere se stessi attraverso il movimento* (Como: Ed. RED, 1991).

M. Feldenkrais, *Embodied Wisdom* (Berkeley: North Atlantic Books, 2010).

M. Feldenkrais, *La saggezza del corpo* (Rome: Casa Editrice Astrolabio, 2011).

E. Fisset, *L'Ivresse de la marche: Petit Manifeste en faveur du voyage à pied* (Paris: Éditeur Transboréal, 2008).

François Delsarte, le leggi del teatro. Il pensiero scenico del precursore della danza moderna, edited by E. Randi (Rome: Bulzoni, 1993).

S. Freud, *Delusion and Dream in Jensen's Gradiva* [1906] (Los Angeles: Green Integer, 2003).

H. Fulton, *Walking Artist Hamish Fulton* (Düsseldorf: Richter, 2001).

H. Fulton, *Keep Moving* (Milan: Charta, 2005).

V. Gautherin, "La Danse", in *Bourdelle*, edited by B. Manara (Rome: De Luca, 1994).

J. Genet, *Il funambolo* [1956] (Milan: Adelphi, 1997).

La generazione danzante, edited by S. Carandini and E. Vaccarino (Rome: Di Giacomo, 1997).

R. Giambrone, *Pina Bausch, le coreografie del viaggio* (Macerata: I libri dell'Icosaedro, Ephemeria, 2008).

P. Giovetti, *Rudolf Steiner, la vita e l'opera del fondatore dell'antroposofia* (Rome: Edizioni Mediterranee, 1992).

J. E. Gordon, *Strutture Ovvero perché le cose stanno in piedi* (Milan: Edizioni scientifiche e tecniche Mondadori, 1979).

M. Green, *Mountain of Truth: The Counterculture Begins: Ascona, 1900–1920* (University Press of New England, 1986).

F. Gros, *Andare a piedi. Filosofia del camminare* [2009] (Milan: Garzanti, 2013).

G. Guniberti, M. Pavan and M. Guderzo, *Il Museo e la Gipsoteca di Antonio Canova di Possagno* (Possagno: Faenza Scientifics, 2012).

E. Guzzo Vaccarino, *Il tango* (Palermo: L'Epos, 2010).

E. Guzzo Vaccarino, "John Cage, il libertador della danza", in *John Cage. Una rivoluzione lunga cent'anni*, edited by G. Fronzi (Milan and Udine: Mimesis, 2013).

H. Hesse, *Romanzi*, edited by C. Magris and M. P. C. Palin (Milan: Mondadori, 1977).

H. Hesse, *Vagabondaggio* (Rome: Newton Compton, 1992).

Highlights in the History of Monte Verità (Museo Monte Verita, 2007).

M. Hochkofler, *Anna Magnani* (Milan: Bompiani, 2013).

B. Hölldobler and E. O. Wilson, *Journey to the Ants: A Story of Scientific Exploration* (Cambridge, Massachusetts: The Belknap Press of Harvard University Press, 1994).

T. Ingold, "Culture on the Ground: The World Perceived Through the Feet", in *Journal of Material Culture*, vol. 9, no. 3 (University of Aberdeen, Scotland, 2004).

S. Jallade, *Il richiamo della strada. Piccola mistica del viaggiatore* (Portogruaro: Ediciclo, 2011).

E. Jaques-Dalcroze, *Le rythme, la musique et l'éducation* (Paris, 1920).

W. Jensen, *Gradiva. Una fantasia pompeiana* [1903] (Rome: Donzelli, 2013).

W. Jensen and S. Freud, *Gradiva* (Pordenone: Edizioni Studio Tesi, 1992).

E. Jünger, "Testa e piede", in *Il contemplatore solitario*, edited by H. Plard, afterword by Q. Principe (Parma: Guanda Editore, 1995).

W. L. Jungers et al., "The foot of Homo floresiensis", in *Nature*, no. 459, 2009.

O. Kaeppelin and R. Barni, *Roberto Barni: sculptures et peintures*, exhibition catalogue (Monaco: Marlborough, 2013).

D. Kamin, *The Comedy of Charlie Chaplin* (Lanham, Maryland, Toronto and Plymouth: The Scarecrow Press, 2008).

H. von Kleist, *On a Theatre of Marionettes* [1811].

R. Laban, *Modern Educational Dance* (Wokingham, Berkshire: Macdonald & Evans, 1948; reprinted 1963 and 1975, third edition edited by L. Ullman).

R. Laban, *The Mastery of Movement on the Stage* (Wokingham, Berkshire: Macdonald & Evans, 1950; reprinted 1967, 1974, third enlarged edition edited by L. Ullman and 1980, fourth edition with introduction by R. Laban).

R. Laban, *Laban's Principles of Dance and Movement Notation* [1956] (London: Macdonald and Evans, 1975).

R. Laban, *Effort: Economy of Human Movement* (London: Macdonald and Evans, 1974), with F. C. Lawrence.

H. Laborit, *Elogio della fuga* (Milan: Mondadori, 1982).

R. Landmann, *Ascona – Monte Verità* (Berlin: Ullstein, 1979).

B. Léal, C. Piot and M. L. Bernadac, *Picasso Totale* (Milan: Leonardo International, 2001).

D. Le Breton, "Camminatori e cammini", in *Pensieri viandanti. Antropologia ed estetica del camminare*, edited by I. Testa (Reggio Emilia: Diabasis, 2007).

D. Le Breton, *Il mondo a piedi. Elogio della marcia* (Bergamo: Universale Economica Feltrinelli, 2013).

A. Le Normand-Romain, *Rodin et le bronze. Catalogue des oeuvres conservées au musée Rodin* (Paris: Éditions du Musée Rodin / Éditions de la Réunion des Musées nationaux, 2007).

A. Leroi-Gourhan, *Il gesto e la parola. II, La memoria e i ritmi* (Turin: Einaudi, 1977).

Lorenzo Bartolini. Mostra delle attività di tutela, exhibition catalogue (Florence, 1978).

F. Losi, preface to M. Feldenkrais, *Il corpo e il comportamento maturo* (Rome: Casa Editrice Astrolabio,1996).

A. Magnien, "Le prugne e le frittelle. Ricordi di gioventù di Auguste Rodin", in A. Magnien and F. Arensi, *Rodin. L'origine del genio*, exhibition catalogue (Turin: Allemandi, 2010).

G. Manzi, *Il grande racconto dell'evoluzione umana* (Bologna: il Mulino, 2013).

Marino Marini. Catalogo ragionato della scultura, introductory essay by G. Carandente (Milan: Skira, 1998).

Y. Matsuyama and L. B. Ribeiro, *Danza classica e contemporanea con la didattica Kniaseff* (Milan: Gammalibri, 1984).

C. McCann, *Questo bacio vada al mondo intero* [2009] (Milan: Bur Rizzoli, 2013).

M. Melucci, *Lezioni di Metodo Feldenkrais* (Milan: Biblioteca olistica, Xenia Edizioni, 2011).

Merce Cunningham, edited by G. Celant (Milan: Charta, 2000).

M. Merleau-Ponty, *The Merleau-Ponty Reader* (Evanston: Northwestern University Press, 2007).

N. Misler, *In principio era il corpo. L'arte del movimento a Mosca negli anni Venti*, exhibition catalogue (Milan: Electa, 1999).

M. Missirini, *Vita di Antonio Canova. Libri quattro* (Prato, 1824; anastatic reprint edited and with an introduction by F. Leone, Bassano del Grappa 2004).

Monte Verità. Le mammelle della verità, edited by H. Szeemann (Locarno: Armando Dadò, 1978).

B. Musetti, "Ceci n'est pas un peintre. Qualche riflessione intorno alla produzione pittorica di Auguste Rodin", in A. Magnien and F. Arensi, *Rodin. L'origine del genio*, exhibition catalogue (Turin: Allemandi, 2010).

C. Negri, *Le Gratie d'Amore* (Milan, 1602).

L. Nepi, "Un Chaplin moderno: Monsieur Verdoux tra Keats Schopenhauer e il fermoimmagine", in *Annali del dipartimento di storia delle arti e dello spettacolo*, XI, 2010.

F. Nietzsche, *The Gay Science* (Mineola, New York: Courier Dover Publications, 2006).

K. Noschis, *Monte Verità: Ascona e il genio del luogo* (Bellinzona: Casagrande, 2013).

Oeuvres de Pierre Camper, qui ont pour objet l'histoire naturelle, la physiologie et l'anatomie comparée, edited by H. J. Jansen (Paris, 1803).

G. Oliva, *Il laboratorio teatrale* (Milan: LED Edizioni Universitarie, 1999).

A. Oliverio, *Prima lezione di neuroscienze* (Bari: Laterza, 2011).

A. Ottani Cavina, "Il Settecento e l'antico", in *Storia dell'arte italiana* (Turin, 1982).

Pablo Picasso. I saltimbanchi, exhibition catalogue (Pistoia: Gli Ori, 2011).

P. P. Pasolini, *Il cinema di poesia* [1965], now in P. P. Pasolini, *Empirismo eretico* (Milan: Garzanti, 1972).

Paul Valéry. Scritti sull'arte (Milan: Guanda, 1984).

G. Pavanello, "Canova e la Danzatrice", in *Canova e la danza*, exhibition catalogue, edited by M. Guderzo (Crocetta di Montello: Terra Ferma Edizioni, 2012).

R. Payne, *The Great God Pan. A Biography of the Tramp Played by Charles Chaplin* (New York: Hermitage House, 1952).

P. Petit, *Funambule* (Paris: Albin Michel, 1991).

P. Petit, *Traité du funambulisme*, preface by Paul Auster (Arles: Actes Sud, 1997).

P. Petit, *To Reach The Clouds: My High Wire Walk Between The Twin Towers* (New York: North Point Press, 2002).

P. Petit, *Credere nel vuoto* (Turin: Bollati Boringhieri, 2008).

T. Pievani, *La vita inaspettata* (Milan: Raffaello Cortina Editore, 2001).

J. H. Pilates, *Your Health*, 1934.

J. H. Pilates and W. J. Miller, *Return to Life through Contrology* [1945] (Bel Air: Christopher Publishing House, 1960).

Plato, *The Laws* (London: Penguin, 1970).

Plato, *Symposium* (Oxford: Oxford University Press, 2008).

K. W. Purcell, *Alexey Brodovitch* (London: Phaidon, 2002).

F. Quadri, *Sulle tracce di Pina Bausch* (Milan: Ubulibri, 2002).

E. Raimondi, "Il coreografo perduto" and "Un teatro terribile: Roma 1782", in *Le pietre del sogno. Il moderno dopo il sublime* (Bologna: il Mulino, 1985).

E. Randi, *Il magistero perduto di Delsarte. Dalla Parigi romantica alla modern dance* (Padua: Esedra, 1996).

M. Reese, preface to M. Feldenkrais, *L'Io potente* (Rome: Casa Editrice Astrolabio, 2007).

S. Risaliti, *Melotti. Catalogo generale della grafica* (Milan: Electa, 2008).

S. Risaliti, "Reminiscenze e ispirazioni di Salvatore Ferragamo", in *Salvatore Ferragamo. Ispirazioni e Visioni* (Milan: Skira, 2011).

F. Russoli and F. Minervino, *Picasso cubista* (Milan: Rizzoli, 1981).

B. Santi, *La collezione Chigi Saracini* (Siena: Banca Monte dei Paschi di Siena, 1998).

L. Scarlini, "Le geometrie dell'anima. I trionfi della moda e la percezione dell'altrove di Salvatore Ferragamo", in *Salvatore Ferragamo. Ispirazioni e Visioni* (Milan: Skira, 2011).

P. Schilder, *Mind, Perception and Thought in their Constructive Aspects* (New York: Columbia University Press, 1942).

W. Self, *Umbrella* (London: Bloomsbury, 2012).

W. Sharp, "Two Interviews", in *Bruce Nauman*, edited by R. C. Morgan (Baltimore: The Johns Hopkins University Press, 2002).

Shoemaker of Dreams. The Autobiography of Salvatore Ferragamo [1957] (Florence: Giunti, 1985).

J. Simons, "Breaking the Silence: an Interview with Bruce Nauman", in *Please Pay Attention Please: Bruce Nauman's Words. Writings and Interviews*, edited by J. Kraynak (Cambridge, Massachusetts and London: The MIT Press, 2003).

R. Solnit, *River of Shadows. Eadweard Muybridge and the Technological Wild West* (New York, 2003).

J. Starobinski, *Ritratto dell'artista da saltimbanco*, edited by C. Bologna (Turin: Bollati Boringhieri, 1984).

I. Teotochi Albrizzi, *Opere di scultura e di plastica di Antonio Canova* (Pisa, 1823; anastatic reprint edited and with an introduction by M. Pastore Stocchi and G. Venturi, Bassano del Grappa, 2003).

H. D. Thoreau, *Camminare* [1862], edited by M. Jevolella (Milan: Oscar Mondadori, 2009).

C. Tomkins, *Vite d'avanguardia. John Cage, Leo Castelli, Christo, Merce Cunningham, Johnson Philip, Andy Warhol* (Genoa: Costa&Nolan, 1983).

N. Tsukui, *Architecture and Urbanism. Cecil Balmond* (Tokyo: A+U Publishing, Bunkyo-ku, 2006).

E. Vaccarino, *Altre scene, altre danze* (Turin: Einaudi, 1991).

E. Vaccarino, *Teatro dell'esperienza, danza della vita* (Genoa: Costa & Nolan, 2005).

P. Valéry, *Degas danse dessin* [1936], now in *Oeuvres*, vol. 2 (Paris: Gallimard, 1960).

P. Valéry, "Degas Danza Disegno", in *Paul Valéry. Scritti sull'arte* (Milan: Guanda, 1984).

W. Van der Will, "The Politics of the Body", in B. Taylor and W. Van der Will, *The Nazification of Art* (Winchester, Ontario: The Winchester Press, 1990).

P. Veroli and G. Vinay, *I Ballets Russes di Diaghilev tra storia e mito* (Rome: Accademia Nazionale di Santa Cecilia, 2013).

C. V. Ward, W. H. Kimbel and D. C. Johanson, "Complete Fourth Metatarsal and Arches in the Foot of Australopithecus afarensis", in *Science*, no. 331 (6018), 2011.

E. O. Wilson, *The Social Conquest of Earth* (New York: Liveright Publishing Corporation, 2012).

J. J. Winckelmann, *Il bello nell'arte. Scritti sull'arte antica*, edited by F. Pfister (Turin, 1973).

H. Windisch, *Das Deutsche Lichtbild* [1929], quoted by C. Phillips, "Twenties Photography: Mastering Urban Space", in *The 1920s: Age of the Metropolis*, exhibition catalogue, edited by J. Clair (Montreal: Museum of Fine Arts, 1991).

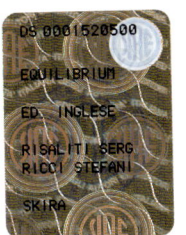

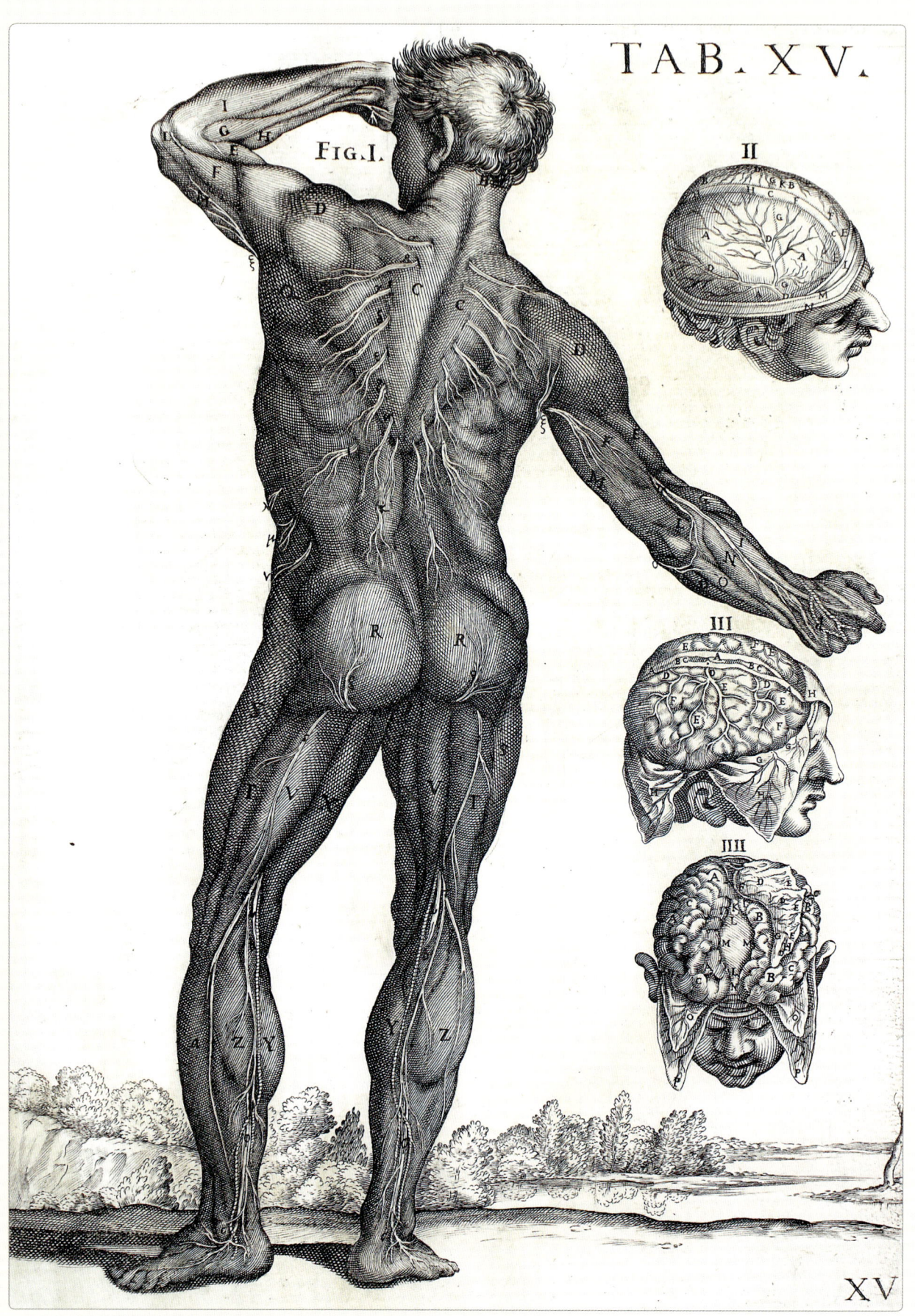

TAB. XV.

FIG. I.

II

III

IIII

XV

RAVIMVS, IN CVIVS DEXTRO LATERE MVSCVLI MOX SVB CVTE RECONDITI, ANTERIORI IN

facie conspiciuntur: in sinistro autem illis resectis obuij sunt, qui in
figura redderetur copiosior, eorum quæ caluaria complectitur ima
characterum index docebit, delineantes.

dextro latere apparentibus proximè succumbunt. Vt uerò præsens
ginem, sectionis ordine proponere incepimus, humi oculi musculos, uti

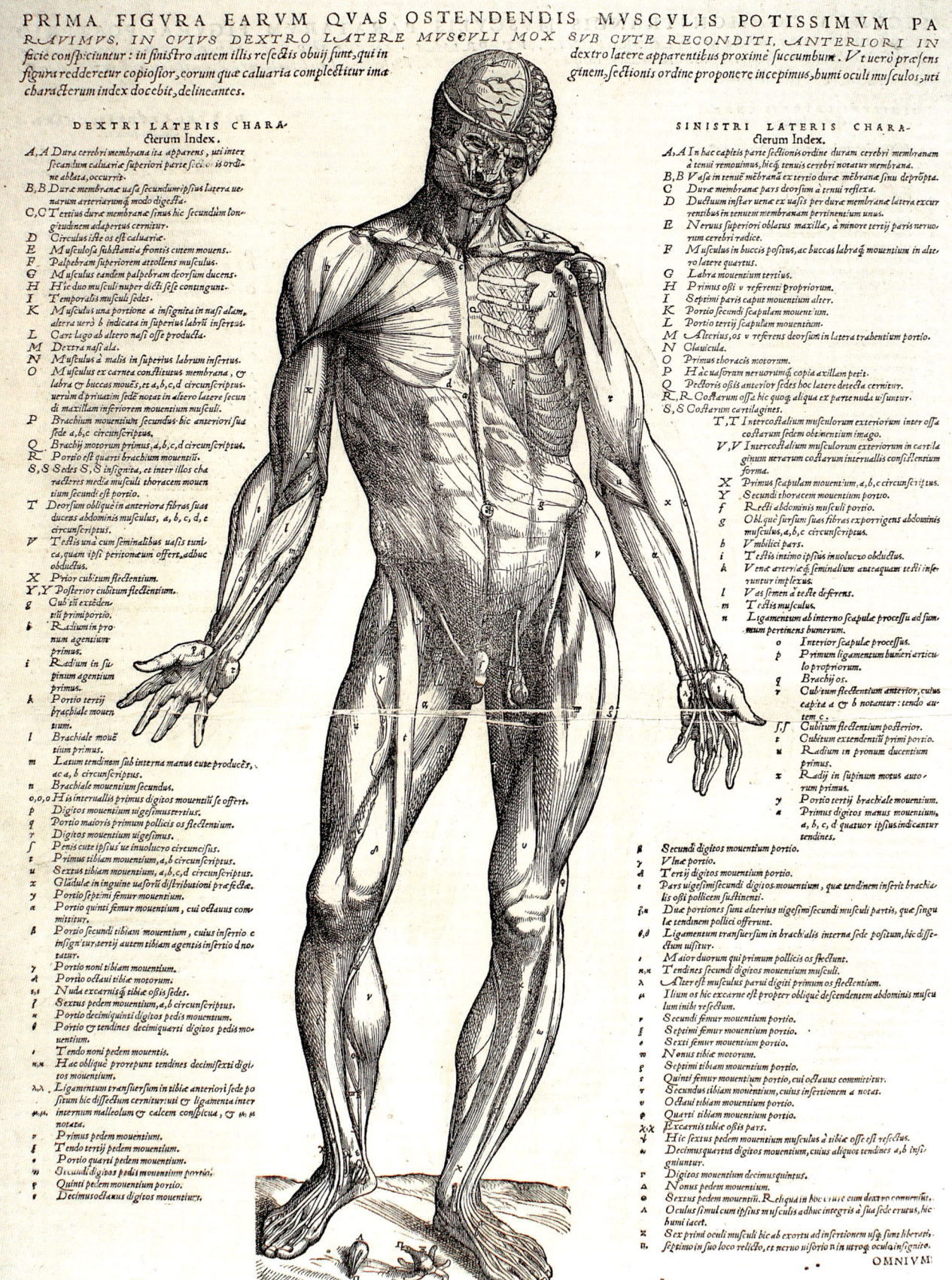

DEXTRI LATERIS CHARA-
cterum Index.

A,A Duræ cerebri membrana ita apparens, uti inter
secandum caluariæ superiori parte sectionis ordine ablata, occurrit.

B,B Duræ membranæ uasa secundum sua latera ue-
narum arteriarumq́ modo digesta.

C,C Tertius duræ membranæ sinus hic secundum lon-
gitudinem adapertus cernitur.

D Circulus iste os est caluariæ.

E Musculosa substantia frontis cutem mouens.

F Palpebram superiorem attollens musculus.

G Musculus eandem palpebram deorsum ducens.

H,I,K Hic duo musculi nuper dicti sese contingunt.
  Temporalis musculi sedes.

L Musculus una portione a insignita in nasi alam,
  altera uerò b indicata in superius labrum inseritus.

M,N,O Dextra nasi ala.
  Musculus à malis in superius labrum insertus.
  Musculus ex carnea constituus membrana, &
  labra & buccas mouens, et a,b,c,d circunscriptus.

P Brachium mouentium secundus hic anteriori sua
  sede a,b,c circunscriptus.

Q Brachij motorum primus, a,b,c,d circunscriptus.

R,S Portio est quarti brachium mouentium.
  S,S Sedes S, & insignita, et inter illos cha-

T Deorsum obliquè in anteriora fibras suas
  ducens abdominis musculus, a, b, c, d, e
  circunscriptus.

V Testis una cum seminalibus uasis tuni-
  ca, quam ipsi peritonæum offert, adhuc
  obductus.

X Prior cubitum flectentium.

Y,Y Posterior cubitum flectentium.

SINISTRI LATERIS CHARA-
cterum Index.

A,A In hac capitis parte sectionis ordine duram cerebri membranam
  à tenui remouimus, hic tenuis cerebri notatur membrana.

B,B Vasa in tenui mèbrana ex tertio duræ mèbranæ sinu depròpta.

C Duræ membranæ pars deorsum à tenui reflexa.

D Ductuum instar uenæ ex uasis per duræ membranæ latera excur-
  rentibus in tenuem membranam pertinentium unus.

E Neruus superiori oblatus maxillæ, à mollore tertij paris neruo-
  rum cerebri radice.

F Musculus in buccis positus, ac buccas labrag̃ mouentium in alte-
  ro latere quartus.

G,H Labra mouentium tertius.
  Primus ossi ν referenti propriorum.

I,K,L Septimi paris caput mouentium alter.
  Portio secundi scapulam mouentium.
  Portio tertij scapulam mouentium.

M Alterius, os ν referens deorsum in latera trahentium portio.

N,O Clauicula.
  Primus thoracis motorum.

P,Q Hic uasorum neruorumq́ copia axillam petit.
  Pectoris ossis anterior sedes hoc latere detecta cernitur.

R,R Costarum ossa hic quoq́ aliqua ex parte nuda uisuntur.

S,S Costarum cartilagines.

T,T Intercostalium musculorum exteriorum inter ossa
  costarum sedem obtinentium imago.

V,V Intercostalium musculorum exteriorum in cartila-
  ginum uerarum costarum internallis consistentium
  forma.

X Primus scapulam mouentium, a,b,c circunscriptus.

Y Secundi thoracem mouentium portio.

f Recti abdominis musculi portio.

g Obliquè sursum suas fibras exporrigens abdominis
  musculus, a,b,c circunscriptus.

h Vmbilici pars.

i Testis intimo ipsius inuolucro obductus.

k Venæ arteriæq́ seminalium antequam testi inse-
  runtur implexus.

l Vas à semen à teste deferens.

m Testis musculus.

n Ligamentum ab interno scapulæ processu ad sum-
  mum pertinens humerum.

  o Interior scapulæ processus.
  p Primum ligamentum humeri articu-
    lo propriorum.
  q Brachij os.
  r Cubitum flectentium anterior, cuius
    capita d & b notantur: tendo au-
    tem c.

ſſ Cubitum flectentium posterior.
  t Cubitum extendentiū primi portio.
  u Radium in pronum ducentium
    primus.
  x Radij in supinum motus auto-
    rum primus.
  y Portio tertij brachiele mouentium.
  a Primus digitos menus mouentium,
    a, b, c, d quatuor ipsius indicantur
    tendines.

g Cui ill extenden-
  tili primi portio.
b Radium in pro-
  num agentium
  primus.
i Radium in su-
  pinum agentium
  primus.
k Portio tertij
  brachiele mouen-
  tium.

l Brachiele mouen-
  tium primus.
m Latum tendinem sub interna manus cute producēs,
  ac a, b circunscriptus.
n Brachiele mouentium secundus.
o,o,o His in interuallis primus digitos mouentiū se offert.
p Digitos mouentium uigesimustertius.
q Portio maioris primum pollicis os flectentium.
r Digitos mouentium uigesimus.
s Penis cute ipsius ab inuolucro circuncisus.
t Primus tibiam mouentium, a,b circunscriptus.
u Sextus tibiam mouentium, a,b,c,d circunscriptus.
x Glandula in inguine uasorum distributioni præfectæ.
y Portio septimi femur mouentium.
a Portio quinti femur mouentium, cui octauus committitur.
β Portio secundi tibiam mouentium, cuius insertio c
  insignitur, tertij autem tibiam agentis insertio d notatur.
γ Portio noni tibiam mouentium.
δ Portio octaui tibiæ motorum.
ε,ε Nuda excarnisā tibiæ ossis sedes.
ζ Sextus pedem mouentium, a,b,c circunscriptus.
η Portio decimiquinti digitos pedis mouentium.
θ Portio & tendines decimiquarti digitos pedis mouentium.
ι Tendo noni pedem mouentis.
κ,κ Hæc oblique prorepunt tendines decimisexti digitos mouentium.
λ,λ Ligamentum transuersim in tibiæ anteriori sede positum hic dissectum cernitur: uti & ligamenta inter
μ,μ,μ internum malleolum & calcem conspicua, ν, & μ
  notata.
ν Primus pedem mouentium.
ξ Tendo tertij pedem mouentium.
ο Portio quarti pedem mouentium.
π Secundi digitos pedis mouentium portio.
ρ Quinti pedem mouentium portio.
σ Decimusoctauus digitos mouentium.

β Secundi digitos mouentium portio.
γ Vlnæ portio.
δ Tertij digitos mouentium portio.
ε Pars uigesimisecundi digitos mouentium, quæ tendinem inserit brachialis ossi pollicem sustinenti.
ζ,η Duæ portiones sunt alterius uigesimisecundi musculi partis, quæ singu-
  le tendinem pollici offerunt.
θ,δ Ligamentum transuersum in brachialis interna sede positum, hic dissectū
  uisitur.
ι Maior duorum qui primum pollicis os flectunt.
κ,κ Tendines secundi digitos mouentium musculi.
λ Alter ist musculus parui digiti primum os flectentium.
μ Ilium os hic excarne est propter oblique descendentem abdominis muscu
  lum inibi resectum.
ρ Secundi femur mouentium portio.
ξ Septimi femur mouentium portio.
ο Sexti femur mouentium portio.
π Nonus tibiæ mouentium.
ρ Septimi tibiam mouentium portio.
σ Quinti femur mouentium portio, cui octauus committitur.
τ Secundus tibiam mouentium, cuius insertionem a notat.
υ Octaui tibiam mouentium portio.
φ Quarti tibiam mouentium portio.
χ,χ Excarnis tibiæ ossis pars.
ψ Hic sextus pedem mouentium musculus à tibiæ osse est resectus.
ω Decimusquartus digitos mouentium, cuius aliquot tendines a,b insi-
  gniuntur.
Γ Digitos mouentium decimusquintus.
Δ Nonus pedem mouentium.
Θ,Δ Sextus pedem mouentium. Reliqua in hoc crure cum dextro communis.
Λ Oculus simul cum ipsius musculis adhuc integris à sua sede eruus, hic
  humi iacet.
χ Sex primi oculi musculi hic ab exortu ad insertionem usq̃ sunt liberati,
π septimo in suo loco relicto, et neruo uisorio π in utroq̃ oculi insignito.

OMNIVM.

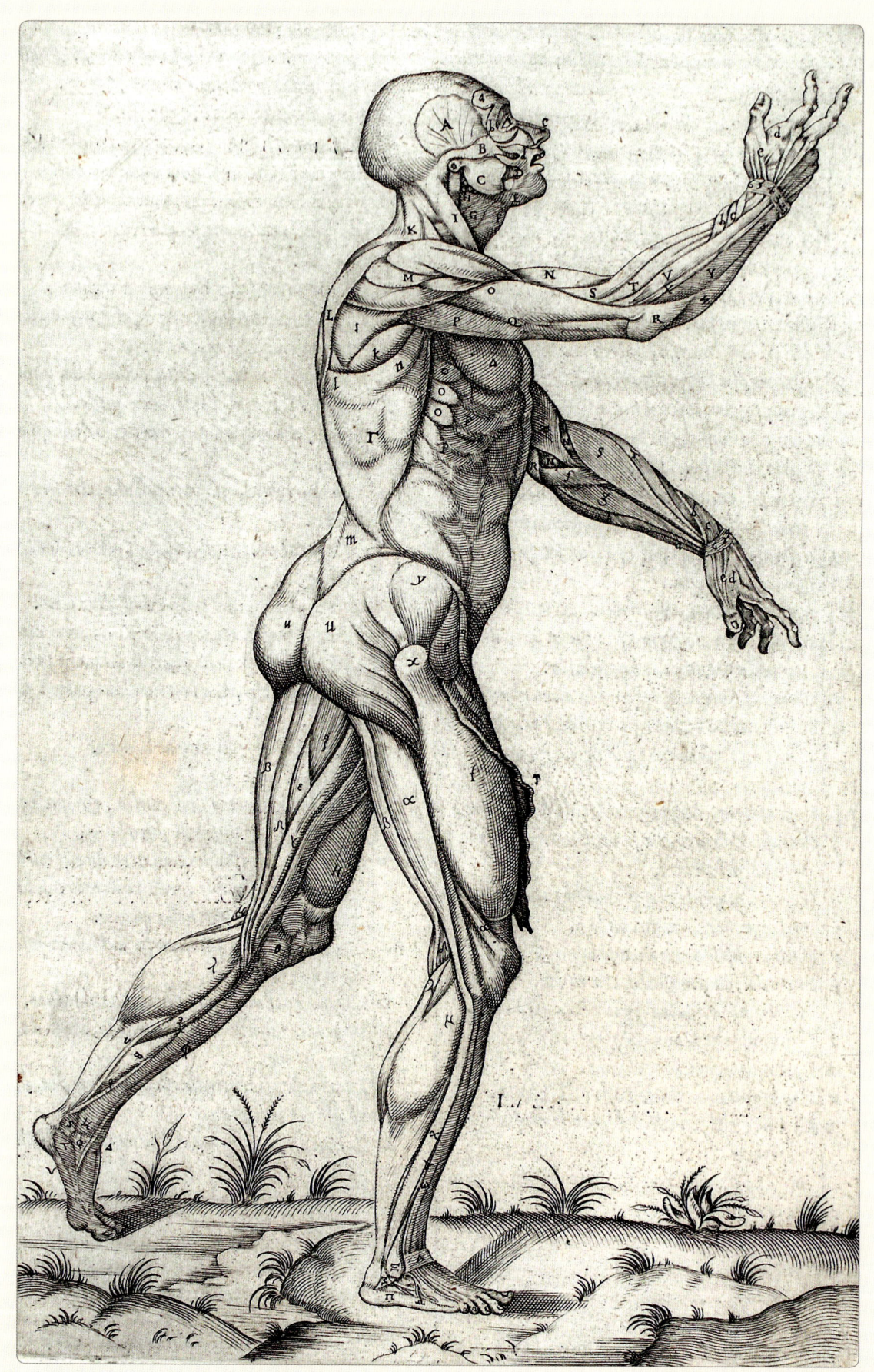

TAB. V.

FIG.I.

FIG.II.

FIG.III.

FIG.IIII.

V

TAB. XVIII.

FIG. I.

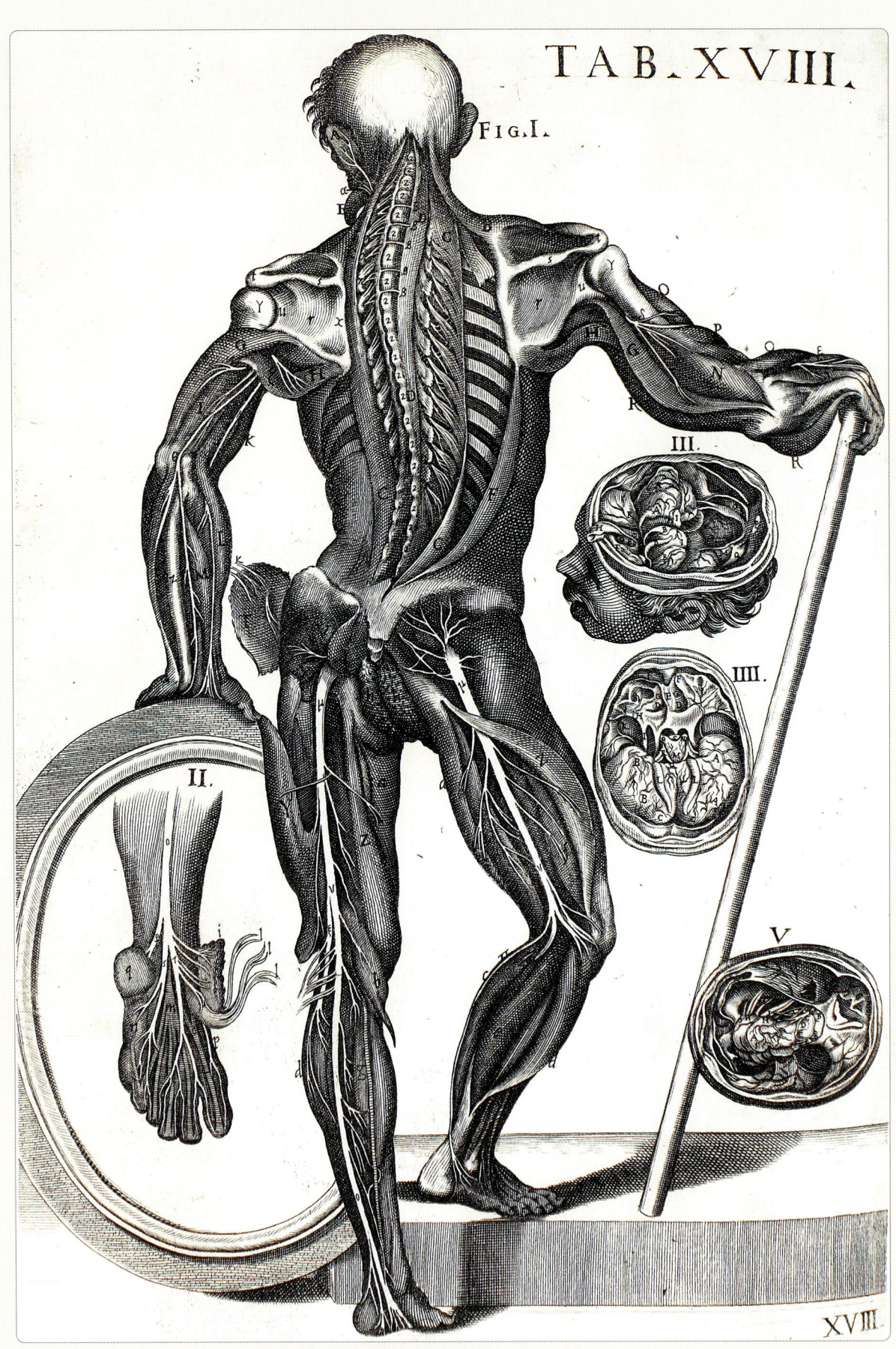

III.

IIII.

II.

V.

XVIII.

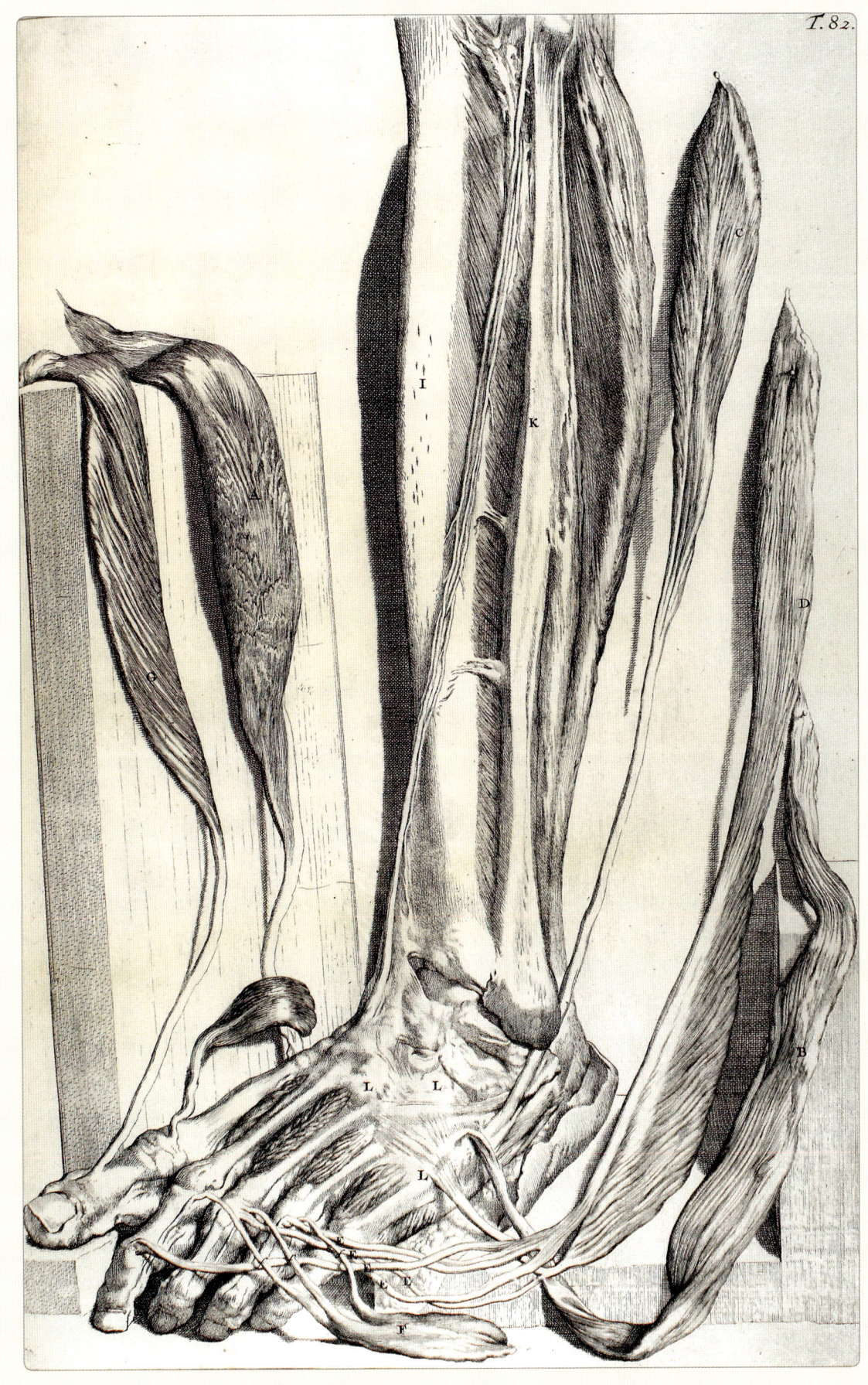

T. 82.

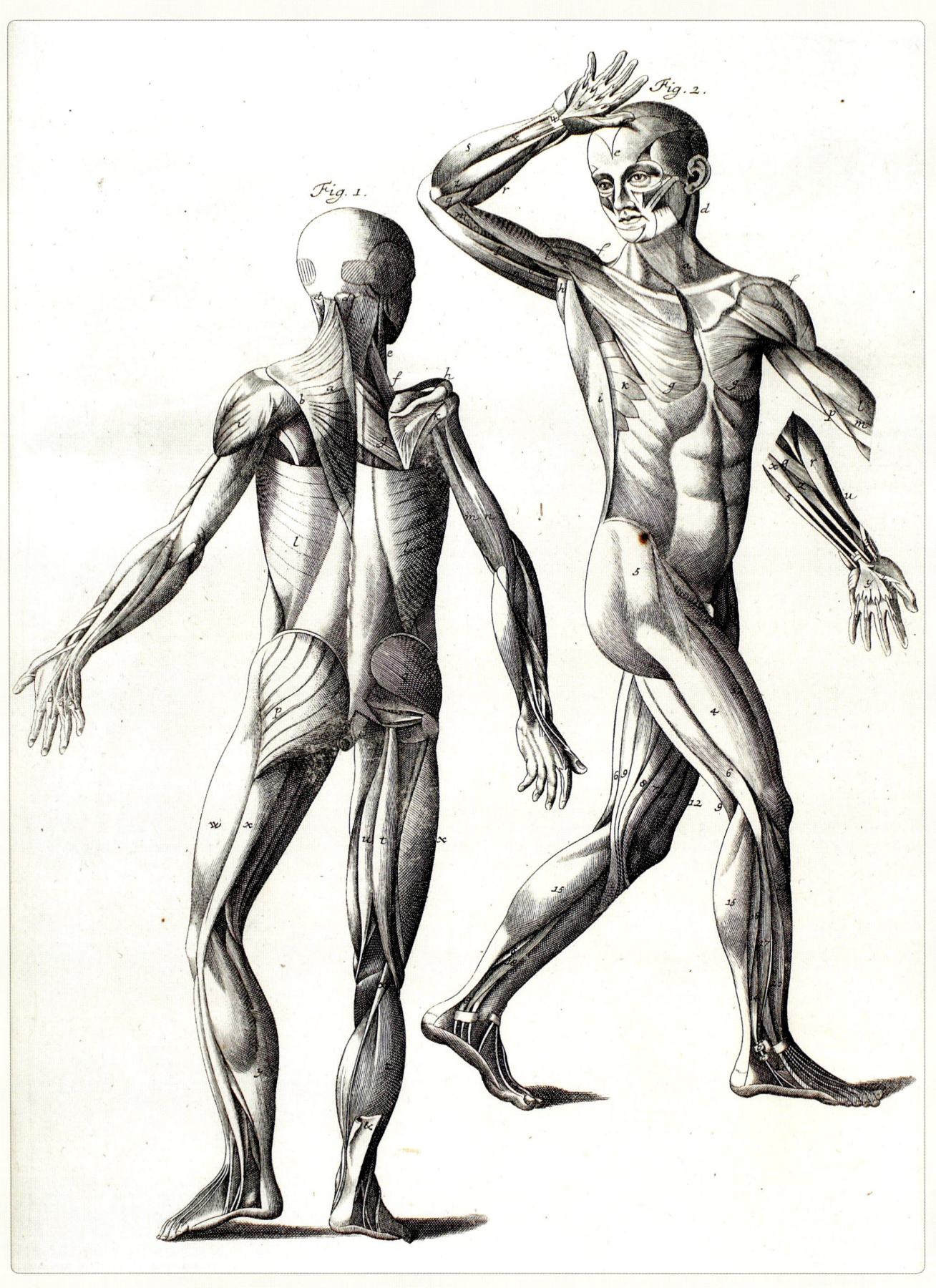

Fig. 1.

Fig. 2.

left and right flaps

Pietro Berrettini (called Pietro
da Cortona), *Tabulae anatomicae
a celeberrimo picture Petro Berretino*
(Rome: A. de Rubeis, 1741),
44.4 x 32.5 x 3.4 cm
Florence, Biblioteca Nazionale
Centrale di Firenze.

inside, left to right

Andrea Vesalio, *Suorum de humani
corporis fabrica libro rum epitome*,
(Basileae: ex officina Ioannis Oporini,
1543), 50 x 35.7 x 1 cm
Florence, Biblioteca Nazionale
Centrale di Firenze.

Juan Valverde, *Anatomia del corpo
umano* (Rome: A. Salamanca and A.
Lafrery, 1560), 31 x 22 x 3.7 cm
Florence, Biblioteca Nazionale
Centrale di Firenze.

Jean-Jacques Manget, *Theatrum
anatomicum…* (Geneva, 1716–17),
2 vols., 44 x 29 x 8.3 cm
Florence, Biblioteca Nazionale
Centrale di Firenze.

Govard Bidloo, *Anatomia humani
corporis…* (Amsterdam: the Widow
of J. Van Someren, the Heirs of J.
Van Dyk, H. Boom and the Widow
of T. Boom, 1685), 53 x 38 x 5.7 cm
Florence, Biblioteca Nazionale
Centrale di Firenze.

that a person's musculature will continue to follow its own schema, determined by his or her self-image: habitual patterns of movement are imprinted upon the nervous system, which reacts to external stimuli according to prearranged, habitual patterns and models, since no other reaction models are available (it still has to learn any). In order to enable the body to modify these patterns, previous models must be removed; one must then teach the nervous system some new patterns, "leaving it free to act or react – not according to habit, but according to the given external situation".[11]

TOUCHING AS AN ACT OF KNOWLEDGE-ACQUISITION

Salvatore Ferragamo's autobiography is rich in tales and images related to feet and the perception of feet. At one point, it even describes a kind of current, an "electric shock" running from the foot he is touching to the palm of his hand.[12] On another page, it invites the reader to reproduce the structure of a foot, so as to better understand and perceive it, by placing a hand on the table, palm down, and pulling the fingers back a little.[13] It is interesting to look at the photographs in which Ferragamo is touching the feet of his customers, as though through this contact he sought to help them better perceive feet and their anatomy. Besides, foot reflexology has drawn upon all the ancient knowledge surrounding feet which civilizations such as the Chinese and Indian had already acquired as early as five thousand years ago. Feet are like a microcosm reflecting the body – a kind of second heart. Doctor William H. Fitzgerald (1872–1942) was the first to promote this knowledge in the Western world.

Moshe Feldenkrais would have smiled at these intuitions – the neurosciences tell us that nerves transmit electric current through the junction between one nerve and another, or synapse. After all, he was a man of science with a training in the martial arts, which make the most efficient possible use of supports, pushes and spirals in order to move up or down from the ground. Given the above premises, touch is precisely what enabled Feldenkrais to communicate with others and their nervous system, and to ultimately confirm his intuitions: touch could make people – both the one touching and the one being touched – more aware of their own movement patterns, by activating the connectivity between hands and body parts. What helped improve this perception was the hand, the hand's touch: *eliciting* a neuro-sensory-motor response, *increasing* the awareness of one's body and self-image, *broadening* motor patterns, *refining* perception and the arrangement of the body in the gravitational field, and *inhibiting* parasitic movements that limit the possibility of performing free voluntary movements.[14]

Feldenkrais was to describe this aspect of his method, related to touch, as functional integration, precisely because it integrates the parts engaged in connections with the movement pattern as a whole. He repeatedly stressed the fact that it was not a matter of treating a body part by means of touch, but rather of enabling it to improve through learn-

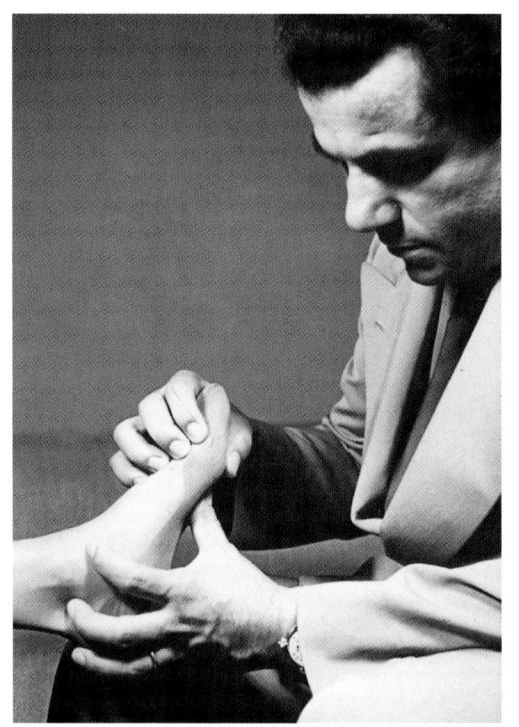

Salvatore Ferragamo touching
the foot of a model during the war.

August Sander, *Three Young Farmers in Sunday Dress, Westerwald*, 1914
Gelatin silver print,
30 x 25.5 cm
Paris, Centre Georges
Pompidou.

ing; of accompanying the other person through touch, as though dancing with him or her, as he liked to say.

Within the framework of functional integration, Feldenkrais also developed some highly effective methods for restoring the sensitivity of the feet through the use of a kind of artificial floor. While making people lie on their back – for the aforementioned reasons – he started employing a thin board of hard wood: facing the soles of the feet, he would hold the board with both hands and move it towards the feet, focusing on one at a time. First he would approach the fifth toe, the little toe, breaking the contact and re-establishing it several times; then he would do the same with the fourth toe, for as many times as was necessary to make it tremble. After that, he would tilt the board in such a way as to touch only the little

toe, then both toes and so on, reaching the third toe, the second, and finally the big toe. Once all toes were touching the board, he would move the latter in such a way as to have only the heel touching it first, and then the toes. He would continue in this manner until he detected a reaction in the ankle joint, or a more regular movement if a reaction had already occurred, and then tilt the board in order to touch the outermost part of the foot. He would switch between the side of the big toe and that of the little toe until there too a twisting of the foot would become perceivable, followed by a softer and more or less normal movement. After roughly half an hour of this slow and regular approaching of the various parts of the foot, the latter would automatically follow the board in all directions, having "re-learned" to use its surface and to adapt its supports to the inclination of the floor.[15] In this case, the effect was to awaken the feet of people suffering from cerebral palsy, yet the method was also applicable to "normal" people who had simply lost some sensitivity in the feet by keeping them too long in shoes unsuited to their anatomy. In just thirty minutes a person's feet and legs would rediscover the explorative state, the tactile feeling of having always walked without any shoes on – a situation in which feet could respond to the stimuli from different supports and textures of the ground (be it sand, marble, concrete, wood or stone).

August Sander, *The Artist
Anton Räderscheidt*, 1926
Gelatin silver print,
27.9 x 21.9 cm
Paris, Centre Georges Pompidou.

August Sander, *Peasant Children,
Westerwald*, 1931
Gelatin silver print,
22.5 x 30.2 cm
Paris, Centre Georges Pompidou.

and surfaces, rags lying on the ground, lines and veins in the concrete in the works of Eugène Atget and André Kertész. Like children, these photographers pushed the explorative and tactile tension to the very limit, bringing the act of walking into play once again – as ballet was to do as well.

"Touching the click button, for the photographer, is like giving a kiss; it is the most direct expression of his love for things".[17] László Moholy-Nagy, one of the leading exponents of Bauhaus, tried his hand both with a series of shots of pavements and with the film camera. He described this attempt to cast a fresh gaze on the surface of pavements as a "psychophysical" approach, which he saw as "a training of the eye, concerning both morals and the senses, a broader kind of work compared to any other means of visual creation. … In this respect, film in the future will be the most important means of space apprehension".[18] This helps us understand even better the parallel journey made by some "new" arts, such as photography and the cinema, which came to explore the moving image from without through a different way of perceiving and touching it, and Feldenkrais' work, which transposed this perception within the body. And given that – as should be clear by now – posture is always set within a social context which influences it, what remains to be evoked is a gesture, the philosophical gesture of a foot that, in the same period, paved the way for the momentous changes of the twentieth century.

At the turn of the twentieth century, certain gestures sprung from the new expressive possibilities afforded by the photo camera – and later the film camera – brought into play a poetics of the ground, of the urban pavement. It was as though an attempt were being carried out to reactivate a kind of exploration of touch and supporting surfaces that might amplify the stimuli from the ground. "What do we know of the street corners, curbstones, the architecture of the pavements, we who have never felt the streets, heat, dirt and edges of the stones under naked soles, never investigated the connections, to see if this might serve as a bed for us?", Walter Benjamin asked himself.[16] The outcome were series of shots of pavements

August Sander, *Schoolboys Celebrate the Kaiser's Birthday*, 1915
Gelatin silver print, 22.1 x 30.2 cm
Paris, Centre Georges Pompidou.

MAKING A FRESH START FROM FOOT GESTURES: THE GAY SCIENCE

From a given moment onwards, the philosopher Friedrich Nietzsche started pushing to the very limits the short circuit which his thought was leading him towards. He realized the need to read the whole history of philosophy from a physiological perspective, heeding the complaints and reasons of the body. The foot itself spoke forth: "My objections to Wagner's music are physiological objections. Why should I therefore begin by disguising them under aesthetic formulae? My 'point' is that I can no longer breathe freely when this music begins to operate on me; my foot immediately becomes indignant at it and rebels: for what it needs is time, dance and march; it demands first of all from music the ecstasies which are in good walking, striding, leaping and dancing. But do not my stomach, my heart, my blood and my bowels also protest? Do I not become hoarse unawares under its influence? And then I ask myself what my body really wants from music generally".[19]

The "whole body" is invoked by the philosopher, who professes a Gay Science of walking, leaping and dancing, since knowledge does not mean sitting in fusty chairs and burying oneself in books: "it is our custom to think in the open air, walking, leaping, climbing, or dancing on lonesome mountains by preference, or close to the sea, where even the paths become thoughtful".[20]

"Touch, contact and the manipulation of living human bodies allow me to understand the books of these extraordinary writers and to apply the science they teach", Feldenkrais stated.[21] It was necessary to access the inside of the body in order to understand its outside.

The Gay Science, this amazing anticipation of a new approach to thought, turned people into poets, scientists, children, dancers and walkers. It reinterpreted Western thought from a physiological perspective, against the great misunderstanding of the body: the

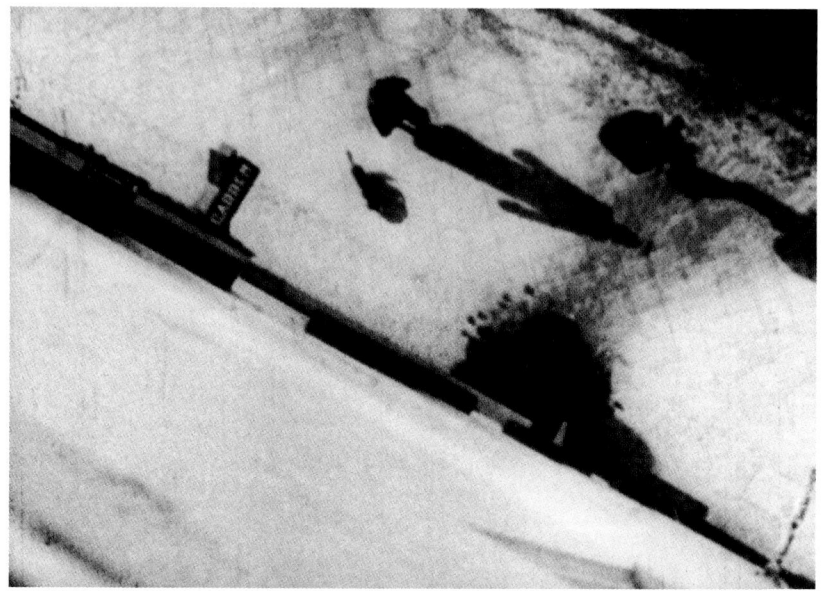

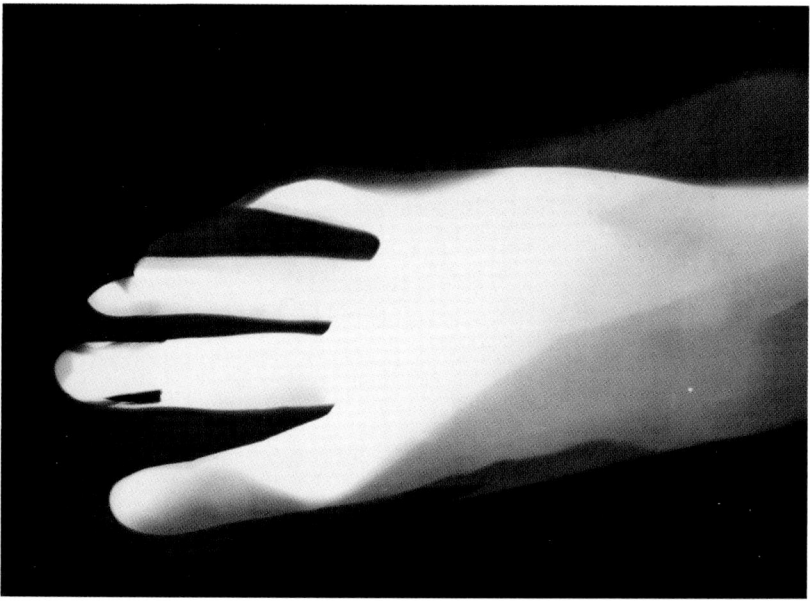

László Moholy-Nagy,
Berlin Still Life, 1929
Black and white 35mm film
Paris, Centre Georges Pompidou.

László Moholy-Nagy,
Untitled, 1925–26
Gelatin silver print,
17.7 x 23.7 cm
Paris, Centre Georges Pompidou.

constant separation between high and low, thought and the seat of thought, head, arms and hands and the rest of the body – the lower part, from the waist to the feet. After all, had not the whole history of metaphysics been a constant attempt to separate high and low, as though to establish, even in symbolic terms, the greatest possible distance between the feet and the head?

Did the rules of classical ballet – arguably the loftiest embodiment of this philosophical idealism – not reflect this very idea of a central perspective in the way they employed the body on stage, making a foot more charming, the more it was raised from the ground? Court dances, just like folk dances, used to be circular: the public would stand all around, in such a way as to gaze at the dancers and their feet from all angles; but when at the French court staged ballet was introduced, it was found that if the public stood in front of the performers, with the latter's feet pointing at them, the feet would disap-

pear.[22] An awareness thus emerged of the fact that if the performers externally rotated their feet and arms, the public would always see them. In such a way, through the arrangement of dancers' positions to suit the central perspective of the viewer – which is to say the gaze of the spectator at the centre: the king, his retinue, and the whole public – their arms and feet were always made clearly visible as they moved back and forth.

As any posture is a more or less successful form of "social adaptation", twentieth-century ballet largely sought to renew itself by perceiving the moving body from within.

Perceiving the origin of movement – where it comes from – and creating skeletal awareness of movement: this is the very principle followed by Feldenkrais. For one's self-image, within the motor cortex, is never visual, but kinaesthetic. We move not according to a visual image – even though we may think so – but according to a kinaesthetic one. And so the parts of ourselves that are not provided with sense perception are not the ones we use – even though sometimes we may think we are using them. No doubt, we look and imitate things, as recent discoveries about mirror neurons have revealed; but the point is how that imitation is transmitted within our perception of movement. This is what makes movement conscious and perceived.

László Moholy-Nagy,
Untitled, 1941
Gelatin silver print, 40.5 x 50.4 cm
Paris, Centre Georges Pompidou.

BETWEEN HANDS AND FEET. CHARLES DARWIN

In his studies on the value assigned to hands and feet by cultural systems, anthropologist Tim Ingold has shown that Darwin inherited the distinction between the two from the past. Darwin's evolutionary theories ultimately stressed three crucial factors responsible for the characteristics of the human species: 1) The anatomical changes that led to an erect posture: the S curvature of the spine, the widening of the pelvis and the full stretching of the legs through the hyper-extension of the hips, ensuring the possibility of standing on two legs; 2) the enlargement of the brain, and especially of its frontal area; 3) the use of hands, starting from the capacity to bring the end of the thumb in contact with any other finger. From this a range of new possibilities ensued in terms of the ability to do things and craft objects.

Following this line of thought, however, as stressed by T. H. Huxley himself in *Man's Place in Nature* (1863), increasing emphasis came to be placed on a sort of physiological distinction (again this word: even behind philosophical systems a physiological behaviour is to be found), according to which, as man shifted to an erect posture, hands and feet came to acquire almost complementary functions in relation to the work performed by the human body: ensuring support and locomotion in the case of the feet; grasping, handling and transforming materials in the case of the hands. Darwin stressed that the act of standing, thereby freeing the hands from their role as ground supports, ensured the separate development of the two. So the theory of evolution was largely to attribute the value and transmission of the use of the intellect to the hands (capable of even crafting shoes), while confining the feet to a state of nature, creating a division between a noble upper part of the human body and a lower part, from the waist down. Paradoxically, it was as though, despite the progress made in walking on two feet, the feet themselves had remained unaffected by the new brain functions.[23]

This is why Nietzsche's gesture, the rebellion of the foot, acquires such crucial importance.

LEARNING TO LEARN: HOW CHILDREN MOVE

How do children learn things? How do they build their connections in that privileged moment when they come to explore a body that is still unfamiliar to them? How do they use their head, the seat of the telereceptors – mouth, nose, eyes and ears – to extend their relations with the outer world beyond the body and find their bearings in space? How do they start expanding their movement within the body, in such a way as to perceive it and then finally set it in motion by crawling,

Eadweard Muybridge, *Animal Locomotion* (plate 747): sequence with a baboon walking, 1887 Florence, Raccolte Museali Fratelli Alinari (RMFA) – Palazzoli Collection.

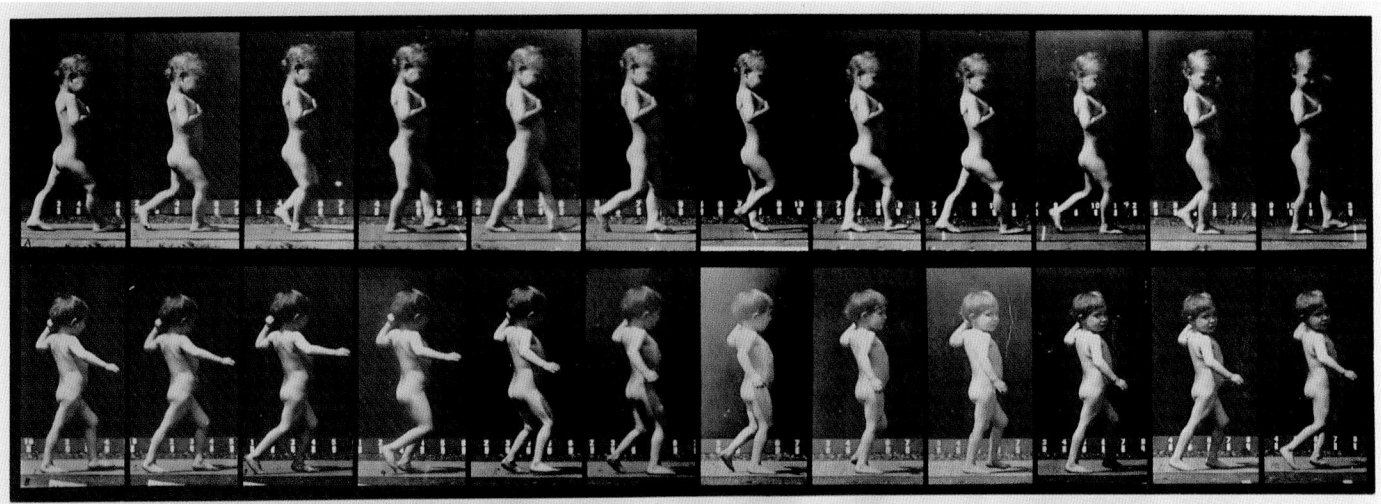

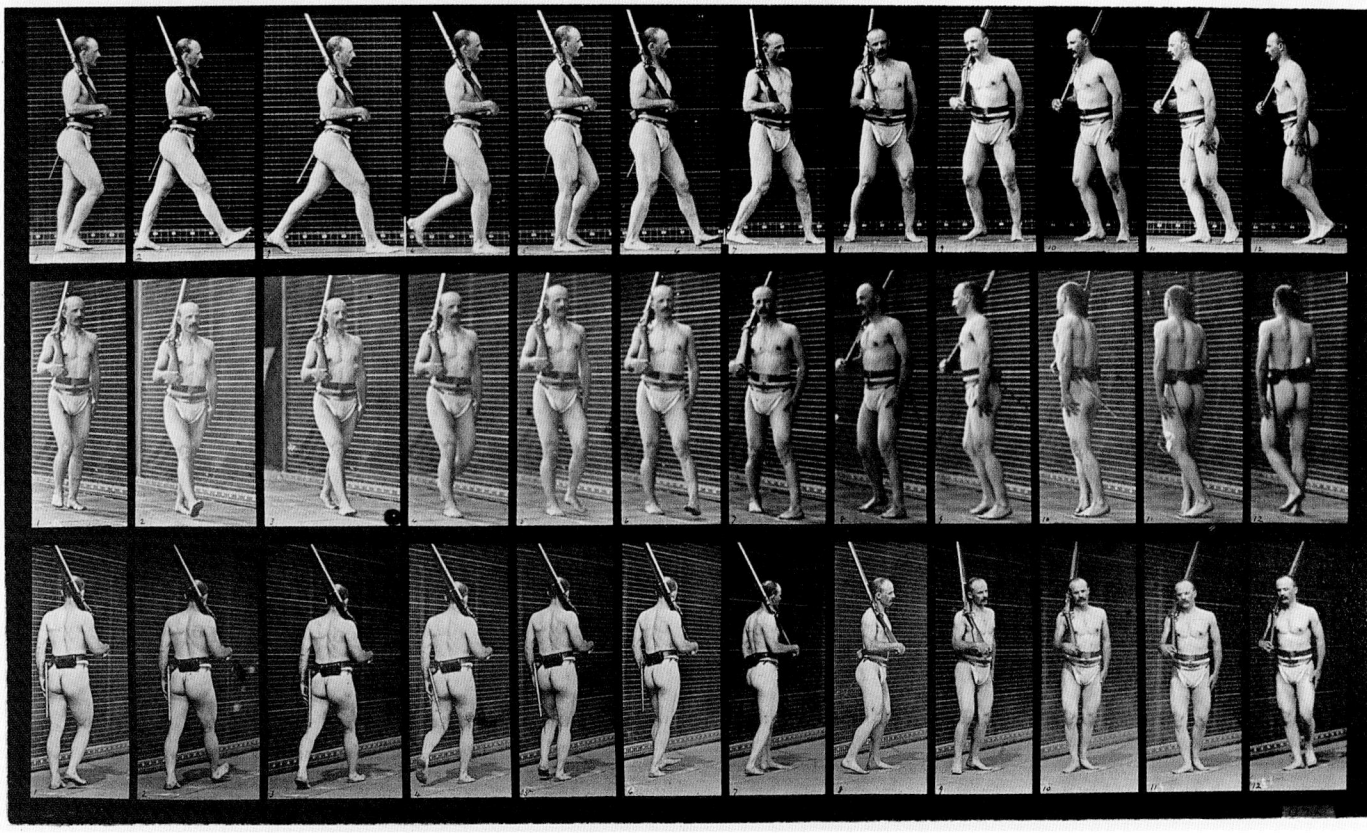

Eadweard Muybridge, *Animal Locomotion* (plate 540): sequence with a naked boy walking, 1887 Florence, Raccolte Museali Fratelli Alinari (RMFA) – Palazzoli Collection.

Eadweard Muybridge, *Animal Locomotion* (plate 355): sequence with a man walking and carrying a rifle, 1887 Florence, Raccolte Museali Fratelli Alinari (RMFA) – Palazzoli Collection.

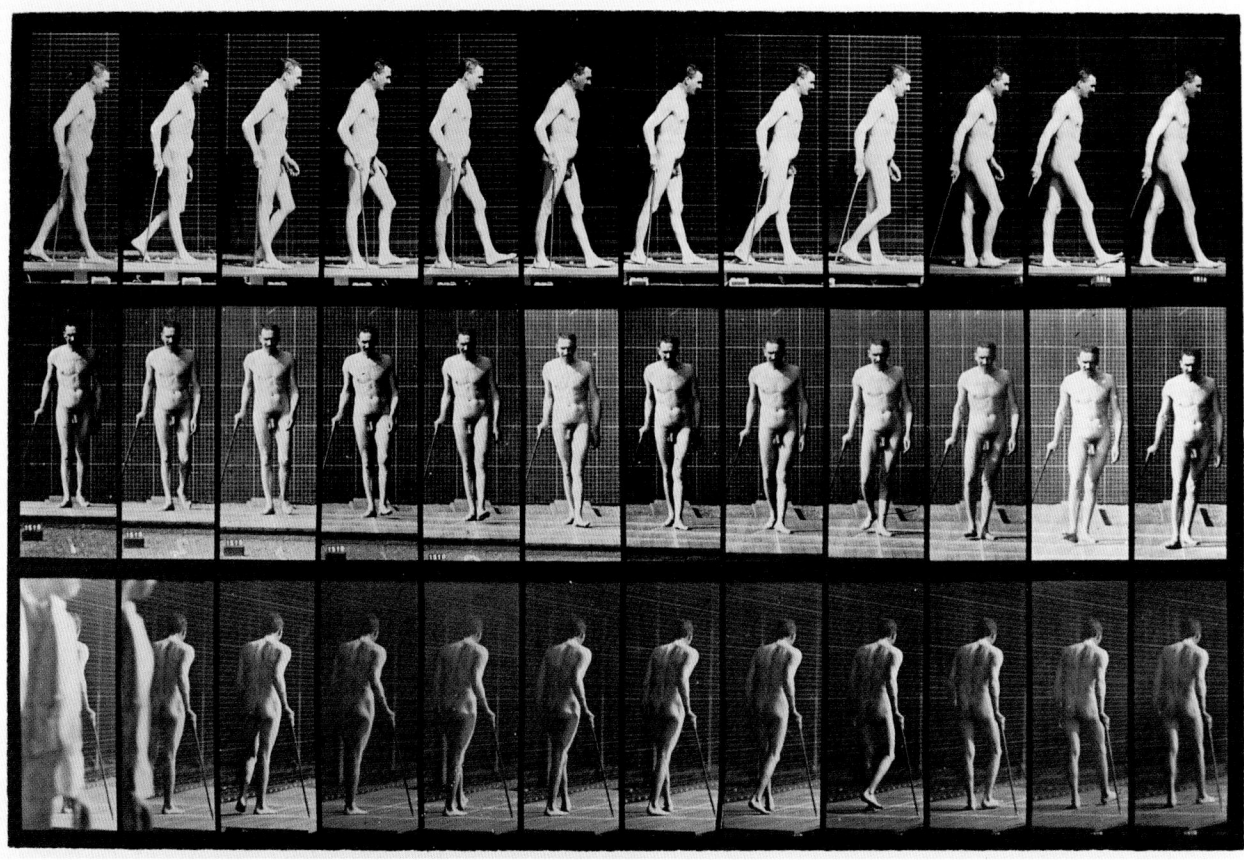

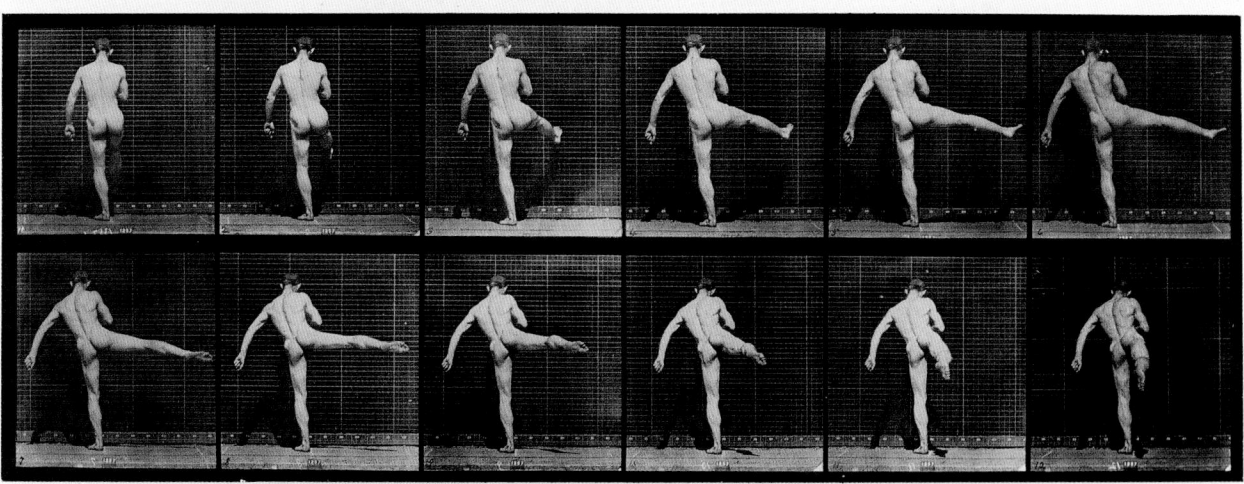

Eadweard Muybridge, *Animal Locomotion* (plate 559): sequence with a naked man walking, 1887 Florence, Raccolte Museali Fratelli Alinari (RMFA) – Palazzoli Collection.

Eadweard Muybridge, *Animal Locomotion* (plate 369): sequence with a naked man moving his right leg, 1887 Florence, Raccolte Museali Fratelli Alinari (RMFA) – Palazzoli Collection.

rolling and toddling about (gestures that almost all cultures tend to abolish in the case of adults, thereby depriving them of the richest learning tool available), until they can stand on two feet and develop their own body schema? Here Feldenkrais had another brilliant intuition, as he came to identify this phase as the process of greatest connective exploration between movement and the fact of learning from movement. And in doing so, he successfully enabled adults to learn anew. Feldenkrais would deconstruct movement through an endless variety of practices meant to foster an awareness of it: dancers, athletes, actors, musicians and even "ordinary" people would use them to retrace those stages in which they had learned how to move. Feldenkrais here lay the foundations of what is described today as organic learning or somatic education.

At birth, Feldenkrais explains, the human brain weighs around 300 grams, roughly one-fifth of its final weight. The brain of a newborn ape has a similar weight, 200 or 300 grams; but this already amounts to two-thirds of its final weight. This is not so with the brain of a human baby, because many connections between neurons, or synapses, have yet to be formed. As we have seen, the first expression of the anti-gravity function is the reflex response of the vestibular apparatus of the ear to abrupt changes of position or the loss of bodily support. At a later stage loud noises too will elicit a similar reaction, again by affecting the eighth cranial nerve.

The newborn, this I-all that is still undifferentiated from the rest of the world, will then start acting, since in order to differentiate himself he must act – and explore: "Feeling with the point of the fingers the contact of a part of the body that will feel in turn the contact of the fingers, he will have the perception of a closed circuit on himself, while the feeling will be open when the body will come into contact with the surrounding world. It needs that, acting on the objects, he gathers in the nervous system the sensory influences that come you through different channels (touch, approves, heard, smelled, etc), but originating from a same object thing that can discover through the action on such object".[24]

In the first two years of his life, a baby will learn by exploring: he will touch, lick, smell, and bite: "Post-natal development largely concerns brain wiring and the isolation of many nerve fibres: what are still waiting to be developed at birth are the fibres running from the brain to the muscles, which are what enable the baby to keep his head up, then to sit, and finally to walk … around the age of 3–6 months, the occipital, temporal and parietal areas will start developing (in the sense that efficient connections between nerve cells will be established). Through the development of these areas, a baby will start to control the muscles of his body (first those of the neck, then those of the arms and trunk, and finally those of the legs), to perceive visual messages with increasing clarity, and to recognize sounds. Until these areas are fully developed, a baby will only control his facial muscles, and will be living in a blurry visual world. Around the age of 9-10 months the frontal cortex starts to develop, together with the long nerve fibres connecting the areas governing the decipherment of sounds and visual or tactile messages, and the production of movements … from the age of ten months to the completion of the second year and beyond, the brain of a child undergoes a profound transformation: his nerve cells produce a huge number of synapses, tiny buttons through which the extensions of neurons, axons and dendrites, enter in contact with the extensions or cell bodies of other nerve cells".[25]

In the course of his development, each child, as if by magic, reproduces and encapsulates all the various phases in the evolution of the species: from the liquid environment he finds himself in as a foetus to his first experience of gravity when he comes into the world – like the first amphibian creatures, who passed from water to dry land, adapting their body to suit new functions. The baby will roll about on his back or stomach, using head movements to switch from one position to the other. He will cross space by crawling and then move

on all fours like a monkey; and once he has gained control of the upper part of his body – the head, neck or even pelvis – his feet will be ready to help him stand. At this stage, he will already have traded this vast reservoir of new experiences with the social tradition of his milieu. He will discover that he is no longer allowed to perform many of those movements, such as licking, smelling, crawling and rolling about. Even the sound production typical of this stage – lallation, that beautiful playing with sounds, with the very origin of sound, as the child discovers the larynx muscles – will gradually disappear, turning into one of the many world languages.

Some twentieth-century artists joyously or desperately experimented with this auditory, physical and sensory yearning to revert to an original mode of exploration, in such a way as to shatter language and free it from the codes governing it. It is only in such terms that we can grasp the meaning of the gibberish in Samuel Beckett's works, the poet Andrea Zanzotto's babbling in Federico Fellini's *Casanova*, the stammering and gestures in Tadeusz Kantor's plays and the practice of making bodies roll about and explore all points of support that is made in contact improvisation.

I will end with the words of Feldenkrais himself, which lead us back to the figure of the artisan-artist: "Even in our culture a number of us succeed in continuing their healthy life process to an old age – an age, that is, where the unhealthy are already dotty and sick. Some of our best and healthiest men (who, by the way, may be hunchbacks or have other deformities) … are the sort of people of whom we think of as artists … be they cobblers or sculptors, composers or virtuosos, poets, or scientists. The outstanding difference between such healthy people and the others is that they have found by intuition, genius, or had the luck to learn from a healthy teacher, that learning is the gift of life. A special kind of learning: that of knowing oneself. They learn to know 'how' they are acting and thus are able to do 'what' they want – the intense living of their unavowed, and sometimes declared, dreams". [26]

[1] M. Feldenkrais, *Embodied Wisdom* (Berkeley: North Atlantic Books, 2010, p. 3).

[2] F. Losi, preface to M. Feldenkrais, *Il corpo e il comportamento maturo* (Rome: Casa Editrice Astrolabio, 1996), p. 8.

[3] M. Reese, preface to M. Feldenkrais, *L'Io potente* (Rome: Casa Editrice Astrolabio, 2007), p. 24.

[4] E. Canetti, *Massa e potere* (Milan: Rizzoli, 1972), p. 11.

[5] Ibid., p. 185.

[6] V. Crapanzano, *Orizzonti dell'immaginario* (Turin: Bollati Boringhieri, 2007), p. 109.

[7] M. Feldenkrais, *Body and Mature Behavior* (Berkeley: North Atlantic Books, 2005), pp. 118–19.

[8] P. Schilder, *Mind, Perception and Thought in their Constructive Aspects* (New York: Columbia University Press, 1942), p. 134, quoted in Feldenkrais 2005, pp. 120–21.

[9] A. Oliverio, *Prima lezione di neuroscienze* (Bari: Laterza, 2011), pp. 70–81.

[10] Feldenkrais 2010, pp. 42–43.

[11] Ibid., p. 5.

[12] *Shoemaker of Dreams. The Autobiography of Salvatore Ferragamo* [1957] (Florence: Giunti, 1985), p. 193.

[13] Ibid., p. 80.

[14] Reported from M. Fusero, Feldenkrais Training Milan-Levico, 2012.

[15] M. Feldenkrais, *Le basi del metodo per la consapevolezza dei processi psicomotori* (Rome: Casa Editrice Astrolabio, 1991), p. 134.

[16] W. Benjamin in G. Didi-Huberman, *Ninfa Moderna. Saggio sul panneggio caduto* [2004] (Milan: Abscondita, 2013), p. 55.

[17] H. Sieker in Didi-Huberman 2013, p. 64.

[18] Ibid., p. 73.

[19] F. Nietzsche, *The Gay Science* (Mineola, New York: Courier Dover Publications, 2006), p. 184.

[20] Ibid., p. 182.

[21] Feldenkrais 1991, p. 13.

[22] Reported from F. Wildman, *Levico's lecture*, Feldenkrais Training 2011.

[23] T. Ingold, "Culture on the Ground: The World Perceived Through the Feet", in *Journal of Material Culture*, vol. 9, no. 3 (University of Aberdeen, Scotland, 2004), p. 315.

[24] H. Laborit, *Elogio della fuga* (Milan: Mondadori, 1982), p. 53.

[25] Oliverio 2011, pp. 39–58.

[26] Feldenkrais 2010, p. 58.

A PORTRAIT OF THE PHOTOGRAPHER AS A TIGHT-ROPE WALKER

WALTER GUADAGNINI

FROM SHOE-SHINER TO ACROBAT

In the last decades of the nineteenth century, photographs became filled with curious figures, unlikely events, and scenes that appeared to disprove – once and for all – the old idea of photography as a mirror of reality (one with a memory, as Oliver Wendell Holmes pointed out): of photography as a necromantic apparition of the world through its self-representation on a chemically treated surface. A series of discoveries, improvements, technological innovations and lifestyle changes gave rise to a new photographic landscape, radically different from that which had witnessed the birth and early development of photography. With the invention and rapid spread of gelatin silver films in the 1860s, the gradual improvements made to cameras, and the research on the rendering of movement – most famously, yet not exclusively, exemplified by the work of Eadweard Muybridge and Étienne-Jules Marey – along with the release of the Kodak camera in 1888 and the growing taste for the manipulation of photographic images, professional and amateur photographers, artists and enthusiasts, at different levels and for different purposes, came to explore new themes, or ones which had hitherto been regarded as marginal. The portrayal of movement and the sudden freezing of time; the breakdown of motion into endless fragments or its transformation into a constant flow of lines; the creation of new spaces through the possibility of presenting new vantage points and of recombining images: these are only some of the pursuits and themes which in the late nineteenth and early

Underwood and Underwood,
Above Fifth Avenue, Looking North, 1905
Gelatin silver print,
24.2 x 18.6 cm
New York, MoMA.

twentieth century grabbed the attention of photographers – both real experts who exploited their profound knowledge to achieve striking aesthetic or scientific effects, and ordinary folk who would use the latest findings in modern technology to take a few snapshots of their family.

The male figure getting his shoe shined in Boulevard du Temple in Daguerre's famous photograph was only half a century away, yet it was as though he belonged to a different geological era in intellectual and technical terms. When he had his photo taken, the man stood still not by choice but because that was the only way in which he could come out on the surface, while all the moving objects and people around him vanished. Now, by contrast, photographs came to be filled with speeding cars, ladies leaping down steps as though they were flying, tight-rope walkers walking across squares suspended in mid-air, acrobats and dancers, men and animals running about as never before. Again, as a reflection of the taste for visual tricks and illusions that has always been a part of photography, we see an elegant gentleman balancing on one leg on the roof of a skyscraper, with eleven men on his shoulders, forming a circle: this open-air circus stands as a counterpart to the no less striking real-life circuses. It was in this atmosphere that for the first time an enduring and widespread interest emerged in the rendering of balance and of everything this term implies in photography: an interest that was destined to grow up until the present day and that has inspired some of the most famous pictures in the history of photography.

ICARUS AND THE PRINCESS

Lewis Hine turned the workman featured in one of the most spectacular photographs of his *Men at Work* series into a mythological figure: Icarus, who had dared challenge the gods to give mankind the gift of flight. Like all his colleagues on

American anonymous,
Man on Rooftop with Eleven Men in Formation on His Shoulders, c. 1930
Gelatin silver print,
34.5 x 27.2 cm
New York, George Eastman House.

Courtesy of George Eastman House, International Museum of Photography and Film, Rochester.

the scaffolding of the emerging Empire State Building, this modern Icarus challenged hundreds of years of building and town-planning conventions by tracing the skyline of a city which instead of expanding from its foundations or over the ruins of pre-existing cities – as had been the case in the Old Continent for thousands of years – was rising anew, as a metropolis destined in turn to spawn new legends, especially through films and photographs. The man at work hundreds of metres above the ground – caught in a precarious balance, suspended in mid-air with the soles of his shoes resting on steel beams – makes an amazing photographic subject. Not the ground at his feet but the living body of the city: there is no trace of nature in this new world – only artificial elements. As a German intellectual wrote in those years: "We live in cities. Our fields are asphalt, our stars are electric street lights, our forests are high tension wires".[1] This period was marked by a constant urge to affirm the present through the past, and technology through nature, as though, faced with the radical innovations of modernity, people sought to take refuge in what was familiar to them.

Hine's photographs celebrate both man's power and his technological achievements, perfectly in line with the global trend in those years to celebrate work. From Roosevelt's America to Stalin's Russia, from Fascist Italy to the still democratic France, each country had its own Stakhanovs and Icaruses, through which to celebrate the indissoluble connection between muscles and machinery, for the establishment of a better world (it will be superfluous to recall that within a few years a very different form of construction was to be carried out, upon the ruins of cities and civilizations).

Leaving these general considerations aside, it is interesting to note that the building of skyscrapers enabled photographers to turn into tight-rope walkers themselves, in order to capture a crucial feature related to the construction not so much of the actual structures, as of their legend. Many pictures from these years portray photographers

at the top of these modern steel towers, with daring poses that stress the image of themselves as witnesses to contemporaneity: a concept and an image that took root in the public imagination in this very period, through the circulation of illustrated popular magazines.

At the same time, as further confirmation of the fact that these developments occurred within a myth-building discourse involving all the people portrayed, aloof young women started appearing on those very same scaffoldings instead of builders. Caught in the same precarious balance – and hence all the more charming – they smile in the magnificent clothes designed by the most sophisticated stylists of their day. The most famous of these models is Lisa Fonssagrives, Irving Penn's future partner, who was portrayed by Erwin Blumenfeld;

László Moholy-Nagy,
Bauhaus Balcony, 1926
Gelatin silver print,
49.5 x 31.3 cm
New York, George Eastman House.
Courtesy of George Eastman House,
International Museum of Photography
and Film, Rochester.

Lewis Hine, *Icarus,
Empire State Building,*
1931.

Lisa Fonssagrives
on the Tour Eiffel
photographed by
Erwin Blumenfeld
for *Vogue France*,
May 1939.
The outfit is by
Lucien Lelong.
© The Estate of
Erwin Blumenfeld.

Alexander Rodchenko,
Female Pyramid, 1936
Moscow, House of Photography,
Rodchenko Archives.

the photographer had been a member of the iconoclastic Dada movement, and was therefore quite used to intellectual – rather than merely physical – balancing acts. The model is seen marvellously anchored to the Tour Eiffel by only one hand and foot, as she gazes below and at the same time lifts up the lower hem of her dress to form a semicircle of moving fabric: a striking image of lightness and a bold counterpart to Hine's Icarus, as removed from it as the clothes worn by the two figures are from one another's.

No less striking, once we examine the two pictures side by side, are the formal similarities they display: the city crossed by the river, the broad stretch of sky, the upward swing of the two bodies. Indeed, both pictures rest on the same idea: the construction of a sort of triangle of modernity consisting of the human figure viewed in relation to its new life setting according to a bold new perspective, by the bold new narrator – the photographer, who alone can convey this novel reality. And for the time being, we can leave class differences aside.[2]

PUDDLES

The stress on the instant snap brought the theme of balance to the fore, making it central to photographic practice and theory. Henri Cartier-Bresson famously stated: "To photograph is, in the same instant and in a fraction of a second, to recognize a fact and to organize rigorously the visually perceived forms that express and signify this fact … to place head, heart and eye along the same line of sight". This claim essentially constitutes a description of the point of balance that is to be attained between the various moments in the photographic process – from the choice of subject and the framing to the exposure time. Indeed, one of the standard images that are used to illustrate Cartier-Bresson's statement of poetics is *Behind the Gare Saint-Lazare*, which is marked by the contrasting tensions between the jumping man, his reflection in the puddle, and the figure drawn on the poster in the background. An example of compositional balance – in terms of forms, first of all, but also of tones – this photograph illustrates the capacity to make the most of the poetics of the instant, of the freezing of time in a moment that is significant in itself. A given place in time, thought and vision here transcends both the enthusiast's prowess and the amateur's haphazardness, attaining a different dimension: the conceptual development of an idea of photographs as images of both space and time.

Alexander Rodchenko,
I Dominate, 1923
Illustration for Vladimir Mayakovsky's
poem *Pro Eto*, photomontage, 1926
Moscow, House of Photography,
Rodchenko Archives.

Friedrich Seidenstücker,
The Puddle Jumper, 1925
San Francisco Musuem of Modern Art.

Henri Cartier-Bresson, *Behind
the Gare Saint-Lazare*, 1932.

Cartier-Bresson was not the only one to explore this kind of image, which clearly was particularly appealing to photographers equipped with the new compact cameras that had become available from the mid-1920s. However, we only need to compare his picture to that taken by a consummate professional such as Friedrich Seidenstücker to grasp the difference between a kind of photography that focuses on events and one that is capable of turning events into a view of the world. Moreover, this also coincides with a transition – or, better, a flow – between the poetics of the avant-garde (of which Cartier-Bresson was also an exponent, via Surrealism and his *anti-graphic* approach) and professional photography, as well as between instant and constructed photography. These crucial passages lent functional definition to the *Neues Sehen* (and besides, in Soviet Russia too the avant-garde frequently put its poetics to the service of propaganda, until the authorities developed a more direct form of communication based on the standard of "Socialist Realism"). Considerable influence was also exerted on fashion photography. It is interesting to note how Martin Munkácsi revisited the theme of jumping over a puddle in 1935, by combining what up until then had essentially been two irreconcilable approaches: that of fashion photography and that of the rendering of motion. Not even Edward Steichen, who had taken some bold steps indeed in order to bring the poetics of "straight photography" into the *Vogue* shrine, had gone as far: what it took to really stir the scene and its protagonists were photographers such as Blumenfeld and Munkácsi, who came from very different cultural milieus – Dada and photo-reporting, respectively.

Constructing the instant and staging improvisation: a kind of fiction within fiction operating on a double level. On the one hand, it opened new paths for a constantly evolving genre; on the other, it looked towards a kind of photography that in those years had already started writing its own history and hence endorsing the use of allusions and citations.

Such practices were to reach their zenith in relation to this subject some twenty years later, when Alexey Brodovitch's pupil Richard Avedon – already famous at the time – created his explicit *Homage to Munkácsi*, thereby turning Cartier-Bresson's extraordinary balancing feat into a Pop icon. Avedon's picture stands as the citation of a citation, devoid of any meaning outside the rhetorical context of the norms of representation and of the genre (in this respect, it would be interesting to consider Avedon's and Wahol's common stress on the surface as that which alone brings out the nature of objects and people, and to consider as well the two artists' background in advertising – but this is not the place for similar considerations).

Martin Munkácsi,
The Puddle Jumper, 1934.
Ullstein Bild – Martin Munkácsi.

DANCING
IN THE DARK

Edward Steichen and Alexey Brodovitch, again: the former photographed Isadora Duncan and the *Isadorable* Maria-Theresa against the backdrop of the Parthenon in 1921,[3] the latter created one of his masterpieces in 1945 by publishing the now legendary volume *Ballet*, which was destined to influence the photographic portrayal of dancers like few other works right up to the 1980s – if not beyond. Steichen was still indebted to the stylistic motifs of an artistic culture rooted in nineteenth-century painting, but about to be swept away by Modernism. Brodovitch, by contrast, "spat in the face of technique and pointed out a new way in which photographers could work":[4] precisely by exploiting the limits of his early attempts at photographing ballet – no flash in poorly lit settings, moving figures and slow films – Brodovitch developed a new way of conveying dynamism and especially continuity in dancers' movement. Likewise, by taking frontal shots of the lights, he sought to respond to the rhythm of the music. The outcome is a form of photography consciously deviating from academic rules (so much so that Brodovitch even touched up his photographs in the printing phase by heightening the contrasts, bleaching or darkening them, bringing out the grains and blurring the edges). The hallmark of this photography is a striking continuity of lines, shapes and forces within each picture and across the pages of the volume, which creates a kind of visual stream of consciousness that is utterly revolutionary. Possibly for the first time, at least in this field, greater emphasis was placed on the idea of the overall balance of the photographic and publishing project than on individual shots; for the first time, dancers were portrayed not in order to illustrate their prowess, or according to the canons of Classical beauty, but rather according to the notion of a "dérèglement de tous les sens", which recalls the Dionysian – rather than the Apollonian – aspect of dancing. Or which, in other words, appears to correspond more to the "explosante-fixe" beauty of the Surrealists than to the Classical beauty conjured up by Paul Valéry in his fine pages on Degas, on drawing, and indeed on ballet.[5]

Certainly, only ballet – along with the circus and gymnastics – has played such an important role in the representation of the theory and practice of balance in the

Man Ray, *Explosante-fixe*, 1934.

history of photography. One can hardly overlook the curious pictures taken by photographers such as Barbara Morgan and especially Gjon Mili, which illustrate a more direct – yet no less fascinating – approach to the theme. Partly indebted to Muybridge, and even more so to Anton Giulio Bragaglia and Harold E. Edgerton, Gjon Mili stands as a kind of counterpart to Brodovitch: by contrast to the latter's constant flow, Mili offers the fragmentation of the body, its endless multiplication through a kind of arrangement that breaks time and space into an infinite number of particles. The technical skill of the photographer and the potential of the camera are here exploited to attain a form of representation that has an intentionally playful quality to it, once again reminiscent of that idea of prowess which – as we have seen – is always implied by the representation of motion and the capacity to visually convey it in all of its iconographic variants.

Once again caught between these two poles – not least as a *trait d'union* – we find Martin Munkácsi, whose series of shots of the two dancers Ramon and Renita was published in *Harper's Bazaar* in 1935. The pictures were magnificently arranged by Brodovitch in the shape of a T expanding onto the following page – with one of the photographs printed full page – as though they were frames from a film: through this striking example of formal balance, the graphic designer experimented with a visual approach he was no longer to take up as a photographer, but which he acknowledged to be most valuable and which was certainly destined to influence the future choices of authors such as Mili.

Gjon Mili, *Gene Kelly in Multiple Exposure*, 1943.

pp. 228–29
Gjon Mili, *Martha Graham*, 1941.

COFFEE CUPS, CATS AND FLYING POETS

With regard to feats and *divertissements* related to the theme of balance, in 1933 a possible path was once again opened by Munkácsi through the amusing picture of two dancers, Tibor von Halmay and Eva Sylt, caught during a coffee break during reharsals. What we have here is the conscious and sophisticated interplay between photography's claim to realism and its equally characteristic inclination towards fabrication: the genuine, highly fragile point of balance between truth and fiction, spontaneousness and artifice. The crucial moment – Munkácsi seems to be saying, echoed in even more radical terms by Philippe Halsman – is not necessarily the one we grasp in real life, since it may also be arranged in the studio. The important thing is for

the outcome to be plausible (and the claim to realism will help make the trick even more fascinating – in a way like in an illusionist's show, where the public knows there is a trick, and yet the important thing is not to detect it and thus to continue to believe in the magic). With his series on "jumps" – actually anticipated by a small, lesser known series of studio portraits by Barbara Morgan – Halsman starts by going completely against the reader's expectations (we should bear in mind that these pictures, just like Mili's, were taken in order to be published in magazines with a wide circulation), since what he presents is an unusual situation: the photographer's subjects – film and entertainment stars – are not usually seen in this kind of pose. He then stresses the visionary quality of his *mises-en-scène*, particularly in the case of figures such as Salvador Dalí and Jean Cocteau, who would not simply pose for him, but would personally contribute to designing the set – knowing well that self-representation is of fundamental importance for making a legend of oneself.

MENS SANA...

The ballet world marks the apex of the attempt to photographically freeze the balance attained – or to be attained – through movement. Another pursuit entirely grounded on the control of the body is gymnastics, which is indeed another favourite subject among photographers. Compared to dancers, gymnasts are more frequently captured in moments of stillness, illustrating the outcome of that work of preparation which comes before the actual athletic performance, and which at the same time reveals the beauty of a body with harmonious, powerful muscles, according to the standard of Classical beauty.

Often, moreover, ballet and gymnastics merge with one another, particularly in contexts where the worship of the body acquires a markedly ideological significance. Starting in the 1920s, the modes of

Philippe Halsman,
Dalí Atomicus, 1948.

Philippe Halsman,
Dream of a Poet, 1949.

representation of the body and its harmony – especially that deriving from the control of movement and the balance thus attained – manifest themselves in the form of plastic poses, in situations that tend to abstract the subject from reality by turning him or her into an ideal and exceptional figure, a model. Countless examples of this may be found in Soviet Russia and Weimar Germany, where the tragic seeds of Nazi culture had already been sowed. The worship of the body and of nudity constitutes the core of Hans Surén's volume *Der Mensch und die Sonne*. Published in 1924, the book sold 61,000 copies its first year, reaching 145,000 in 1936. The work features a series of pictures of naked men and women performing sport and dance exercises in the open, or posing to display the perfection of their bodies.[6] This iconographic construction of the race myth overturned the original meaning implied by certain forms of nudism and naturism, which expressed the yearning for a society founded on freedom, the removal of taboos related to the body, and the rediscovery of nature as a way of oppos-

ing the nightmarish mechanization of the world (illustrated in the same years by the film *Metropolis*).

Meanwhile, in the Soviet Union gymnasts and dancers were also portrayed in settings halfway between the Edenic and the theatrical with the four exhibitions on the "Art of Movement" held by the Choreological Laboratory of the Russian Academy of Artistic Science (RAChN) between 1925 and 1928 – one of the first utopian projects of the post-revolutionary years to engage art historians, dancers, photographers, philosophers and scientists. This stimulating project, however, soon gave way to the gradual restrictions imposed by the regime, starting from the ban on nudity: as is always the case with regimes, the eroticization of the body came to be perceived as something that could undermine and weaken the population, both physically and morally. Still, the Soviet Union too used bodies as a means of propaganda: the bodies of athletes and gymnasts filing in parades as human pyramids were especially chosen to illustrate the rise of the socialist *homo novus*, whose individuality – including physical individuality – is always to be put to the service of society.[7]

Despite all this, one area remained in which it was still possible to admire and photograph acrobatic feats without any ideological implications: the circus world, a no man's land free from opposing yet equally false forms of rhetoric. Tight-rope walkers and trapeze artists fill the pictures by photographers ranging from Umbo and Rodchenko to Ted Croner and the very young Stanley Kubrick, who in the late 1940s played with the principles of modernist photography and reportage in an exemplary photo shoot published in *Look*. In several ways, this work stands as a counterpart to the pictures taken in Europe by the great Robert Doisneau, who also combined realism and artificiality in the representa-

Stanley Kubrick, *Circus (Trapeze Artists)*, 1948
Museum of the City of New York.

Arno Rafael Minkkinen,
Hite, Utah, 1997
Courtesy Photography +
Contemporary, Turin.

tion of gymnasts training indoors or outdoors. Besides, if we were to search for an antidote to the idealized and ideologically charged photographs of the Nazi, Fascist and Communist regimes, the best place to look would be the bodies of young men and women strolling hand-in-hand down the streets of Paris in 1934: to paraphrase Benjamin (who in those years was reading Atget), pictures capable of "disinfecting the musty air" created by the various Riebickes and Pragers.

THE UNBALANCED

Starting from the 1960s, the theme of balance, as developed in the former half of the century, no longer seemed to capture the interest of more conscious photographers. Sport, gymnastics, ballet and the circus were largely confined to genre photography – and increasingly published in special-interest magazines. These pictures no longer reflected the ability or desire to capture the defining features of contemporary society, or even an interest in the language of photography itself, of the sort that had marked the crossroads between instantaneousness and artifice, primal vision and novel vision, over the course of the 1920s and 1930s.

The only – physical and conceptual – space in which the theme still found a raison d'être and continued to convey an original visual approach, with interesting outcomes, was the human body. Unlike in the past, however, the body came to be defined in terms of its precariousness, primarily as a means to reflect upon the identity of individuals and their relation to the world. Based on these premises, it is clear that the photography in question is chiefly artistic in nature, that it explores both the world and its modes of representation, both subjects and language.

We need only consider here authors such as John Coplans, Arno Rafael Minkkinen and Francesca Woodman, who – albeit in different ways and with different results – focused on the verification of identity through the perception and investigation of the body. They would photograph themselves not so much to affirm their presence in the world, as to evaluate their modes of being in the world as bodies. For Coplans, this meant filling every centimetre of the surface with his own physicality, pushing its display to the limits and even recording its decline. For Minkkinen it coincided instead with an attempt to measure the world starting from himself and the body, as a genuine way of extending its size and shape by setting it in contrast to other existing forms – so much so that the establishment of this relation often brought

Elliott Erwitt, *Paris*, 1952.

top
Robert Doisneau, *The Brothers,
rue du Docteur-Lecène, Paris*, 1934.

right
Francesca Woodman, *Untitled,
Providence, Rhode Island*, 1976
Gelatin silver print,
13.3 x 13.3 cm each
Foto Forum – San Francisco
Museum of Modern Art. Courtesy
of George and Betty Woodman.

out extreme, precarious or apparently impossible forms of balance, only made possible by the presence of a camera to record an event materially arranged for it. All in all, this is still a positive vision, which accepts precariousness as a mode of relating, if not to others, at least to the world and oneself. A different approach is illustrated by Francesca Woodman, whose research represents a constant attempt to affirm an identity that is both multiple and non-existent, concealed behind a mask or displayed through a far from erotic nakedness. Woodman would operate in everyday settings, or devise new ones – as in the case of her series of shoots from 1976 in which some movements (possibly dance steps) in an empty room are set in the virtual space created by geometrical figures drawn upon the surface of the photographs by the artist herself. This brilliant way of expressing the ambiguous relation between a condition of balance and one of imbalanced occurs again – arguably with more sophisticated intentions – in two compositions in which precariousness takes the shape of a kind of conjuring act, with a spherical shape moving between arm and

hand: further evidence of the fact that this theme and its symbolic implications deeply touched the artist.

Erwin Wurm has turned the self-reflective tensions of the 1970s into a clever game with his public by inviting people to stand in the most unlikely poses for a shoot that would immortalize their having been sculptures, if only for a minute. The most succinct and all-round picture of the contemporary approach to the idea of balance, however, is arguably provided by Jeff Wall through the peculiar mode of constructing images that has marked his career ever since his debut. It is also worth noting that the underlying principle, for both photographers, is the idea of the loss of balance as a trigger for comical effects, in line with a long-established practice – common in the theatrical and cinema world – which has not been widely explored in the field of photography. The theatre of the absurd staged in *The Stumbling Block* truly encapsulates the relation between reality and fiction, which has always been a hallmark of photography: it plays with the conventions of instant photography, offers perfect recon-

Jeff Wall, *The Stumbling Block*, 1991
Slide in lightbox, 229 x 337.5 cm
Courtesy of the artist.

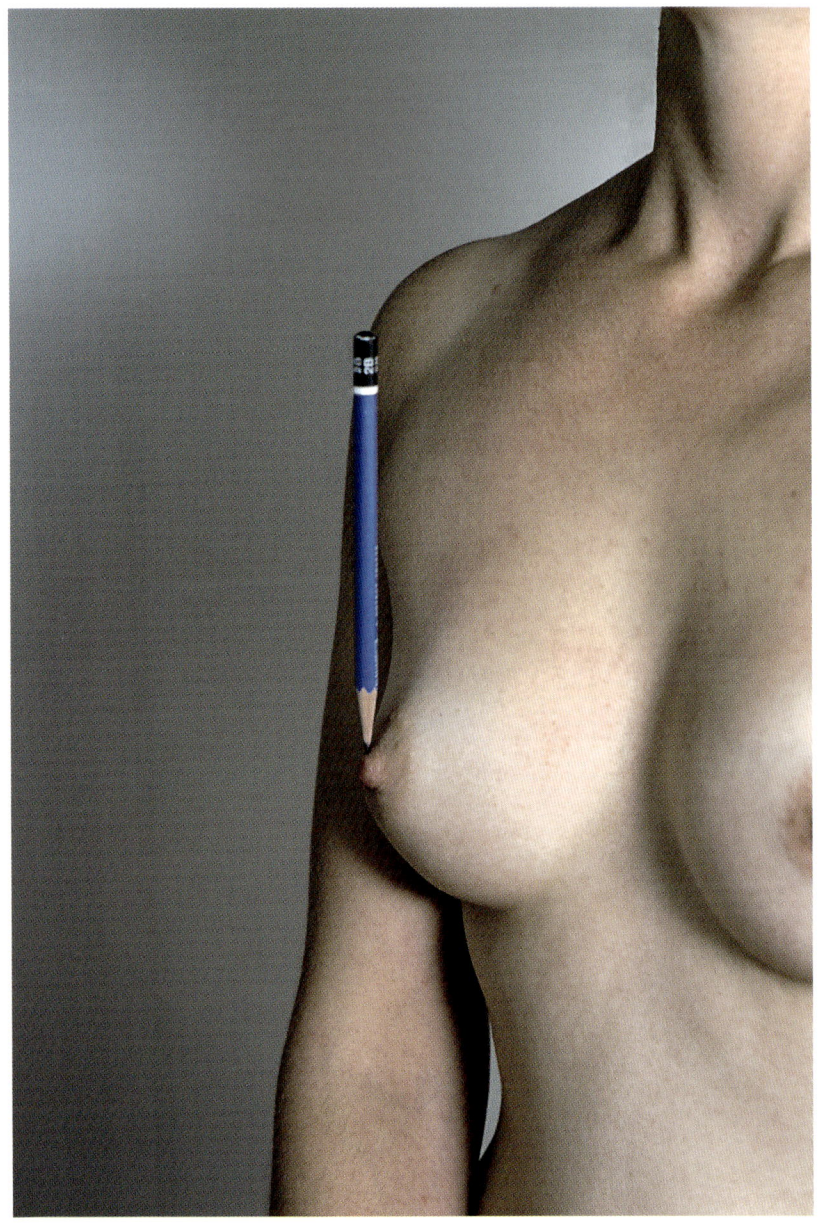

Steven Pippin,
A Naked Singularity, 2007
Photo on aluminium, 80 x 60 cm
Courtesy Galleria Astuni, Bologna.

structions, and sets a (concealed) compositional balance in contrast to the (displayed) loss of balance on the part of two of the three protagonists of the scene. Ultimately, this is precisely what photography would appear to be exploring today when it engages with the theme of balance, namely the impossibility of attaining the latter by ordinary means, by those physical means passed down to us by tradition: it is as though athletes and gymnasts were tumbling down together with the businessman on the right, and all the models leaping over puddles were falling together with the young woman in the foreground. What is left is the artist's desire to craft a perfect, albeit unlikely, image. For the only possible balance in art today is the awareness of its loss and the willingness to restore at least its semblance through a reflection on the one concept that, right from the start, has marked the history of this theme: the relation between the passing of time and the need to freeze it at a given instant. As in this shot by Steven Pippin, where the tip of the pencil has been sharpened to the point that it stays balanced on the model's breast.

[1] H. Windisch, *Das Deutsche Lichtbild* [1929], quoted by C. Phillips, "Twenties Photography: Mastering Urban Space", in *The 1920s: Age of the Metropolis*, exhibition catalogue, edited by J. Clair (Montreal: Museum of Fine Arts, 1991), p. 209.
[2] The saga of contemporary tower-dwellers continued in the post-war period: the best example is no doubt Marc Riboud's ironically affectionate series devoted to the "painter of the Tour Eiffel", which was published in *Life* in 1953. In the more specific field of fashion photography, Peter Lindbergh was to pay a tribute to Blumenfeld in 1989.
[3] In 1929 the Greek photographer Nelly was to take up the themes and stylistic features of this celebrated photo-session with her famous series devoted to the Hungarian ballerina Nikolska.
[4] Quoted in K. W Purcell, *Alexey Brodovitch* (London: Phaidon, 2002), p. 132.
[5] Aside from the names already mentioned, among the countless classical ballet pictures that were taken between the 1930s and the 1950s it is especially worth mentioning: André Kertész's 1938 series devoted to American Ballet, which was revolutionary in its open-air setting; the photographs that Georgy Petrusov took at the Bolshoi in the 1940s and 1950s; and the photographs by Paul Himmel collected in the 1954 volume *Ballet in Action* – deeply indebted to Brodovitch. Also noteworthy is the series that Ellen Auerbach devoted to Renate Schottelius, again in the early 1950s.
[6] See W. Van der Will, "The Politics of the Body", in B. Taylor and W. Van der Will, *The Nazification of Art* (Winchester, Ontario: The Winchester Press, 1990).
[7] See the essays by A. Loginov, N. Misler and B. Groys in *Nudo per Stalin*, edited by L. Anisimova and P. Khoroshilov (Rome: Cangemi Editore, 2009).

THE PRESENT'S FRAGILE EQUILIBRIUM

ARABELLA NATALINI

Acrobats, maenads, dancers, animated skeletons, walkers, hovering bodies, traces of absent figures, abstract figures, whole and broken down figures... Multitudes of forms suspended in their peculiar and momentary balance live the time and the space of art. Our individual paths are accompanied by a quest for equilibrium; even more so, this quest accompanies artists, who often move along unstable boundaries, poised between sense and nonsense, past and future, and are capable of showing us the fragile balancing of the present only when they accomplish that risky further step that moves away from it. And if this quest occupies a fundamental amount of space in the figurative arts throughout the twentieth century, from the end of the Second World War to the present, it has increasingly focused on one way of moving that characterizes our species and, at the same time, distinguishes and identifies the single individual: the way a person walks. Great artists have undertaken new paths, over the past decade, creating poetic and unexpected works in which the relationship between the body and space is time after time redefined by way of the gait. And so the yearning to leave traces of one's own existence, of one's transit through this world, has achieved further steps in unorthodox directions. Let's try to follow a few of these.

1.

A man walks in the desert. We see him approach us, his figure gradually becomes more distinct until the movie camera frames just his face, while the sequence of steps he takes are hidden from our view, and we are only allowed to hear them.

Bill Viola, *Inner Passage*, 2013
(tribute to Richard Long)
Colour HD video on plasma
display mounted vertically
on wall, stereo sound,
155.5 x 92.5 x 12.7 cm, 17:12 mins.
Performer: Blake Viola
Photo: Kira Perov.

This is followed by a quick succession of images in movement – a "sampling" of water, fire, eruptions, plants, pathways and people – at the end of which the initial figure reappears, this time with his back to us, moving away towards the desert, in the direction from which he had come.

The work in question is *Inner Passage* (2013), the recent tribute paid by Bill Viola, an artist universally known for his pictorial videos, to Richard Long, the undisputed father of British Land Art.[1] An acknowledgement that reminds us of how around the second half of the 1960s "the art of the landscape", or "of the land", gave rise to a rethinking of the making of art *tout court*, transforming the traditional representation of nature into a direct and personal aesthetic experience.

Indeed, the appropriation of the territory radically changed the perception of the landscape, which from being the classical subject of painterly representation became the space and material that made up the work itself. Spurred on by the desire to contrast the growing merchandising of art, some artists abandoned the institutional sites of galleries and museums to instead move to and operate in open spaces, often remote and at times uncontaminated. This yearning to go beyond boundaries was common to both American and European experiences, which adhered to the same need while using substantially different means on the two continents. In England, in particular, this aspiration harked back in an unprecedented way to the romantic tradition, to a more intimistic attitude, to a need to experience nature directly, which discovered in the body, and in walking, its ideal instrument.

In 1967 Richard Long produced *A Line Made by Walking*. The title unhesitatingly reveals the contents of the work: a line traced while walking, the footprints left by the feet and the weight of the artist's body on the grass, "an action that is engraved directly upon the site".[2] Long, unlike his American colleagues such as Smithson, De Maria and Oppenheim, does not use diggers and bulldozers to intervene in the landscape; rather, he uses his own body as the sole instrument. The simplicity as well as the radical innovation of a work made without the help of tools put the artist's body and mind in direct contact with nature, thereby transforming them into tools through which to leave ephemeral traces, which are then given back to our gaze only by the photograph that is taken of them.

Despite the fact that Long has in most cases conducted his "explorations" by himself, on some occasions he has been accompanied by his friend Hamish Fulton. Both men have wondered how to express and communicate the experience of walking: unlike Long, Fulton leaves no trace in the landscape, as his interest is totally drawn to the process of crossing the territory. Although this experience will never be fully represented, the "walking artist", as Fulton likes to describe himself, brings back from his long journeys magnificent photographs with meticulously accurate captions. Hence, exhibitions of his work are mostly made up of photos, documents and wall drawings, often developed in diagrams that recreate a sort of experiential map of the places he has travelled.

2.

There is perhaps no "natural way" of walking for the adult.
Marcel Mauss, Les techniques du corps*, 1935*

The value of Bruce Nauman's "walks" is profoundly different. Towards the late 1960s Nauman realized in his own studio a series of pioneering photographs focused on the encounter between the body and space, and on the physiological and emotional effects of time passing.

In *Walking in an Exaggerated Manner around the Perimeter of a Square* (1967–68), the artist walks while following the perimeter lines traced previously on the ground. By accentuating the movement of the hips, he constantly tests his own equilibrium, producing a wavering movement one step after another. The same way of walking is reproduced backwards, with the constant search for balance that re-

calls the gait and the concentration of a tight-rope walker. By staging the reiteration of the experience for a prolonged period of time, Nauman fuels an elevated tension, in both the performer and the viewer, where the chance for error and that "something might change" – positions and attitudes at the limit of the possibilities of the muscles – evidences the artist's interest in avant-garde dance, based on ordinary movements as well as on the awareness of the body and space.

The subsequent video, *Slow Angle Walk (Beckett Walk)* of 1968,

was made at a 90-degree angle, following the artist as he walked around a track, while he repeatedly lifted his leg high before him and, as it landed, lifted the other leg high in the air behind him – long-kept positions and balancing actions fraught with suspense. "I thought of them as dance problems without being a dancer, being interested in the kinds of tension that arise when you try to balance and can't. Or do something for a long time and get tired".[3]

The interest in avant-garde music and dance is further complicated by a visual equivalence with the stories of Samuel Beckett. The Irish writer's influence is revealed in the subtitle as well as in the barren *mise-en-scène*, where futile and repeated activities are described in a circular form of reasoning, alluding to a subject trapped by the circumstances.

The attention towards circularity, reaffirmed and emphasized through the use of the loop, is also likened by the artist to a certain idea of music: "The circularity is also a lot like La Monte Young's idea about music. The music is always going on. You just happen to come in at the part he's playing that day. It's a way of structuring something so that you don't have to make a story".[4]

In spite of this, Nauman does not work on improvisation; rather, he realizes a series of well-defined actions. As shown in exemplary fashion by a preparatory drawing, *Beckett Walk Diagram II*, the artist painstakingly defines the direction of the footsteps, the position of the figure and the space that can be trodden. Within this grid, which is not just spatial but temporal as well (the duration of the action corresponds to exactly that of the tape on which it is recorded), reiteration becomes the means to put the balance and the physical and mental resistance to the test.

Unlike taking meticulously planned steps in a delimited space, following a path means choosing a direction, whether or not knowingly. It can mean approaching a destination, but also, at the same time, moving away from other things, other people.

Bruce Nauman, *Walking in an Exagerrated Manner around the Perimeter of a Square*, 1967–68
Black and white 16mm video, silent, 10 min.
Courtesy Electronic Arts Intermix (EAI), New York.

Bruce Nauman, *Slow Angle Walk (Beckett Walk)*, 1968
Black and white video, sound, 60 min.
Courtesy Electronic Arts Intermix (EAI), New York.

Marina Abramović ventured down her own path around the late 1960s and she still hasn't stopped. A long walk that has been marked, however, by a rupture: her separation from Ulay, artist and partner in life, with whom she had realized extreme performances, testing the individual's physical and psychological limits. Within this frame, the two artists performed many works in which they tried out the complex relationships of interdependence between a couple. This was the case in *Death Self*, when they breathed in and out of each other's mouths until they exhausted the oxygen supply and both collapsed to the ground, or *Rest Energy*, in which between them they held a bow and arrow in tension by balancing the weight of their bodies. Ulay held the arrow at the point where the notch was fitted into the string, while Marina, standing opposite him, held the bow. If either one of them had given in, or if there had been an oscillation beyond the unstable threshold of equilibrium, the arrow would have struck Marina's heart: only their ability to maintain constant tension by continually making tiny adjustments to the re-balancing of the forces at play – a capacity sustained by the training of their muscles and fuelled by their reciprocal faith in each other – could avoid collapse and its consequences.

Years later, when their relationship was coming to an end, the two artists decided to venture down one last path together and then separate for good. In 1988 they left for China where they were to take long walks along the Great Wall of China. Moving from two opposite ends, Ulay from the edge of the Gobi Desert, Marina from the Yellow Sea, they walked for three months, travelling 2,500 kilometres: they met only to embrace and say goodbye forever. A farewell ceremony at the meeting point of two opposite paths: the title of the performance was *The Lovers*.

Marina Abramović and Ulay,
The Lovers, The Great Wall Walk,
1988/2010
Two-channel video, colour,
16:45 minutes
Based on the 1988 performance
90 Days, the Great Wall of China
Courtesy of the Marina Abramović
Archives and Murray Grigor.

Marina Abramović, *Shoes
for Departure*, 1991
Amethyst, 26 x 50 x 20 cm
Paris, Collection Enrico Navarra.

The reference to walking has marked other works by the artist from Montenegro; among these, *Shoes for Departure*, 1991, invites us once again to start moving. Very few but significant directions are given to the public: "With naked feet enter the shoes. Eyes closed. Motionless. Depart". Contrary to what they usually serve as, the shoes used in the performance, solid, heavy ones, do not favour walking, but act as a burden, the imperative of insisting upon the place where one is, an inhibition of going as a real movement, which frees the energies for a journey of the mind. To "put yourself in someone else's shoes" means shifting your point of view, identifying in another person's position, empathizing. And here, perhaps, someone else's immobilizing shoes – inhospitable and anonymous ones – by preventing us from exploring the real world, inhibiting every escape from the place we find ourselves in, force us to try imaginative experiments, to reposition ourselves not so much in the world as with respect to ourselves and others. The equilibrium we can find in the "indefinite time" we are invited to wear the shoes is the indefinite time of life, in which we shuttle between different angles. Time is indefinite, not infinite, and therefore precious.

3.

Camper describes very exactly how the best shoes must be made, but he certainly could not make one.
In my country a common man … will not refuse to apply the term art to the performance of a ropedancer.
Immanuel Kant, Critique of the Aesthetic Judgment, *43*

Perhaps it should come as no surprise that when Immanuel Kant – in a work that constitutes one of the foundations of Western aesthetics – introduces the notion of "art" for the first time in order to distinguish it from nature, science and trades, he mentions shoes and tight-rope walkers (ropedancers) first. Two exemplary references, which in the German philosopher's eyes serve to show the insufficiency of knowl-edge and the use of rules for artistic production, and, at the same time, refer back to walking, to gait, to balance.

Petrus Camper, the scientist mentioned by Kant, was one of the most famous naturalists in eighteenth-century Europe, the author, among other things, of a treatise entitled *On the Best Form of Shoe*.[5] In the passage recalled here Kant means to say that, however much it is necessary to have a scientific knowledge of the anatomy of the foot and of walking, it will never be enough to know how to make a shoe. As for all the other activities that legitimately claim the name "art", something that cannot be referred back to science will have to be added to this necessary technical-scientific condition, to knowledge: that something is a gift, talent, an ingenious element. Otherwise it would be sufficient to learn the rules that preside over the realization of a certain product in order to become artists. The example of the tight-rope walker, which immediately follows that of shoes, is particularly suited to dispelling any doubts on this point: who would be willing to walk on a rope suspended above the void, armed only with the scientific knowledge of the reasons why he or she can do so? Physics and anatomy may explain why the tight-rope walker remains in balance, but the tight-rope walker who doesn't want to fall must have acquired something that cannot be boiled down to the order of knowledge. Something that isn't knowledge, but that is rather a more intangible non-knowledge, and that will be crucial to the success of his art. If this element of non-knowledge is necessary for every art (that of the shoemaker or the tight-rope walker), its role will become increasingly important for the production of "beautiful" works of art, the ones that nowadays we refer to as "art" *tout court*. Kant called them, without giving in to "intemperance", works of "genius".

Hence, the rules for *producing* "the best shoe" will not be enough to actually produce it: something else is needed, something which is hard to define. Nor will it be enough to represent a pair of shoes, no less to turn them into a work of art. Since the time when Martin

Heidegger placed a painting by Vincent van Gogh of a pair of farmer's shoes at the heart of *The Origin of the Work of Art* (1936), shoes seem to have become a privileged benchmark for many reflections on art: from art historian Meyer Schapiro's dispute with Heidegger to Jacques Derrida's intervention; from Jacques Lacan's interpretation to that of the Marxist Fredric Jameson, of the philosophers of analytical art, and the installation of an exhibition on that specific theme.

Lying at the margins of this controversial chain is an element that Kant, perhaps unintentionally, had immediately and quite naturally associated with his thoughts on the production of shoes: walking in balance on a tight-rope. It is precisely in tight-rope walking that we can clearly see how balance is not a static result that is obtained once and for all: for each living being – and, we might add, also for those half alive "things", those "quasi-subjects" that make up a work of art – a similar condition of staticness would be tantamount to death, to inexpressive petrification, to a return to an inorganic state. Equilibrium is, rather, the constant search for stability within a dynamic and living imbalance. For all animals, whether or not human, "balance consists of the coordinated play of organs and muscles, according to the unfolding of rhythmical concatenations of different breadths, fixed in a regular order".[6] But we know that it is only in human beings that this concatenation of natural, essential operations for adaptation and survival can take on another dimension, recursively operate on itself and complicate itself on different symbolic and expressive levels. That is why, as Marcel Mauss believed, there is no "natural" way of walking for us.

The study enacted by *Animazione*, a video made by Daniela De Lorenzo in 2008, is precisely a reflection on "rhythmical concatenations". Inspired by a famous chronophotograph by Eadweard Muybridge showing a child whose motor problems prevent him from walking erect, the artist asked a performer (the dancer Ramona Caia) to mime the movement of walking on hands and feet and, above all, the wide grin with which the child seems to want to distance from himself and from those watching him his unsuccessful attempts to get up.

The video begins with a shot of the empty space, slowly approaching the intersection between the wall and the floor, between vertical and horizontal, between quadrupedalism and bipedalism, nature and culture. Now appearing in our visual field is Caia who, with hands and feet on the ground, slowly lifts her head and mimics the wide grin that distinguished the efforts of Muybridge's child. The performer moves to the side, keeping her body at a right angle, her feet touching the crucial intersection between wall and floor, crossing her legs and arms with a wavering, lithe gait, which is at the same time extremely similar to a monkey's. By way of a sort of artistically controlled regression, De Lorenzo thus stages that zone of indifference between the natural "concatenation" that preserves balance in all animals, and the further symbolical elaboration that is proper to the human animal, an interweaving that is never definitively extricable between animality and culture, childhood and maturity, artifice and instinct. It is here – in this zone of indifference, of indecision, of oscillation that belongs to every art – that a contact surface between our contemporary awareness of always being interwoven with artifice and the deep anthropological roots that connect us to and separate us from the rest of the animal world seems to emerge in this video: "Acrobatics, exercises in balance and dance to a great extent express the effort of subtracting oneself from normal operational concatenations, the search for a creation that will break the daily cycle of positions in space. Freedom is produced spontaneously when one dreams of flying, at the very moment during slumber when the resting of the inner ear and the muscles creates the exact opposite of the everyday scenario. In a different way, during a vigilant state, the spectacle of the acrobat also represents an act of freedom, a sort of challenge to operational concatenation".[7]

Thus, it is easy to understand the fascination of many late-nineteenth and twentieth-century artists with the circus world, with ac-

pp. 246–47
Jacques Lipchitz,
Mother and Child, 1912
Pencil on paper, 32 x 22.6 cm
Prato, Museo di Palazzo Pretorio.

Georges Rouault, original etching on wood from *Cirque de l'étoile filante* (Paris: A. Vollard, 1938),
45.5 x 34.5 x 5.2 cm
Florence, Biblioteca Nazionale Centrale di Firenze.

forced to walk on two feet; we feel like returning home walking on our hands, doing cartwheels…".[11]

Alexander Calder began his artistic experiments by inventing a small-scale circus, *Circus Calder*, populated by human and animal figures made of metal wiring, rubber, cloth and other found materials. A circus (that could be carried around in a suitcase) whose performances charmed the Parisian avant-garde in the late 1920s. If we consider the fact that upon returning to America Calder collaborated with Martha Graham, and that, meanwhile, he began working on his famous aerial sculptures named *Mobiles* by Marcel Duchamp, we can see that he emblematically synthesized the interest in and exploration of the world of the circus, balance, dance, which were rooted in the late nineteenth century, albeit influenced by the crucial and irreversible encounter with Mondrian's abstraction. *Stabiles* – this time it was Jean Arp who invented the term – are instead a series of self-bearing sculptures, complementary to the hanging ones. The search for equilibrium radiates outwards: from acrobatic to environmental (suffice it to recall the *Mobiles* moved by the surrounding air), from that of kinetic sculptures endowed with their own engine and the one that, like in *Stabile-Mobile* (1973), questions the cohabitation of heterogeneous balances.

But not even the remarkable wealth of experiments performed by Calder succeeded in exhausting the suggestions that the world of acrobats offered to the twentieth century. Fausto Melotti is proof of this. In a moving "poem" written by Jean Genet for the young acrobat Abdallah Bentaga, the "death-like" dimension of "everyday walk-

robats and saltimbanques (from Picasso to Lipchitz, from Klee to Calder, from Severini to Melotti…): it is not just the discovery of a "mythology alternative"[8] to Classical mythology that attracts them, but the clown-acrobat's agility at challenging gravity, the metamorphosis that enables him to achieve "Puck's wanderings on wings".[9] Similar considerations could be made for dance, that "poetry of the arms and legs" where the "material, gracious and terrible, … comes alive and is beautified by movement".[10]

What struck the nineteenth-century critics and chroniclers was the attempt to use the body itself to overcome the body's limit, to dissolve everydayness, daily movement into winged walking: "O great buffoons, miraculous leapers, we feel humbled, upon seeing you,

Ugo Mulas, *Cirque Calder*, 1963–64
Fotografie Ugo Mulas © Eredi Ugo
Mulas. Tutti i diritti riservati
Courtesy Archivio Ugo Mulas, Milan –
Galleria Lia Rumma, Milan/Naples.

pp. 250–51
Alexander Calder, original lithograph
from *Gouaches et Totems* (Paris:
Maeght, 1966), 38.5 x 29 x 2 cm
Florence, Biblioteca Nazionale
Centrale di Firenze.

right
Alexander Calder,
Stabile-Mobile, 1973
Varnished metal, 80 x 80 x 80 cm
Florence, private collection.

Fausto Melotti, *Equilibriums*, 1971
Gold and enamel on Plexiglas
base, 49 x 43 x 15 cm
Courtesy Galleria Christian Stein,
Milan.

ing" should be produced in favour of a total transfiguration of the self into one's own glorious image balancing on a steel wire: "Why dance tonight? Perform leaps and somersaults under the spotlights eight metres from the floor, on a wire? Because you need to find yourself. Prey and predator at the same time, you escape and you seek yourself. Where were you before you entered the ring? Sadly dispersed in your mundane gestures, you did not exist. … But you touch and grab yourself only for an instant".[12]

That's when this "total transfiguration of the self" can lead to the "death" of the acrobat's body as the erasure of his figure: in Melotti's *Invisible Acrobat* of 1980 (see here on p. 58), all that remains are the thread-like traces of equilibrium, while the acrobat, having achieved the extreme in his art, can become invisible. If we needed to emphasize how the artist comes to the twentieth century and sees himself as a tight-rope walker, an acrobat poised between control and contingency, technique and spontaneity, balance and lack of it, sense and nonsense, it would be enough to remember that "invisible acrobat of the twentieth century" were also the words actually used by Italo Calvino to describe Melotti.

The work of another great artist represented here also moves along the balancing line. Starting from his famous squaring off of the canvas (*Geometric Drawing*, 1960), Giulio Paolini has for many decades focused on the very fundaments of art, realizing works poised between the visible and the invisible. In a continuous overturning of gazes (suffice it to recall *Young Man Looking at Lorenzo Lotto*), with an inclination towards spatial fragmentation and the multiplication of images, Paolini moves along the crest that connects past and present, the everyday and the history of art, breathing life into ambiguous and fleeting temporality. His self-portraits, in which the face is more or less hidden, offer clues to his refined and complex poetics. In particular, in *Carte Noire* (1999–2000), Paolini portrays himself as a tight-rope walker advancing carefully in a uniform space, whose coordinates are

Giulio Paolini, *Carte Noire*, 1999–2000
Silk-screen, coloured pencils and
collage on black paper, 163 x 223 cm
(nine framed elements
53 x 73 cm each)
Turin, collection of the artist.

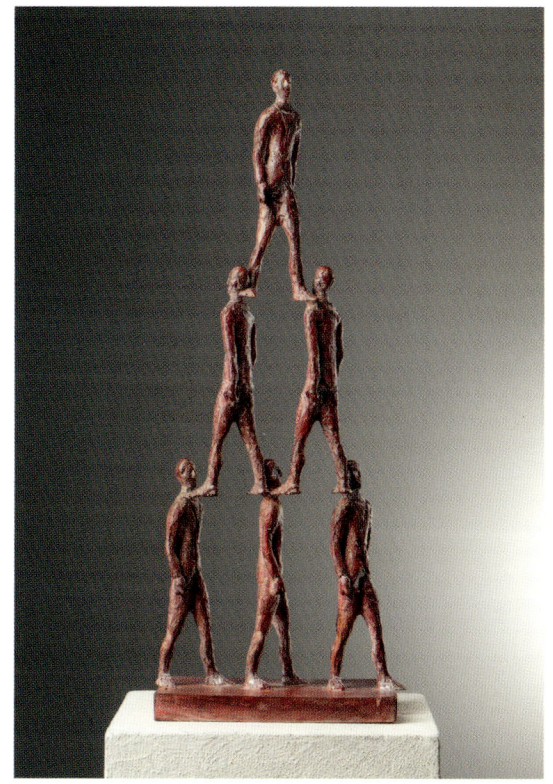

Roberto Barni, *Enterprise*, 2010
Patinated red bronze,
56 x 22 x 12.5 cm
Florence, collection of the artist.

Roberto Barni, *Joke*, 2013
Patinated bronze, 145 x 23 x 23 cm
Florence, collection of the artist.

Roberto Barni, *Continuous*, 2001
Patinated bronze, 19 x 39 x 63 cm
from 1 to 7 + 1 artist's proof
signed A. P.
Florence, collection of the artist.

Roberto Barni, *Razor 2*, 2003
Bronze, 71 x 18 x 50 cm
from 1 to 6 + 1 artist's proof
signed A. P.
Florence, collection of the artist.

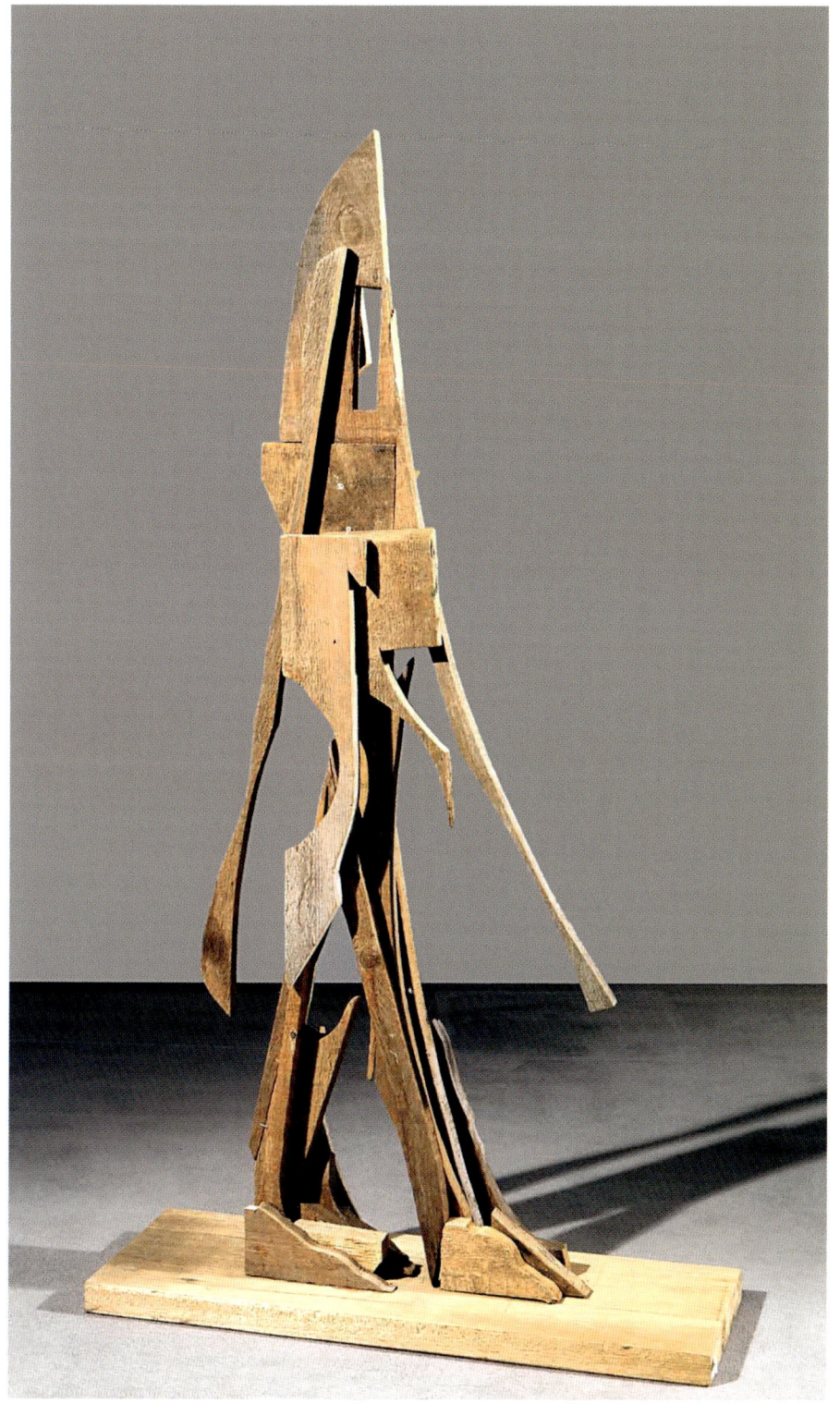

Mario Ceroli, *Untitled*, 2002
Russian pinewood, 144 x 75 x 25 cm
Private collection, courtesy of
Tornabuoni Arte, Florence.

exclusively established by the intersection between two diagonals. This indeterminacy is only articulated by the figure of the artist's suspended body, surrounded by eight images that depict just as many of his works. Paolini's points of reference are past and contemporary works offering not spatial but temporal coordinates; as such, they are subjected to changes in taste and to a multiplication in points of view that increasingly articulate and complicate the fragile equilibrium of the present.

Also dated to the late 1950s and early 1960s are the early works of Mario Ceroli, another well-known representative of Arte Povera. Akin to Paolini, Ceroli has dedicated countless tributes to the art of every age (from Michelangelo to Alberto Burri, from Giorgio de Chirico to Dante), but his sculptures, installations and large set designs are imbued with a radically different ambiguity. His works, often characterized by a strongly static appearance, are indeed connoted by a particular spatial research that places them at the crossroads between two and three dimensions, on the boundary between painting and sculpture. His untitled work from 2002 shows a man walking, a walker with a "composed" gait, albeit broken up into geometric planes and elements. Although the sculptural element is there, further underscored by the presence of the pedestal, what we see here is the exploration of a form of balance that is not that of the tight-rope, as it does not stem from staying in balance, but from the narrowing of the sculptural three-dimensionality refuted and preserved via the juxtaposition and intersection of two-dimensional planes.

The sculptures by Roberto Barni showcased here are also filled with walking figures bearing other figures upon their heads or shoulders (*Joke*, 2013; *Enterprise*, 2010), or upon a razor blade (*Razor 2,* 2003). In their impassive anonymity they defy the laws of gravity, standing in unstable balance upon the pre-

carious meaning of life that belongs to each one of us, on the crest between the rootedness in tradition, the community and solitude, alluding to the way in which we stumble along on the narrow paths of life and history. In *Continuous* (2001), two male figures, their movement virtually endless, waver as they step on the opposite ends of a swing, resembling a curved staircase consisting of slender stakes and voids. This bronze work cannot help but bring to mind the small pen drawing *Perpetual Swing* with which Aby Warburg, always painfully grappling with a precarious psychological equilibrium, synthesized the conception of artistic activity. Warburg's "science without a name" – art history that emerged from its traditional confines and from the usual dichotomies – revolved around the construction of an "aesthetics of the dynamogram", of the figural and plastic form that takes on that force that is time. This is why we find a multitude of "oscillating schemes", "weights", or the representation of the artist as someone who "dances" or "debates" in precarious equilibrium on "a perpetual swing".[13]

[1] However much the term may commonly be used to also describe his work, Long is not fond of this label, which he sees as being closer to the experience of American artists and their broad use of technology.

[2] F. Careri, *Walkscapes. Camminare come pratica estetica* (Turin: Einaudi, 2006), p. 112. In the same volume the author uses a wealth of material to clearly focus on several key moments in the history of the "artistic walk".

[3] W. Sharp, "Two Interviews", in *Bruce Nauman*, edited by R. C. Morgan (Baltimore: The Johns Hopkins University Press, 2002), p. 253.

[4] J. Simons, "Breaking the Silence: an Interview with Bruce Nauman", in *Please Pay Attention Please: Bruce Nauman's Words. Writings and Interviews*, edited by J. Kraynak (Cambridge, Massachusetts and London: The MIT Press, 2003), p. 337.

[5] When the complete works of the Dutch anatomist Petrus Camper (1722–1789) were translated into French and published posthumously in the early nineteenth century (*Oeuvres de Pierre Camper, qui ont pour objet l'histoire naturelle, la physiologie et l'anatomie comparée*, edited by H. J. Jansen, Paris, 1803), they were already widely known to the European scientific community.

[6] A. Leroi-Gourhan, *Il gesto e la parola. II, La memoria e i ritmi* (Turin: Einaudi, 1977), p. 334.

[7] Ibid.

[8] J. Starobinski, *Ritratto dell'artista da saltimbanco*, edited by C. Bologna (Turin: Bollati Boringhieri, 1984), p. 42.

[9] Ibid., p. 58.

[10] Charles Baudelaire's words in Starobinski 1984, p. 82.

[11] Théophile Gautier's words in Starobinski 1984, p. 58.

[12] J. Genet, *Il funambolo* [1956] (Milan: Adelphi, 1997), p. 120.

[13] Cf. G. Didi-Huberman, *L'immagine insepolta. Aby Warburg, la memoria dei fantasmi e la storia dell'arte* (Turin: Bollati Boringhieri, 2006), p. 167. "All the Warburgian temporality seems to be built around rhythmical, pulsating, suspensive, alternating or panting hypotheses" (ibid.).

Mario Ceroli, *Groma*, 1990
Wood, bronze, Belgian black marble,
Portuguese pink marble,
h 111 cm (Ø 50 cm)
Rome, collection of the artist.

TRANSMITTING THE EXPERIENCE OF EQUILIBRIUM

edited by EMANUELE ENRIA

Staying in balance or seeking physical and, in consequence, inner balance is one of the truest and oldest experiences that we are still allowed to have in the contemporary world, owing to the fact that it is linked to man's primordial condition, i.e. that of getting up on his feet and walking. The writings of Salvatore Ferragamo, artisan and shoemaker, his studies and creations, his shoes, have inspired us to conduct an in-depth study of this intriguing and complex subject. The words of Wanda Miletti Ferragamo, Salvatore's wife, about her husband's great interest in feet and his discoveries in relation to the arch of the foot, find a contemporary dimension in those of James Ferragamo, the famous artisan's grandson, who explains how the knowledge he inherited from his grandfather is a priceless heritage to be cultivated and maintained. In this section, the testimonies of several well-known personalities, from the alpinist Reinhold Messner to the wire-walker Philippe Petit, from the writer Will Self to the architect, engineer and artist Cecil Balmond and to the ballet dancer Eleonora Abbagnato, help us to understand how equilibrium is at the very heart of their experiences and, we might add, of our own as well. Through these interviews we come to understand that the cornerstone of their equilibrium, even when it involves heights that are unheard of for most of us, as in the cases of Messner and Petit, always lies in a profoundly artisanal element, which seems to be the real secret not just of their near-perfect relationship with balance, but of their essential philosophy of life. Hence, they reveal how the very idea of equilibrium is actually much broader than we might believe. Not merely something having to do with the stability of a body, a building, a technique; what comes into play is the relationship with the materials, how we perceive them, feel them, how we get our bearings in space. And it has to do with our mental balance as well. Each one of these figures tells us how they transmit this experience, by means of what might be a gesture that is danced, suspended, walked and designed. Cecil Balmond's Renaissance-like approach to architectural design, which joins an ancient knowledge of numbers to the most highly evolved dynamics of algorithms and spirals, stimulates and invites us to create forms from within, not without, if we wish to avoid ending up being frozen inside rigid, predetermined formulas or models. Including those of a shoe. Salvatore Ferragamo's great insight was the bounce of gravity on the ground, and how the arch of the foot, if we allow it to, will absorb it quite naturally: an insight that is still being transmitted to us today through the experiences of Jerry Ferragamo, his nephew. As the writer Will Self tells us, ever since our bodies were swallowed up

by the city, which no longer reflects our physical shape, but rather that of cars and the technology we use to get around and to communicate, we can even wander about like sleepwalkers in the urban space, and thus hand over our equilibrium to the artifice of technology. The art of Philippe Petit, the poet of the sky who walked on a wire suspended at the top of Notre Dame and the Tour Eiffel in Paris, the World Trade Center in New York, stems, after all, from a small gesture that even circus performers often overlook: walking. "Give your skin enough time to understand", he advises in his *Treatise on Tight-rope Walking* to those who are learning to walk the wire barefoot. He actually sounds a lot like a shoemaker. Not accidentally, he claims that he too is an artisan in every detail that, on the whole, constitute the art of the tight-rope walker. The great Reinhold Messner, whom we are accustomed to seeing as a hero of epic climbs, tells us that his equilibrium as it relates to mountains became complete and conscious when he chose to live like a mountaineer. And, lastly, étoile Eleonora Abbagnato talks to us about feet, but not just that. Feet that are constantly being shaped, day after day, reinforcing the way they lean on the ground and are then elevated *en pointe*. She also tells us about her experience with Pina Bausch and William Forsythe, who both focused on developing new relationships with balance.

The philosopher and writer Walter Benjamin dedicated some of his most illuminating pages to the fundamental moment of transition between the nineteenth and the twentieth centuries. In one of his most famous writings on the figure of the narrator, Benjamin stressed the risk of the unrecoverable loss of experience. We mustn't overlook the fact, he tells us, that the first storytellers recounted tales of experiences which were then passed on by word of mouth: "farmers and sailors were the first master storytellers, and the superior school of storytelling was craftsmanship". Because true storytelling implied a sort of practical utility that was also related to living. It was a way of learning about peoples' journeys and the finer details of a profession, proverbs and the rules of life. As such, the storyteller became the person giving advice to his listener. When artisans started to disappear, storytelling began to change as well (paving the way for the novel). There was a decline in the communicability of experience "sewn into the fabric of life as it is lived, which becomes wisdom", in Benjamin's words. But it is only through this type of storytelling that equilibrium can be transmitted. The impression is that of finding, in the stories told by these figures in regards to their art, much of that same transmission of wisdom.

WANDA MILETTI FERRAGAMO, JAMES FERRAGAMO, A HISTORY OF EQUILIBRIUM ACROSS THE GENERATIONS

EMANUELE ENRIA Salvatore Ferragamo called himself a shoemaker. Today we consider him as one of the major figures of twentieth-century fashion. What is extraordinary about his shoes is not just their aesthetic appearance and the uniqueness of the materials he used to make them, but above all the actual making of the shoe, behind which lies an expert and a connoisseur of the foot's anatomy. He dedicated his entire life to maintaining in the shoes he made the equilibrium that the foot and body preserve in the natural act of walking. As Salvatore Ferragamo's wife and the Honorary President of the company, could you tell us some of the key points along this path and the results that have been achieved in footwear?

WANDA MILETTI FERRAGAMO I'm fascinated by the word equilibrium, because you find it everywhere, in everything we do, in our actions, in our way of thinking. Without equilibrium a person would be empty. It is what puts a stop to excess. I heard my husband pronounce this word countless times.

When Salvatore and I got married he was 42 and I was just 18. I learned a lot from him, even as far as his work was concerned, what was substantial about his work. One of his greatest lessons focused on the equilibrium he bestowed upon the shoes he made and, as a consequence, upon the human figure. He believed that the weight of the body falls vertically on the arch of the foot, so he added a steel support, the shank, between the insole and the sole to give the plantar arch stability. His clients understood that the shoes he made featured a stability and a reliability that couldn't be found elsewhere. He patented his invention, but then, as we all know, after a few years patents expire, becoming public domain. This is why today every pair of shoes has a shank, Salvatore's invention.

The interviews were conducted in Florence, Palazzo Spini Feroni, 22 April 2014

EMANUELE ENRIA When Salvatore died in 1960, the artist left the international scene. You are the heir to your husband's values, and you decided to helm the company with the help of your elder daughter Fiamma, and then with that of your other children. How did you cope with that moment?

WANDA MILETTI FERRAGAMO There are many who ask me that question, but now that so many years have passed I can safely say it was quite easy. I remember that all the workers came to my husband's funeral. They shook my hand and said to me: "Wait and see, Signora, we'll help you out, we can do it, rest assured". Since then I've kept these words inside, I've never forgotten them, and they've come true.

Luckily, I already had Fiamma by my side who had worked with her father a year. Salvatore wanted her to quit school, Fiamma was in her second year of secondary school and instead wanted to go on studying. He would say to me: "You're her mother, you have to convince her because what she learns from working beside me will be of use to her in the future". So, rather reluctantly, Fiamma left school.

After my husband died, she felt that her decision had come as a blessing, because what she had learned from her father would serve her well.

I'd just like to add one thing: how important a perfect union is between two people who get married. If one of them passes away, the other still holds all his or her secrets, advice, all the experiences they shared. So a wife or a husband can continue to transmit all the good things he or she has learned, absorbed. This is what happened to me, because when Salvatore passed away we had been married twenty years; but I had already understood the work he did, I knew his secrets. So I was able to transmit these values to my children, who have treasured them. All of us have made every effort to carry forward and reinforce the basic principles of his activity. We have been able to gather together and instruct within the company people who are keen on his work. If Salvatore were still here with us today, he would be happy to see his name written in every corner of the world. And to know that his work lives on.

WANDA MILETTI FERRAGAMO has been at the helm of the group since 1960, the year of the death of her husband Salvatore, the founder of the company of which she immediately took over the presidency. At first alone, and later with the help of her six children, she managed to successfully overcome the immense problems of a difficult legacy such as the one she was left by her husband. It was under her leadership that the company took the "big step" forward – from a single type of product, footwear, to the ready-to-wear and total look market. She is currently the company's Honorary President.

EMANUELE ENRIA Salvatore Ferragamo was your grandfather. You represent the third generation of the Ferragamo family and you're in charge of the leather department and of the creation and production of women's shoes, originated by the founder. When your grandfather was alive, shoes were entirely handmade and made-to-measure. The principle of the fit on which all of Salvatore's work was based was applied to an artisanal productive model. After him, how did the company manage to grow without falling short of the founder's values?

JAMES FERRAGAMO My grandfather was a genius, not just because of the creativity of his models, the selection of the materials and the ability to make a product that was lightweight, but also because he paid close attention to the fit, to the foot's equilibrium inside the shoe. He was one of the first to consider this problem, a pioneer in making shoes that were, of course, beautiful and original, but that could also be worn, that weren't just an object of desire for his clients, famous actresses, at the expense of comfort. To achieve these results he researched a lot; in Los Angeles he studied the anatomy of the foot. When I was just a boy, I remember hearing lots of stories told by my grandmother that helped me to know him well, since I didn't actually have the chance to meet him. I learned a lot about his yearning to create special products, ones that didn't just take into account the form but also the client's well-being, his or her body's equilibrium as well.

After he passed away, a lot of people collaborated to keep his principles alive, starting from Jerry Ferragamo, one of Salvatore's nephews who had worked at his side for more than a decade. Jerry made every effort to transform a product that was born for bespoke production into an industrially manufactured shoe. He worked on a project that made it possible to offer a hundred different fits for one model: an almost bespoke product but conceived on an industrial scale. This idea was key at a time of historical upheaval, when industrialization was becoming increasingly important and the number of "career women" was rising.

My aunt Fiamma, Salvatore's eldest daughter, instead took over the creative department, unquestionably counting on the huge archive of products left behind by her father as a source of inspiration. But she needed to create shoes that would reflect a way of living that was undergoing deep changes. There was continuous tension toward a balance between the creative side, helmed by Fiamma, who was trying to place unique and innovative products on the market, and the technical side, represented by Jerry, who wanted to defend the laws of fit and comfort. Sometimes, when you were passing by in the corridor, you could hear them arguing, although their opinions were always for a constructive purpose: to produce shoes that reflected the equilibrium of the foot, and that were at the same time innovative.

Fiamma became famous for Vara, a shoe she created in 1978, which continues to symbolize a shoe designed with the spirit of offering a bespoke product but on an industrial scale.

EMANUELE ENRIA How does today's company handle Salvatore Ferragamo's demanding legacy, having chosen to continue to produce and keep its headquarters in Italy, in Florence and the surrounding area?

JAMES FERRAGAMO My grandfather's research into making a shoe with a perfect fit is the reason why the company chose Florence: because of the opportunity the city offered to collaborate with highly skilled artisans, who could create shoes according to his standards. The shoes the company makes are still artisanal products, which need to bring together many professionals, not just those who make the upper, but the ones who work on the inner parts as well, which in our case comprise a steel shank to support the arch of the foot, like a bridge that joins the heel and the sole and distributes the weight of the body on the two parts in equilibrium. The shank is still one of the basic component parts of our shoes. But this is just one of the things that Salvatore Ferragamo passed on in the creation of footwear. The shoes we still make on our conveyors are the fruit of the work of artisans who, through their experience, try to respect these criteria by combining them with cutting-edge technology.

The market and client typology are evolving all the time. Our aim is to update what Ferragamo was in the past and what Ferragamo is today. Even if our great creative has left us.

JAMES FERRAGAMO is a member of the third generation, the son of Ferruccio Ferragamo and grandson of Wanda and Salvatore. Born in Florence in 1971, he got a degree in Marketing and International Business from the Stern School of Business, New York University. After working for Saks Fifth Avenue, in 1998 he entered the family company, at first as General Merchandising Manager. He is currently Manager of Women's Leather Products.

CECIL BALMOND, THE DYNAMICS OF EQUILIBRIUM

The interview was conducted in London, Balmond Studio, 13 December 2013

EMANUELE ENRIA "Nature, the supreme architect, from whom man has borrowed and adopted many of his ideas, has created the human foot with an arched shape because – as any architect can confirm – an arch can support more weight than a flat surface. The function of the arch of the foot, however, is not simply that of supporting a static weight, like the arch of a church portal, for it must support our weight in motion." These are Salvatore Ferragamo's words. It would be interesting to know how you envisage the relationship between the arch of the foot and the architectural arch.

CECIL BALMOND You could think that the foot is an arch, which it actually is. But it's the most complex body part in my opinion. And when you think of the arch of a bridge, or the Roman arch, they are beautiful things but they are static. They are purely in compression. The foot has three layers of tensions that move in counter tension: when the forces come through, part of the joints move up and others move across to hold them back. So, it's like a massive orchestration of tension and compression in the foot. No architect can make it like that, at all.

When I did my bridge in Portugal, I was not thinking of the foot, but now that I post-rationalize it, I can see how it is a metaphor for the foot in a way, because it looks like an arch, but actually it's one arch and another one displaced. The displacement causes all sorts of issues because there's no direct transmission of compression – you have to have tension. But also, the arch is very flat. In architecture, when the arch comes from this size, to get flat, it is like the foot. Tensions are coming in, and actually it's not an arch any more. So, in a way, it's acting as the foot does, in a complex interaction of tension and compression. Spatially mapped, not in any direct opposition either. It raises the whole question of equilibrium. Because equilibrium is a huge concept. It's the basis of the whole universe held together. At one hand – the galaxies, the stars, everything… in a dynamic equilibrium that is moving but held together thanks to the centrifugal force. And there is another kind of equilibrium, a spatial equilibrium that is a complex of balances. Now, the normal idea of balance would be very reductive: you have an

opposition. This to this. And you balance. Good to bad. But that's static equilibrium. Dynamic equilibrium is when the two oppositions start breaking into parts, and each of these parts would also be a movement. A dance. Equilibrium, therefore, is for me a massive orchestration of parts being held together by a composition, be it static or dynamic. And, of course, the idea of the dynamic will be implicit to the best static composition.

So, we take a very simple thing. The classical sculptures. Michelangelo. At the beginning of his career, when he was still a young man, you find the *contrapposto* movement: you have the leg coming straight down, and the other leg back down with the arm pulled out against this. Now, if you see my Orbit, the big tower I did, what is that but a *contrapposto* movement? You get one thing going up, but then against this movement there is a huge spiral in space in the other way. It is a sort of balance. But if you are around that, it's all moving and you are not sure where this balances, because the whole thing is moving. Michelangelo, in his later career, was doing exactly this. In his earliest *Pietà*, Christ's body is over that of the woman; the problem is how you balance the weight of a heavy man on a delicate woman. That's why, in that early *Pietà*, the woman is standing up and Christ is lying on her arms. Otherwise, in the *Rondanini Pietà*, the woman is moving and she is holding Christ from dropping. What is actually balancing everything is what Michelangelo did: the small counter movements running through the marble, which you can find especially in the unfinished works. The sculptor creates these *contrapposto* movements that give equilibrium to our eye... and the thing becomes full of life. Equilibrium is a basis for life, it's a balance in the way we perceive space as you

move through. So, we feel secure but a little at risk: when we move through a forest, or a mountain, or listen to music, or make a piece of architecture, we feel safe at the end of it but we know we have experienced a degree of risk. Risk is the movement that can disbalance you. Risk is the instability that an artist introduces into his work. Risk is the trick of good creativity that achieves balance through imbalance and asymmetry. Ultimately, risk is the trick of equilibrium itself. And the foot is doing this all the time, as we walk and as we run. We underestimate the fact that if we're standing it's thanks to our feet. Because we have one structure that, without the foot on the ground, would be nothing. All transmission of flow is constantly going from and to the foot, which is the biggest dynamic of the body (apart from the brain working in virtual space). And it is an extraordinary example of equilibrium created by nature.

EMANUELE ENRIA In recent decades, dance has arguably been the art that has best represented the concept of balance on stage. A parallel can be found between dance and architecture: both have freed themselves from traditional notions of space and form. We could mention choreographers such as Merce Cunningham or – nowadays – William Forsythe. You once claimed that dance is nothing but the algorithm of movement. Would you care to explain this concept in relation to your own work better?

CECIL BALMOND The dynamic root of a movement is an algorithm. The most primitive algorithm was in the early ritual dance. What does algorithm mean? A rule, a repeated sequence. The primitive algorithm only repeats the same sequence. So you go in the dance around the fire, in the same steps, making the same body movements and the same

songs. And chanting, all the ritual chanting, is also an algorithm, a repeat of the same formula of the words. So, if you take Laban with his dance movements, or Merce Cunningham, or Martha Graham and her early works, and even break dance – it's all algorithmic, in that there is always improvisation going on the base motif. Bach did the same thing. You have the fugue, four notes, or just four sequences, and then it is just repeated, repeated on different levels... The trumpet, the bass... Dance is the same thing for me. It's an algorithmic move. In the end, it's a rapid repetition of small movements, four parts, five parts, and when you make it all together, you overlay the whole thing, if you look at the performance and map the dances and moves over a period of, say, 40 minutes, you'll see the overall patterns being repeated. One interesting thing is when you see Indian classical dance. The woman is moving her head and the whole story is in her hands; if you put lights on each of the fingers and film it in darkness, and afterwards you listen to Bach, in 5 or 7 minutes you will find that Bach is being repeated in the movement of the dance.

If you analyse the Indian woman dancing, you'll see that her body is being held very firm: she keeps the navel at a fixed centre and her movements create a series of circles around it – the first circle centres on the navel, then the second circle, third circle, round the body and chest, then the elbows and then the fingers. It is all held within a ratio, so the centre is still but everything around it is moving.

EMANUELE ENRIA In some of your projects, including the ones with Daniel Libeskind, you have made wide use of the spiral shape. The spiral is an efficient movement that is also used in the martial arts and

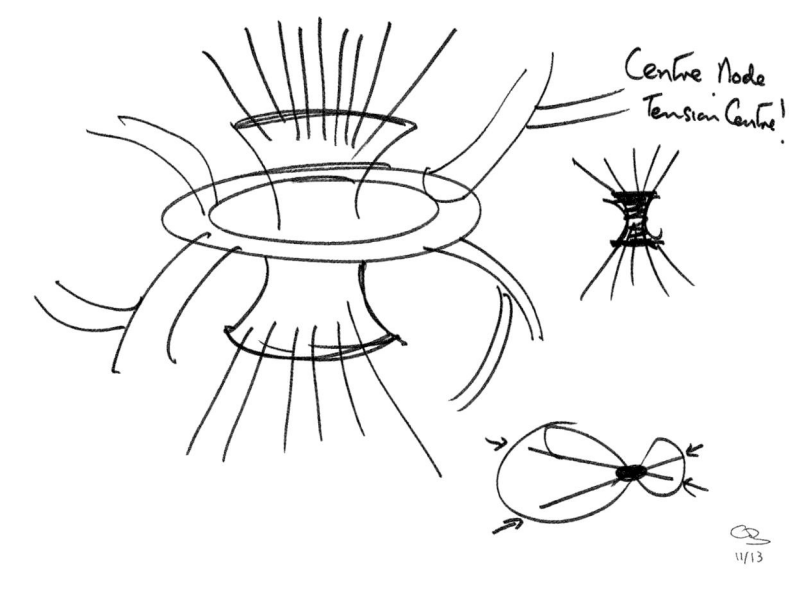

Centre Node
Tension Centre!

Orbits

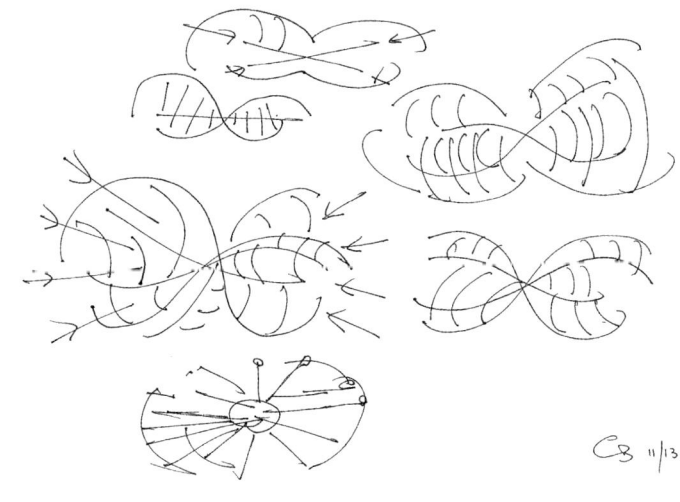

CB 11/13

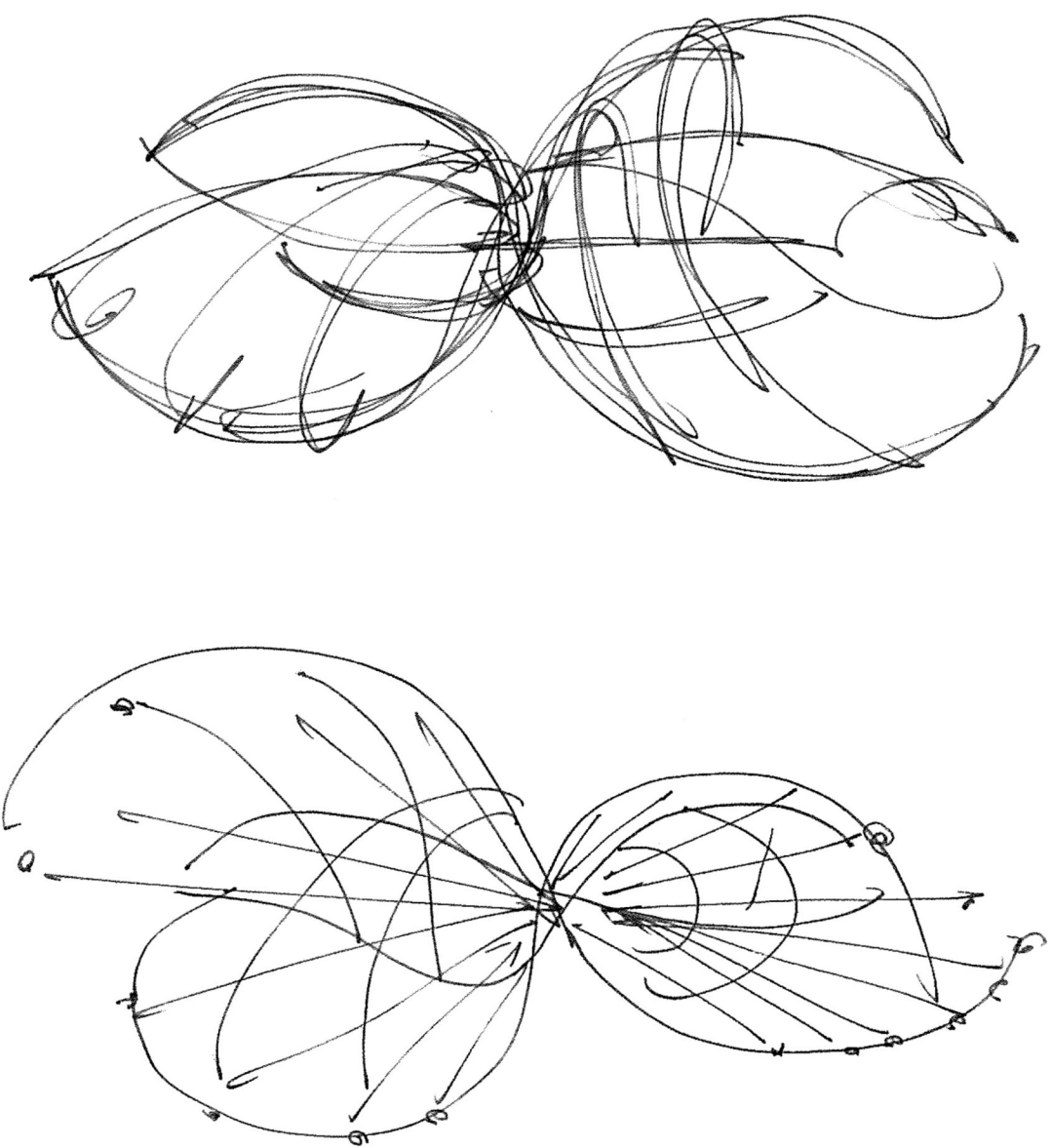

CB 11/13

in contemporary dance, as it enables the expansion of movement across space by exploiting the centre of gravity. Nature itself is made of spirals: in minerals, galaxies, the DNA, and so on. You've explained that there are two spiral models, logarithms and Archimedes' spiral, and that both have a fixed centre. In architecture, from the Egyptian pyramids down to Le Corbusier's work, the spiral was often used to explore the golden ratio. You've stated that you are fascinated by new, non-linear spiral models. Could you tell us more about these?

CECIL BALMOND The spiral movement, from a classical perspective, is just one widening spiral, but it's not uncontrolled. There is always a control on this spiral. It is not just free – it looks free, it looks like an open discovery, but it is controlled. Natural spirals are always controlled. The centre is fixed, but there's a movement around. The mathematical spiral is just a complete wrapping, like a string.

I have been interested in what I call the moving centre, because it's another metaphor for equilibrium. The centre moves, everything is just moving around, and displacing. You see, the centre is holding everything, but the centre itself moves. Ultimately, it's a new idea of energy. Moving energy. The classical idea was that energy was contained – you couldn't just add or destroy it. I don't believe that, I think energy creates its own energy. I am interested in a whole new concept of one moving centre. It releases all fixity and you are free to explore.

EMANUELE ENRIA As you have stated, ever since the dawn of architecture, geometry, proportion and ideas of balance have been at work in the conception of form. Balance is not static but dynamic. This also applies to the postural oscillation of our body. Our brain must constantly balance the information it receives from within and without the body. How, then, do you envisage the dynamic of balance in relation to all of these aspects?

CECIL BALMOND I think, as I am talking of this freedom to move and create, that there has to be some kind of control. And I use numbers a lot because they are a kind of control. They represent a definite value. You can give them a position in space: for instance, if you look at early dance manuscripts, you can see people were marking the dance movements through certain numbers. The oldest intellectual language that we have, and the only universal intellectual language that everyone understands is numbers. A basis of intellectual arguments and agreements is a number. But, unfortunately, we consider the numbers merely as calculators and this means nothing. In early Greek times, when Pythagoras lived, they had a sense of the number being a position and creating ratio, creating a dynamic. We have completely lost this idea. To us a number is just a cipher. For me, a number is a position marker, a symbol, but above all it is potential. It's got potential. So if I put number 4 and 6, and join them, am I joining just a gap of 2? Maybe 6 is one and a half 4… and so on. I can create movement by marking a position in space, in a computer, with numbers, giving value to the numbers. By giving potential to the numbers. I give the numbers charm. They have their own life. So they can start moving around and dancing in space, according to some design I give them. And then I wait to see what happens, and enter again; it's a kind of ongoing choreography. Numbers are very intriguing and they lead to the most amazing constructions; numbers give shape. You can get the spiral we talked about by a series of numbers, by taking a rectangle, and drawing a side 1 and a side 3, then a big rectangle with 3 and 5, and then 5 and 8, and as you get these rectangles growing, if you connect the edges, you get a spiral. And nature works this way, based on Fibonacci numbers. I have been fascinated by that: it's the most amazing set of numbers, because they combine algebra and geometry. Numbers are arithmetic, their continuum is algebra, and in between arithmetic and algebra there is geometry. The Fibonacci numbers pulled the arithmetic and the algebra into the geometry. And what is geometry? Position, movement, line, space. A never-ending fascination to see how numbers can be, how numbers create space. I call them a new material. I use them as a material. As a sculptor would use bronze, or something like that. I can use concrete steel glass, but I can use numbers to create the forms as well, when I want to.

I designed the Serpentine Pavillon in 2002 – it's a series of lines moving in space. What is it but a dance, for sure? It goes from one number to another number, which is the algorithm, it repeats, but then it moves, and jumps and jumps… it all comes together, and it's a beautiful building of lines and velocity crossing… it's a drama. But it has infinite balance at all levels.

EMANUELE ENRIA I have heard you say again and again that architects all too often tend to envisage forms as though they were already built from the outside, as abstract impositions upon space. This is why you stress the importance of envisaging forms from within, through the creation of individual geometries. Form is dynamic: it has something to do with the configuration of space as connections. And it is the rhythm of these connections that gives rise to denser resonances, to the meaning of archetypical structures. How do you work on these connections?

CECIL BALMOND Ultimately, I think I try to bring to people the archetype of response to form and shape. For instance, I think we only have certain constructions in us, which are a spiral form, a folding form and a branching form. We also find them in microphysics. In modern architecture as well as in the Rome of the old days, there was always a human proportion in classical architecture, from the bottom of the building to the foot, to the body and then the head to the top. The classical world had 3, 5 or 6 floors as for the body of buildings. But when you make it up to 8 floors, or you build a big bridge, 220 metres, where is the human index? A bridge can easily become a mechanical conduit – you just go from A to B. How then can you give the narrative to space?

So I try, in my work, to feel space as being compressed or elongated or closed around you. Or, as on the bridge in Coimbra, I try to play on the feeling of going through. As you step forward, your horizon is lost and you get the impression that you are moving from length to lateral and that the sights are moving with you. As you keep moving through, the idea of the crossing has gone and you are now orientating the bridge into your own space. Why? Because I am playing with what I believe are certain archetypes that are complex phantoms. They are not explicit but they are there and if I can bring the human beings to negotiate,

to be aware in a subliminal way of the space they are immersed in, they are participating in that very space – that is what's important to me. At the same time, this jumping of scale is easily understood and perceived, since it is in all of us. Our molecular structure is like that. I believe that the synapses as well (when we come to understand the brain one day) are what I call fractal – the same signature but on a different scale. I have a deep intuition of these primitive archetypes of force at work in the human beings. So, when you walk through the building, or on the bridge, or leaf through the book I designed with colours (I use colours as structure as well), the archetypes of feeling come out to resonate with you. I call it: the recognition of deep structure. This pattern I am referring to is not something trivial, it is actually this equilibrium we are talking about. There are deep resonances that somehow balance your psyche. And they're rich, they're open, they're always moving. They fold, spiral, branch. I think, like Carl Jung says, that a collective deep unconscious exists. It's the same with dance. The literal movement is one thing, but what really happens is that it resonates something deeper within you. So, I do believe that this idea of the archetype is crucial to driving and powering everything. And I am happy when the actual shape disappears and you can actually feel its form.

EMANUELE ENRIA To conclude, I get the impression that within your way of conceiving balance colours too play a role. What is it?

CECIL BALMOND In my work, I feel that form and shape is not enough, that colour is fundamental. As far as I am concerned, I have to create my own grammar of invention, in which the body is a shape, the mind is the form and colours are the soul. Colour is absolute force. Colour brings out the soul of the work.

I did the Kunsthaus in Rotterdam – it was a very provocative work of shifting the equilibrium in four different places. Not one building, it is four buildings, all different. Yet, it is one building. And I was thinking: what is the colour for this building? Even though I had a red line on it, for me blue was its colour. Why? I cannot explain. What I do is a sort of deep meditation that allows me to actually perceive the colour, which is the form you can't see behind the shape. Colour gives life to form, it's a metamorphosed tension that gives ultimate equilibrium to a piece of tectonic work. Or a book, an object.

CECIL BALMOND is a structural engineer, artist and writer of international renown. He was born in Sri Lanka, where he trained as a civil engineer. He currently directs the Balmond Studio in London, which he founded in 2012. He is the author of *No. 9* (1998), *Informal* (2002) and *Element* (2007). His most recent works include: the ArcelorMittelOrbit, a 130-metre-high tower conceived together with Anish Kapoor for the 2012 Olympic Games in London, and the footbridges in Coimbra (2006) and at the University of Pennsylvania (2009).

REINHOLD MESSNER, INSTINCTIVE EQUILIBRIUM

EMANUELE ENRIA In *Solo. Nanga Parbat*, the book in which you tell the story of your complicated relationship with this peak, which has affected your entire life, there's a picture of your feet and frozen toes. You yourself say that: "The loss of several toes limits my sense of equilibrium". How did your sense of equilibrium change from that moment on, and how important is the use of one's toes for a mountain climber?

REINHOLD MESSNER When I speak of equilibrium, I'm speaking of inner equilibrium, which matures when a person is capable of knowing himself. I started climbing mountains as a child. At the age of five, I climbed my first 3,000 metres with my parents, which was possible because, as we lived in the mountains, I had been walking in forests and climbing rocks from the age of two or three. I needed to relate to balance to avoid falling. When I later became a rock-climber, whether or not the climbing was successful depended on balance, in its physical form.

When you're engaged in a hard climb on a mountain wall that drops straight down, balance is created with the flow. The flow means that a person almost soars, forgetting everything else. If I didn't forget the abyss below me, if I didn't forget my day-to-day concerns, I wouldn't be able to climb up so high.

When I climb a wall, even if it's only 1,000 metres high, all I have before me are two or three metres of wall, I can't see anything else. I try to move and as soon as I feel, by moving a foot or a hand, that I can't hold my balance, then I change my movement, I try to move with the other foot or with the other hand. This happens thousands of times a day: I try and fail and then succeed in it; forgetting everything that's outside, I climb the wall without thinking about my equilibrium, without thinking about the risk of falling, as if the rest of the world didn't exist. I achieve an inner and outer equilibrium, and I climb.

When I became aware of the fact that I wasn't climbing like before after the loss of some of my toes, that I was less keen on this activity, as soon as I realized that I would never again be able to climb to the same heights as before, I retired. I'm still doing some rock-climbing, I did

The interview was conducted in Bolzano, Messner Mountain Museum, 17 December 2013

some after 1970: forty years have gone by since I was a good rock-climber.

However, during this second part of my life I have tackled the highest mountains in the world, not just the ones that are 8,000 metres high, but the great mountains of America, Alaska, the Andes, the Antarctic, New Guinea, the Caucasus. In places such as these, equilibrium is something completely different: you climb with very thick shoes for insulation, you don't need to be in the best physical shape or have the best capacity for equilibrium. What counts is above all your passion and ability to withstand pain, because it's cold, and at high altitudes your head aches. What especially counts is your inner balance. This is what I had learned before the age of 25, and what I could use in this second phase of my life.

Afterwards, in the third phase of my life, I travelled to the biggest deserts: the Antarctic, Greenland. In those places I couldn't even fall. I could sit in the middle of the desert, or the Antarctic, and nothing would happen. However, being outside the civilized world was a much stronger feeling than being on a wall of the Dolomites. I needed to know how to be even calmer than before, despite the fact that I was in a world that was no longer human.

EMANUELE ENRIA   The Messner Mountain Museum, with its five complexes in five different locations in the Alps, tells the story of your relationship with mountains, as well as with the populations of India, Tibet, Nepal. Of that invisible world we're talking about, there would be no story without the memory that is preserved of it. How would you define your memory?

REINHOLD MESSNER   In those museums, I don't just tell the story of my life, but also of the relationship between man and the mountain. Having had the opportunity to climb many mountains and go on many expeditions, I was able to "get inside" those of others. To imagine, for example, what happened on Mount Cervino in 1865, because I've experienced similar emotions, the same as those of the alpinists before me. In this museum most of the objects can be touched, in the exact same way that I was able to get to know the mountain through the sense of touch: the crumbliness and moisture of a rock, both signs of danger. Thousands of moments lived that are etched in my experience and memory, the harder the problem that needed to be solved, the more deeply. In an emergency you don't need to be able to reason, reasoning is too slow. What really counts is instinct. In my opinion, instinct for man is just as important as intelligence when it comes to creating and surviving. We descend from animals and, over millions of years, we have had experiences thanks to our instinct, from which we then developed intelligence, language and communication. This doesn't mean that our animal dimension is no longer of any value.

EMANUELE ENRIA   In your adventures you came into contact with the civilizations of Tibet, Nepal, India. What did you learn from their approach to walking, both horizontally and vertically?

REINHOLD MESSNER   In Tibet, people don't climb mountains, they move about on the plateau, at a high altitude, up to 5,500 metres, taking their yaks and sheep with them. When these Tibetans leave for a week they use the expression kalipé, which means walking slowly. In Nepal, instead, they say namasté. In Germany, alpinists use a martial word, almost of victory. But this word gives me chills, so when I'm with someone in the mountains I use those of the Nepalese or the Tibetans.

There's a difference between mountaineers and alpinists. The former have been living in the mountains for thousands of years and have learned to survive there. They've gone as far up as where their flocks can still find food. But they haven't gone any farther. They didn't need to, and they respected the mountains: the glaciers, the flower beds, the rock walls, as if it were "the beyond" and could not be touched. It's a relationship with a divine dimension. Monotheist religions come from the mountains: Moses came down from Mount Sinai with the Ten Commandments, Mohammed meditated under a peak, in a cavern of Mount Hira, and said that it was there that he had received the vision of Allah; Buddha meditated on the Himalayas. These are all stories in which knowledge comes from above. This was also true of the Alps 3,000–4,000 years ago. Today, the approach to the mountain has changed, it is much more profane, more touristy.

I've become a hybrid, I'm an alpinist but I'm also a mountaineer. I live a self-sufficient life with my family in the mountains. I bought a derelict maso (mountain house), which was worthless, and succeeded in fixing it up. And together with other families, we lead this self-sufficient lifestyle, which is as important to me now as climbing mountains was before. Most people see me as being the mountain climber who climbed Mount Everest, who crossed the Antarctic. But these adventures mean less to me than the life of the mountaineer. When I spent time with the Tibetans, Sherpas, Nepalese, Bunzas, Banthus, all across the world, I had a simple relationship with them because I could understand them. We were on the same wavelength; I didn't look at them as though they were inferior to me, because the things they did, I did at home as well.

EMANUELE ENRIA   Everyone remembers your "manifes-

to", *The Murder of the Impossible*, the essay you wrote in 1968 against the abuse of artificial means by mountain climbers to climb even higher to the detriment of the understanding of one's natural limits. You've been trying to come to terms with this idea of a limit all your life. Where does the approach to verticality begin?

REINHOLD MESSNER It's very interesting to be able to talk today about the essay I wrote when I was just 22. The attitude of those who do mountain climbing nowadays has changed radically: everywhere young people speak of "no limits". My philosophy starts from an opposite vision. To my mind, mountain climbing has grown and evolved with just two words: possible and impossible. Limit does not mean that we go against all limits; rather, it means recognizing one's own limits. If I use all the equipment, all the tools that are available now, there are no limits in terms of height, because I can use a helicopter to climb up, I can build a cable-lift opposite a big wall. Only if I'm willing to cut down on the auxiliary aids can I understand the limits of my possibilities, without going further: this is the most important experience one can have in the mountains.

The mountain has nothing to share with heroism if man accepts the challenge. The heroic phase of mountain climbing took place during the Fascist era when the values of heroic mountain climbing were exploited. It is the exact opposite of what the mountain climber can get out of his activity: recognizing that nature is infinitely greater than us, even more unpredictable, and that in the man-nature game, man is the only one who can make mistakes, while nature never does. We don't set out to conquer a mountain, we go there to experience the mountain, to understand that we are small, imperfect and filled with fear. I'll tell you a story about something that happened in the Dolomites while we were climbing over the Pilastro di Mezzo of the Sass dla Crusc. At a certain point in our climbing we were trapped, 600 metres above the sloping mass of loose rocks, above the base of the wall: I was in front, my brother was more or less 40 metres below me, around us there were very few intermediate loops to slip the rope through, with very small pegs and bolts that might not have held up. I could neither go up nor down, nor to the left or the right, and after about a half hour maybe, I realized we had two choices: to jump, and at least have a chance, or both of us would fall off the wall. But jumping was risky too, because even if we had succeeded, there was the risk of falling into the abyss because my brother was hanging below me. So I decided to continue climbing, because the only thing that could happen was for us to fall. So in my attempt to climb upwards I had some hope of making it. Because of the emergency situation, I focused, mustering up more energy and equilibrium than I actually had, and I managed to pass. But once I made it I realized it had been a miracle. Until then I had always climbed as though I were the young Siegfried of the German legend, thinking: others die, while we who do the craziest, hardest things, don't. But now things were clear to me: all it took was a minor event and it would be fatal for us all. We needed to stay below the limits of what we had dared to do until then.

We can learn to come to terms with these limitations: instinct helps us to get out of most dangers in the mountains, and by so doing the mountain climber gets more experience. But if I instead use tricks to get around the difficulties, what will I learn? Nothing. I want to have an experience with myself and with the mountain where I see the ice, recognize the weather, the type of rock. And along this path, I'm interested in the psychological aspect, in understanding what man is made of.

EMANUELE ENRIA There's also the silence inside mountains. What is your relationship with these silences, and how has it changed since you started climbing solo?

REINHOLD MESSNER Solitude and silence for us Europeans are two very difficult dimensions, but we often experience them even in major expeditions. We don't talk because we can't. When you climb up to 8,000 metres you breathe faster and you don't want to talk to anyone else, but you understand each other without talking. But there's silence and there's solitude, because, in any case, there are two of us, not just one; it's a sum of solitudes that are easier to bear, a sum of silences.

As I get older, when I spend months by myself, I realize that, in truth, we are alone on this Earth, and that death is a solitary affair. We all die in solitude. These expeditions, like the one in the Gobi desert, are an example of what will happen when I die. I believe that dying is no more than entering a world where there is no sound, no space, no time.

REINHOLD MESSNER began to climb mountains when he was just five years old. Since 1969 he has embarked on more than a hundred journeys to the mountains and deserts of the whole world, and has told the story of his adventures in many books. He was the first person ever to climb all 14 "eight thousands", i.e. the world's mountains that exceed 8,000 metres. He has even crossed the Antarctic, Greenland, the Gobi and Taklamakan deserts. In 2011 he completed his project of the Messner Mountain Museum, dedicated to mountains and comprising five different locations.

WILL SELF, WALKING: PERCEIVING SPACE AND THE SENSE OF EQUILIBRIUM

EMANUELE ENRIA   In 2013, the London Underground celebrated 150 years since the first underground journey took place between Paddington and Farringdon on the Metropolitan Railway. The exhibition showed so clearly all the social changes that happened thanks to this new form of transport in the city. More people, women especially, could finally move easily from an area to another one, go to the centre or out of the centre. Before and after and through this invention, walking has always been a primary form of motion – but even walking was a social process. You call this relation between humans and technology *Man Machine Matrix*. How has the relation between walking and the technological changes in a city like London influenced the way of walking along the century?

WILL SELF   The beginning of the twentieth century was a very important period in terms of the impact of modernity and of technology on all aspects of human life. In that period a certain kind of space changed, and the way in which people moved around the world changed radically because of mechanized transport. The way in which people perceived the space within which they were moving changed as well. In the early nineteenth century, people took most journeys by walking. But very soon, the introduction of streetcar, underground and rail systems, and then of the car dramatically changed all that. People were no longer living in a city that was defined by their own physicality. If we observe the way in which people moved around before the introduction of mechanization, we can easily notice that it didn't matter if you were riding a horse, or if you were walking or if you were in a carriage pulled by horses: your muscles were still responding to movement, you were still physically involved and your body was still registering movement through space. Therefore, the entire environment was mediated by the muscular, because muscles were still involved in the way you felt it. Even when you are on a ship,

The interview was conducted in Florence, Biblioteca Nazionale Centrale di Firenze, 10 January 2014

or a boat, your body is registering the motion of the water. What we need to understand about the cities that we inhabit is that they were conceived in relation to our physicality.

Coming to the twentieth century, we see that architecture began to be defined by what the machine is capable of. This is what I call *Man Machine Matrix*. And as a consequence, our very perception of space started to shift in such a radical way that space itself seemed to not be the same anymore. It was no longer necessary that people walked the kind of distances they would have needed before: this inevitably introduced a significant alteration in the perception of their bodies in their environment. And perhaps most significantly of all, an alteration in the way they thought.

Let's go back to Jean-Jacques Rousseau: he said that people think at their walking pace, meaning that the natural speed of thought in some way relates to walking. We think at three miles per hour. This is something that all sorts of preindustrial cultures share somehow: the idea of the pace, the 4/4 beat of walking. And we can find it in poetry as well, in all of English literature. Probably the most significant and influential poet of the nineteenth century, William Wordsworth, was a walking poet. His metre and his scansion, the very rhythm of his verses are determined by his walking pace. His first great walk was actually a walk to Jean-Jacques Rousseau's grave, from England to France. So even in that period walking was perceived as something different, something strange. In a way, Wordsworth's long walk represents the last great actualization of the act of walking before it was lost, in some sense.

EMANUELE ENRIA   In this approach to walk you mention Psychogeography, which eventually became a book and a column in *The Independent*. Historically, this word refers to the French Situationists at the beginning of the twentieth century and to figures like Guy Debord. Previously, a writer like Honoré de Balzac, drawing his huge project of the *Comédie Humaine*, introduced the description of how people used to walk and move. Then the figure of the *flâneur* appeared, masterfully described by Baudelaire and Walter Benjamin. What were your references to design your approach to walk the way you have been doing up to now?

WILL SELF   Psychogeography exists now as a practice, it is a way of engaging with the contemporary intricacy, a way of doing several things in relation to the contemporary city. It is a way of making the city discoverable and understandable: so it affects thought. It is a way of physically coping with the reality of the contemporary city, which is a toxic, poisoning environment. And it is a way of affecting emotional release and physiological well-being within that kind of context. But psychogeography is also a political act. It is an act against a certain kind of construction of space. Against the way we are required to live in space, specifically because of commercial imperatives. So it is a practice, it is something that you do, and it is firmly connected to walking. Let's go back to England, to Wordsworth, and to Thomas de Quincey, particularly to his *Confessions of an Opium Eater*. In the early nineteenth century London was inhabited by about half a million people – it had reached such a size that it was possible for one to be completely anonymous in it. It was possible to move through the city with no real chance of seeing anybody you knew or being recognized. This was a key psychological moment in its development, and it's related to the physical and the body: it was the moment in which people became alienated from their environment. In a very profound way. The alienation from the environment is not an alienation from the forces of God or of Nature, it's an alienation produced by human beings when they became a mass within the city.

Then came the French, specifically Baudelaire and his *Painter of Modern Life*. Baudelaire defined the concept of the *flâneur* – an artist who, through moving around the anonymous city, around the city of the mass, is able to grasp the spirit of his time. And this is just because of his free floating movement, because of his lack of confinement within one specific goal or one specific social construction. If we go back, for example, to the mediaeval city, people might move around the city but they always belonged to a profession, or a guild, or a class. The *flâneur* is free of all of this, he enjoys an idealized point of view within the city that cuts across specific commercial and class interests. Later, this idea was taken up by the Surrealists, particularly by Louis Aragon with *Peasant of Paris* and by André Breton with *Nadja*; they took the idea further and defined the psychology of the *flâneur*. More than Baudelaire, they were responding to the introduction of mechanized transport that was taking place at the time; they both appreciated the impact of mechanized transport because they thought it enabled people to take up particular kinds of perspectives, but they were greatly concerned with walking, since walking places the body in opposition to the mechanical. Both Aragon and Breton were interested in what they called, with a French word, the *frisson*, a sort of shiver, a physical response to instants of perception within the city. It is walking, it is moving around on foot that allows this perception to rise, as opposed to the growing alienation created by

mechanization. So, we might say that the *flâneur* was a collector of *frisson*.

The *frisson* was picked up by Guy Debord and Raoul Vaneigem and translated into Situationism, a development of Surrealist ideas shifted on a political plane. According to Debord, the city has become completely fractioned by the *Man Machine Matrix*. We are no longer able to experience the city except through commercial imperatives. We travel the city to work, to buy something, to be entertained: every perspective that you have along the way is monetized, has an economic value. Therefore, you cannot see your urban environment naturally for what it is. The Situationists conceived this idea of the *derive*, the drift, the aimless progress through the city, which can be undertaken in a mechanized form of transport – sometimes they used taxis or underground – but mostly is done by walking. They walked through the city in an aimless way in order to set free from what Debord called the "spectacle". It was a way to escape from this confinement, a way to actually attack the very idea of the modern city.

Now, readers might think: well, this is crazy! I walk around the city, I am not obsessed with money, often I look at things, I see the city for hours a year. Actually, they should stop and consider: do I really allow myself to feel the possibility of complete aimlessness, of being decoupled from all the concerns and responsibilities and objectives that I have in this urban context? Do I really know where I am, or have I been told? Do I really understand the relation between the different physical features that underlie this city – its hills, its rivers, its environmental peculiarities, its unique flora and fauna, and the building themselves? Do I really understand how these two levels relate to each oth-

er? Could I get from the city to the country? Say, would I be able to physically walk from the city to the country?

The first time I walked from the centre of London, where I live, to the country, I wasn't even sure, at a psychological level, whether this was possible. Whether it could be done. Why was I not sure that it was possible? Because I knew nobody who had ever done it. And nobody had told me that they had walked, that they had used their physical body to traverse the distance between the centre of London and the green fields outside. So I didn't know whether it was possible. That gives you an indication of how decoupled, disconnected we are from the physical reality of the places where we live.

London is a large city, perhaps forty or fifty kilometres from the sea. There is a river that runs through and reaches the sea, about 50 kilometres away. What I did, one day in 1980, was to get into my car and drive to the end of the river, just to see what it looked like. In a way, this was the very beginning of the psychogeography that I do, which is a search for the physical reality of where we are, in order to understanding where we are. What I do is different from what Debord and the Situationists did, in that I am interested in teasing the *Man Machine Matrix*. I am interested in satirizing the way we live now, making fun of it through my walks. That's why I started what I call "airports walks".

I generally walk from my house in central London to one of London's airports, I fly to another city and then walk from the airport into the city centre. So, it is as if I had walked between the two cities. Usually, when it's a long fly, I walk on two consecutive days, sleeping on the plane; at other times, if the destination is in Europe, I would do it in one day. What I discovered during my first airport walk

is that our bodies and our feet and our legs understand distance in a different way from our heads. So, if you walk for a day, and then fly, and then walk for another day, your body tells you: this must be only one place, because I have walked for two days. And your head tells you: no, we flew 3,000, 4,000 kilometres in the middle, we were on the plane. And your body says: no, I think I know about this since I am the one who walks.

The first time I did this, I walked from my home in London to Heathrow airport, flew to JFK airport, then I walked from JFK to Manhattan, and when I reached Manhattan it felt as if the Thames and the Hudson River had been joined together. Long Island had been pushed inside London: so, by the simple gesture of walking, I joined the two countries together. The Atlantic had disappeared. I became intoxicated by this experience and started to do it more and more.

Throughout the years of these airport walks, this phenomenon of contemporary city dwellers who do not really know and understand where they are has increased exponentially because of the introduction of Global Positioning Satellite technology. A tiny little device in their pocket gives people their exact location in space; this way, you might think that thanks to the GPS people finally know where they are – but actually it is not so, and we are frighteningly near to the Debordian "spectacle", since the small thing hidden in your pocket is a commercial object used for commercial purposes. Shortly, when you know where you are, other people know where you are within the context of a set of commercial imperatives. The crucial point, alas, is that you still *don't know* where you are. What you find in conversations with ordinary people is that they rely on Global Positioning to tell them

where to go since they have become unable to find their way based only on the orientation of their own body. They are told electronically and digitally about their position, and they have less and less understanding about where they are.

The most interesting example of how this technology impacts on orientation comes from taxi drivers. A licensed London taxi driver, in those famous black cabs, had to know the name of every single street, and every single notable building within a ten-kilometre radius of Charing Cross, in the centre of London. So you take these 10 kilometres around and they knew everything, it was a sort of map in their mind. A group of neurophysiologists, some years ago, studied the brains of London taxi drivers and discovered that their posterior hippocampus was bigger than that of ordinary people. The part of the brain responsible for memory and orientation was three or four times bigger just because of this constant and intense mental mapping. But now when you arrive at London's airport and call a taxi, you see that they all have a Global Positioning Satellite. They are losing their own mental mapping and starting to rely on it. Taxi drivers studied as long as two or three years to learn this information, and now they are throwing it away. That's an extreme example, but actually this is what's happening with all of us. And if you're getting used to GPS and are sensitive to your body, you will soon begin to understand how you are losing your ability to orient yourself properly in the surrounding space.

EMANUELE ENRIA You use to say that we should feel the landscape's contours as we might those of a body. What do you mean by that?

WILL SELF We can think about how we understand where we are in a city. There's a kind of equilib-rium. There is an equilibrium between an understanding of the plan, the map of the city, and our actual innate sense of direction, which is a natural sense (i.e. where north is) or orientation within a space, an understanding of distance, which is our exteroception (exteroceptors are scattered on our bodies and are correlated to our peripheral sensibility, both subjective and objective). So, when we look at our hand travelling away from our face, or when we perceive how far away things are, well, this is an extension of our own body, in a natural sense, it is our ability to project things into the environment. As we move about the city, we should – if we are autonomous, and if we are walking as an act of will – constantly be updating our sense of direction or orientation and our sense of distance, since we are physically involved. We are just as physically involved (and aiming and striving for equilibrium within the urban context) as a dancer is in a dance. Dance can be regarded as a stylization of this process we are all engaged in.

But now let's look at the way we actually move in the city, particularly, in cities we are not familiar with. The sense of disorientation we feel when we are arrive in a foreign city is especially deep: we arrive by plane, we are probably jetlagged and experience that awkward loss of equilibrium, like when you come out of the cinema. And we take a taxi. Now, taking a taxi, I believe, is a mistake. What happens when you take a taxi in a foreign city is: you are hiring someone else's equilibrium. You are hiring their orientation. You are hiring their understanding of the cultural and social milieu of this new environment. You think you are just asking them to take you from the airport to a hotel or to a meeting, but actually you are doing much more than that. You are engaged in a profound loss and you don't even attempt to understand. All you have to do when you arrive in a foreign city, in order to feel more comfortable, is take the bus: because even the act of taking the bus, since it won't take you precisely where you need to go, will force you to develop your own autonomy. And to begin to orient yourself within the city. And recognize what's out there.

Let's go back to airports: airports are actually designed to be dull. At times, they reveal little moments of architectural *jeu d'esprit*, but everything is mostly contextualized by the ambulatory movement of travellers, which are kind of exaggerations themselves of the act of walking. The airport is, in a sense, the purest articulation of Guy Debord's society of the spectacle. It is a commercial free fair. Let's refer back to Walter Benjamin's writings on the Parisian arcades and the movement of the walker through them: well, airports are the radical expansion of shopping arcades – so radical that they are seamlessly joined together by the similarly neutral experience of flight. Yes, because when you fly, everything is designed to desensitize your experience, everything is done to make you feel the environment of the plane neutral, dull, centred on a kind of axis… For instance, the introduction of the screen in the seat in front of you for international flights. Perfect. Absolutely perfect. And your seat: the tiny commercial space that you have rented for these few hours has been turned into a simulacrum of your domestic entertainment or work environment. So, now, it is possible for people to move from a situation in which they are working in front of the screen to one in which they are travelling in front of the screen to one in which they are experiencing leisure in front of the screen. Furthermore, within these separate, discrete screen-envi-

ronments there are usually mass trans systems in which you can see the world and the itinerary of your flight through the screen.

So you can see that all of this effort – architecture, engineering systems – has been designed to reduce the problem of disorientation that is caused by mass international rapid transport. We have lost our equilibrium and all this technology is designed to create a new form of equilibrium. But this time it is an equilibrium mediated by the *Man Machine Matrix*.

EMANUELE ENRIA How do you think we can naturally preserve our own capacity to perceive space and equilibrium?

WILL SELF Lévi Strauss, the French anthropologist, said that distortion in scale, in other words the relative size of objects, always sacrifices the physical in favour of the intelligible. What he means by that is once you start playing around with the sizes of things, you alter representation and its awareness, you do it in order to comprehend things more, to understand them. But what you lose is their physical immediacy. The clearest example of this is a map. A map is an alteration in scale of an object. Why do you do it? You do it in order to understand how to get around the object itself. But, using the map, you lose the physical immediacy of the object. We all know this experience: we are walking around an unfamiliar city, we are looking constantly on the map, we are inside the map, we are not in the territory but in the map. With the use of GPS technology, all this gets even more extreme. What happens when we walk? What happens when we walk in these environments that are specifically head and eye designed? The walking automatically improves your orientation, improves your physical equilibrium, puts you back in touch with the

physicality of the environment. The classic example for that is Los Angeles. Everybody says: You can't walk in Los Angeles. It's impossible to walk in Los Angeles. Nobody walks in Los Angeles. Ok. A few years ago, I walked from my house in London to the airport, I flew to LA, and I walked for eight days in Los Angeles. Right around the city. And then I got back on the plane. I was in Los Angeles for eight days without ever getting in a car or any other form of wheeled vehicle. Just walking. It's really easy to walk in Los Angeles. One of the most walker-friendly cities in the world. Why? Various reasons: the weather is good, it's very comfortable, sidewalks everywhere. And the architecture, unlike European suburbia, is highly individualistic: on the whole, through Los Angeles' suburbs each building tends to be different from the others. So, it isn't exhausting for the eye. Why then the common perception that it is impossible to walk in Los Angeles? Because of the cultural construct of the car. Because people believe in the car, and wishing to believe in the necessity of the car they have to deny the possibility of being able to walk. I'm sure that most people in the West are actually in a condition of addiction to the car in particular, which has the most significant impact on their inability to get in a natural equilibrium. Most people use cars to always take the same journey, over and over again. If you could analyse it, probably more than 90 per cent of journeys on motor vehicles are repetitive journeys.

How does walking differ from these technologies I have discussed? Unlike screen-based technologies, walking is 360 degrees. Unlike screen-based technologies, walking is in real time. It is impossible to edit the experience of a real walk. You can only endure it. And, of course, it is completely

mediated by the body. Any movement away from equilibrium or into equilibrium is sensed profoundly. All movement through space, all movement through time, is experienced at the most profound physiological level. In most cases, urban environments are not designed to be walked any more. In this sense I was saying that Los Angeles is quite a walker-friendly city but people feel that walking cities should be different.

In the virtual world, neurophysiologists have studied the impact of receiving a new text message, or an email, or moving on to a new website: each of these introduces a little rush of adrenaline into the body. Each of these creates a stimulus response. So, if you like, if you want to think of this in a virtual working environment and within the *Man Machine Matrix*, we are constantly, at a metabolic level, moving out of equilibrium and back into equilibrium, as our physiology readjusts. So, we are in this constant stimulus. But this is different – because it is static – from how we must be in an environment that is just physically mediated. We are throwing back on our level of perception. This is why walking is such a powerful tool, and why people fear it so much. When I speak to people about my airport walks, they always say the same two things to me – always, always the same two things. What do you do with your luggage? And, in the case of London, did you walk along the motorway? Always the same thing. So, why do they say these things? Who would be so foolish to carry a big suitcase for 25 to 30 kilometres? Or pull behind him one on those little wheels? You would have to be an idiot to do such a thing! So, the question of what do you do with your luggage is a question that doesn't need to be asked. Clearly, if I do that sort of thing, I don't carry very much luggage with

me... well, that's it, that's the answer. So what is the question about? The question is about the person who asks it. What the person who is asking is trying to say is: I am incapable of moving around the world without a little house of my own that I take with me. That is a simulacrum of where I am. Everybody travels with quite a lot of luggage. Why? It's to achieve equilibrium. The luggage is balance. As people move away from their domestic environment, they feel unanchored, they feel a loss of equilibrium. They feel they are rocking about in the world. The suitcase is a counterbalance. Against the weightlessness of their existence away from their own context. So, the question about luggage is a question about how to cope with the loss of equilibrium: psychologically, my luggage will retain me as a parachute in equilibrium.

Second question: did you walk along the motorway to the airport? Again, not a question that needs to be asked. In the case of London, it's maybe 25 to 30 kilometres from the centre of London to the airport. First of all, it is illegal to walk on the motorway; secondly, it would be very dangerous: your average life expectancy on the side of a motorway is 17 minutes! So, again, this is not a question that needs to be asked. Why is it asked? The question says: I don't know where the airport is. It's not a question, it's a statement. The question is a statement about the ignorance of where they are. They are saying: I do not know where I am. And I do not know how to get to the airport. I would only know how to get to the airport in a vehicular means of transport. This gives you the indication of how radical it can be to walk. What a radical, what a revolutionary act it is.

You are reintroducing the idea of physical equilibrium into the environment that most rejects it. Yes, it's fine to go often to the beautiful hills, and walk around them – that's allowed. But to walk to the airport is to attack the Matrix at its core.

WILL SELF was born in London. A novelist and journalist, he writes for *The Independent*. His column *Psycho-Geography*, describing his famous "airport walks", has been turned into a book. Italian editions have been published of the following books by this author: *Cock and Bull*, *Great Apes*, *Dorian*, *The Butt*, *The Quantity Theory of Insanity*, *Dr. Mukti and Other Tales of Woe*, *London. Appunti da una metropoli*, and *Umbrella*.

ELEONORA ABBAGNATO, ON TIPTOE

The interview was conducted in Milan, Milan Triennale, 24 February 2014

EMANUELE ENRIA When you were eight you were already dancing on your toes. A dancer experiences the effort of having to practice day in and day out, of shaping the foot and the body. What can the feet of a ballet dancer tell us? And how does one learn to dance on one's toes?

ELEONORA ABBAGNATO For a ballet dancer, the relationship with one's foot is of crucial importance. I still remember the first time I put on pointe shoes. Perhaps I was too young to do so, but I insisted with my teacher. I remember feeling terrible pain at first, but I also remember an amazing sensation: because the foot is held tight inside the shoe, and you feel as though your whole body is being elongated. In the end, it's easier to stay *en pointe* than with your feet flat on the ground. In time, with constant exercise and the help of the choreographers, you learn to keep working on your foot. It's very important for a dancer's equilibrium. For a dancer, the foot is a crucial part of the body: and sometimes it becomes an obsession. When my daughter was born, the first thing I did was to check her arch! I flexed her feet. I was already doing the same thing with my cousins when I was very young. It was a game. I learned to pay attention to the arch at a very young age. That's where you see a ballet dancer's character.

And when I teach I end up instinctively looking at my pupils' arches. It would seem that if you don't have an arched foot you can't dance, but that's not true. Actually, with exercise, you become much stronger on pointe shoes, you learn to work with your toes. I much prefer a foot that's been worked, that's muscular, to a natural foot. A ballet dancer's character can be very different starting from this very detail: because when you have a very pronounced arch, you have a hard time staying in balance on your toes. Instead, if you learn to use all your toes when you're young, in other words, besides using your arch, this will be of use to you in the future. It will enable you to obtain certain variations in your repertoire of movements *en pointe*.

I really like ballet dancers' feet. Of course, they're not so beautiful to see: they inevitably have calluses, blisters, all the signs of routine work, but you get used to it. Your feet might bleed even when you're still a young dancer, and this still happens to me when I'm dancing an important role, when

I wear my pointe shoes all day long. But today we know how to take better care of our feet. They used to say that in the olden days ballet dancers would roll cutlets around their toes to relieve the pain. Now we have double layers of leather that give us some relief from the burning and pain. Constant exercise on your feet is marvellous. When you start to work out, the first parts of the body you warm up are the arches, the joints, the ankles, the toes. Choreographers like Pina Bausch teach you how to work your toes on the ground. In contemporary dance, working in close contact with the ground is an amazing feeling because everything starts with the feet.

EMANUELE ENRIA What do your ballet slippers mean to you?

ELEONORA ABBAGNATO Ballet shoes are our most important tool. When we set off for our performances, our tours, we have thousands of them in our suitcases, and, along with the ballet shoes, everything else that's needed, because every ballet dancer adapts her ballet shoes to the shape of her foot and to the strength of her ankle. I, for instance, use a cutter to make a slit under the sole, to break them in half. They have to

be soft but, at the same time, the plaster has to be strong because if it's squishy it bothers me. And then I use a special varnish inside the shoes. I always have it with me. And I use elastics for the ribbons because I have problems with my tendons. Lastly, the shape is important: I always want my ballet shoes to fit my feet perfectly – that's why I soften them up a lot. I never put on new ballet shoes before going on stage. It would be too hard to use them, technically speaking. You have to remember that the pointe is made of plaster, so it's hard, and not so easy to stand on.

EMANUELE ENRIA As compared with the early twentieth century, the way of walking has changed. Nothing more than dance has succeeded in telling us that this is so: from classical ballet to the Ballets Russes and the first avant-gardes, with Isadora Duncan and Martha Graham, who encouraged a return to dancing with your feet on the ground; they were followed by Pina Bausch's famous walks and the choreographies of Roland Petit. There's a transformation in the concept of equilibrium in modern and contemporary dance. Can you tell us something about your rela-

tionship with these different equilibriums that have influenced the history of dance?

ELEONORA ABBAGNATO In my career I have met the greatest contemporary masters. My encounter with Pina Bausch was especially important. Her dance theatre calls into play the relationship with equilibrium and with theatrical freedom: the classical ballet dancer is transformed, stimulated to be herself, to express all that she can on the stage. In Bausch's *Rite of Spring* the relationship with the ground is of crucial importance; indeed, the performance begins with the dancers on the ground, so that they can use their hands, face, feet to become familiar with it and the surrounding space. In this particular ballet, there's a very difficult solo dance, Electa's piece for the finale, which is completely danced on the ground. It's very hard to keep your balance in that condition. But it was the strong point of the performance and of the role itself, the exploration of the animal in a woman. We would go on stage free and happy to give it our all. Meeting Pina Bausch gave me confidence, I was only eighteen and had just entered the Paris Opéra's *corps de ballet*. Bausch chose the dancers herself, for each of the roles, even if they were very, very young. She had in fact chosen me at first for the part of the young, innocent girl; after ten years of dancing that part she called me for Electa, the strongest of them all. It was a magnificent experience, which I will remember forever. Not to mention the balance, the equilibrium it gave us: because in any case every morning we would work wearing our ballet shoes, and we'd still do our classical ballet lessons. The rest of the day I would find myself dancing barefoot on linoleum or on the bare ground. This is what's really interesting about it: you completely change the weight of your body and your sense of balance. I would also like to mention my experience with Forsythe – someone who was totally

inspired by classical ballet, by Balanchine. The first ballet I danced by Forsythe was *Vertiginous*, a tribute to Balanchine. Forsythe has brought classical ballet to an extreme and continues to do so today. We worked together at the Opéra, in a piece called *Pass/Parts*, and for the first two months of the workshop we made every effort to absorb his style, which is based on the extreme violence of the movement, an almost bestial energy. With Forsythe, you don't lift your leg, you stretch it as much as you can. That's where the power and the energy that you see in his choreographies on the stage lie. When you work with him you immediately understand what he wants. He works on all the diagonals, starting from the centre and moving towards the head and feet according to the Laban method, which inspired him. He leads you to explore the décalé, which is the word we use for it in dance: what it means is not remaining on the weight of the body, but outside it instead. Everything is outside the body. When you work with Forsythe everything is extreme. This is really hard because you can end up pulling a muscle. But every time we've worked with the force and confidence he transmits to you we've never injured ourselves. I recently played the part of Nureyev's *Sleeping Beauty*, for instance, which is completely based on balance. The first act, *Adage of the Rose*, is wholly an equilibrium, based on the centre of the body. While dancing, I kept thinking that I wasn't bringing my body to an extreme, and I was thinking about what Forsythe had taught me. I was thinking about the modernity Forsythe has brought to classical ballet and kept telling myself that my leg should go farther up, so I would lift it even more. My experience with Martha Graham's technique didn't last long. I danced one of the parts she created when I was very young. And that's another, completely different style of dancing. It's all based on contraction and release, on very particular movements. It was interesting precisely because I was young, and since I wasn't familiar with my body yet, it made me aware of new movements.

EMANUELE ENRIA In the late nineteenth century Honoré de Balzac added a story about how people walk to his *Comédie Humaine*. Eadweard Muybridge did just the same thing with photography, which anticipated the observation of the body in movement before the movie camera. What is your relationship with walking?

ELEONORA ABBAGNATO We ballet dancers are always very self-critical, and we have a very special way of walking. Even today my friends say: I saw a young woman waddling like a duck, she must be a ballet dancer. I don't walk with my feet turned out, but I do have very straight posture, sometimes too much so. But I think that walking is what makes a woman beautiful and elegant. We ballet dancers have fun watching other women walking. Shoes unquestionably influence the way you walk, even if all ballet dancers walk the same way. Models have a way of walking that's different from that of ordinary people. I don't find walking with one foot in front of the other, moving one's hips that much, or else lifting one's feet so high to be especially attractive. That's why I have so much fun when I go to fashion shows. I enjoy seeing how every model has a different way of walking. But maybe models make fun of ballet dancers, because we walk too straight and stiff when we're out on the street!

EMANUELE ENRIA Of course, the magical moment in your career was when you were nominated étoile at the Paris Opéra. Who were your masters, and how would you sum up the differences between the French and the Russian schools?

ELEONORA ABBAGNATO I was nominated étoile last March, about a year ago. I danced in the *Carmen* with Nicolas Le Riche, a great ballet dancer who is now leaving stage. I really wanted to be nominated étoile, and I expected it to happen when I danced the part of Carmen, because Roland Petit really made me work hard. He wanted the utmost perfection in that role. The Carmen I watched when I was young was that of Baryshnikov and Zizi Jeanmaire, who inspired me. The part was supposed to be that of a strong woman, rather like myself, although not so much on the outside. I have blonde hair, I may look angelic, but I'm not. I have always seen Carmen wearing a black wig, which Petit preferred because he doesn't like blondes. Also on the evening of my nomination he kind of "modified" me to give me more character. It was an important day for me and for my career with the Paris Opéra, where I had arrived at the age of fourteen to study at the ballet school, and after many years was dancing the part of the étoile anyway. In my career I have had the opportunity to dance with the Russian theatre, in Balanchine's *Jewels*. It was really plunged into the history of classical ballet, I felt like I was experiencing the days of the true classical ballet dancer. Their school goes way back, they're marvellous, but only in the classical repertoire. We're more advanced in the modern repertoire, and we're more physical, more accustomed to this dance style.

ELEONORA ABBAGNATO, a native of Palermo, began to study dance when she was just four. At the age of fourteen she was admitted to the École de Danse at the Paris Opéra, directed by Claude Bessy. In 1996 she entered the *corps de ballet*; in 2001 she became *prima ballerina* and in 2013 étoile. She has worked with such world-renowned classical and contemporary choreographers as Roland Petit, Pina Bausch and William Forsythe.

PHILIPPE PETIT, MIND AND BODY IN EQUILIBRIUM ON THE WIRE

The interview was conducted in New York, Cathedral Church of Saint John the Divine, 4 April 2014

EMANUELE ENRIA The first step is like a signature. I really like the way you describe this in one of your books: "I place my left foot on the steel rope. The weight of my body rests on my right leg, anchored to the flank of the building. I still belong to the material world. Should I ever so slightly shift the weight of my body to the left, my right leg will be unburdened, my right foot will freely meet the wire. On one side, the mass of a mountain. A life I know. On the other, the universe of the clouds, so full of the unknown that it seems empty to us" . Can you tell us something more about that first step?

PHILIPPE PETIT The first step for me, as a wire-walker, is a point of no return. At the World Trade Center, for example, I had put one foot on the wire and my other foot was on the building. And then I had to shift my weight. This is really the essence of balance: to shift my weight so my right foot will become light and will come to the wire. So, on the high wire, the first step, while holding the balancing pole and detaching myself from the earth, is very important physically, psychologically, and poetically. Physically, because I detach myself, I am balancing on the wire. So I have to be completely focused. Psychologically, because I have to make sure, before I take the first step, that I know I will take the last step. Successfully. I have to make sure that I am going to carry my life across. I never gamble my life – I love it too much. Poetically, then, because I am a human being but when I detach myself after the first step, I become a bird. Half man, half bird. In life, generally speaking, the first step, the first time you do something or the first step of a journey, of a project, is very important. It's a decision. So, when I teach students or when I talk in my lectures, I very often take the image of the first step, which you can translate into anything you are doing.

EMANUELE ENRIA When, in the early twentieth century, the Ballets Russes were brought to Paris, this was an extraordinary moment for visual artists like Picasso, composers like Stravinsky, writers like Cocteau and dancers like Nijinsky, who found inspiration in dancing and used it as a means of working together. For the spectacle *Parade*, Picasso designed those figures of a clown and tight-rope walker that also feature in many of his paintings. In your view, what is the

tight-rope walker's role in the contemporary world?

PHILIPPE PETIT It's had a crucial role in the nineteenth and twentieth centuries. Just think of Jean Cocteau, the Ballets Russes, just think of Jean Genet and the many philosophers, painters and writers who were inspired by balance. And by the figure of the wire-walker. And since I am a wire-walker, it is a bit difficult to say what a wire-walker does to the century in which we are living. But there is one thing to be said: that almost all the time, with very rare exceptions, the wire-walker comes from the circus. And the circus has a very big tradition, I love the circus, but sometimes I find that this tradition of costumes and ways of moving, the choreography, the wire-walker, the funambule, are quite repetitive. And I consider myself almost lucky, since I learned by myself and I had to fight how to learn to balance, and then I began writing books about it, starting with *Le traité du funambulisme*.

EMANUELE ENRIA Is it true that you performed on the occasion of Pablo Picasso's ninetieth birthday?

PHILIPPE PETIT Actually, I met Picasso on the occasion of his ninetieth birthday. He invited me to perform on the wire on the little square of Vallauris, his village. And he told me, after the performance, that I reminded him of the acrobats he painted during his Pink Period. He was also in love with the circus, and the travelling gypsies, the people who moved around with their arts. It was a great honour, for me, to be able to perform for him. When there is such a juxtaposition of my world and the world of a great artist, well, it is really very important to me.

EMANUELE ENRIA I find there is a touch of craftsmanship to your work: the choice of the material to use for the rope, to which you have also devoted several pages of your treatise. Would you tell us about your relationship with materials?

PHILIPPE PETIT All my life I have been much more of a craftsman than an artist. I started as a craftsman, building things, doing things with my hands, experimenting with materials. Acquiring tools, so that I could become an artist. So, as a high wire-walker, the first tool you need is a wire, and at the very beginning I didn't know what kind of wire to get. I knew I had to acquire some engineering knowledge and learned everything about the wire: how to tension it, how to position and stabilize it. It's a rigging science, the wire. Now, when the wire is ready, how do I know if it's a good wire to walk on? I am not just going to walk on it, I have to see and check everything. So, what I do is touch and feel it first, of course. And I pull it, to see the vibration; sometimes I bang on it with a bar of metal and I hear its music. Because a cable is like a cable of a musical instrument. Like the guitar, the piano. Or like the harp. So I kick in the wire, I hear the sound, maybe I tighten it a little bit more, like what you do on the guitar or on the violin. Until I have the right sound. So you can tell that many of my senses are getting together to choose the right wire. Once I walk on that wire, all my senses are growing tenfold. For example, my vision: when I am on the wire, I can see almost behind me. When I am walking on a wire the audience is on the side; so, if I see, for example, out of the corner of my eye a dog coming out, I know I am going to see a dog crossing me and I should not lose my balance. While on the wire, I have a very strong sense of smell. Then touch, of course: my feet on the wire and my hands on the balancing pole. I can feel the vibration, I can feel the place where the wire is tied better and then where it's a little bit soft. So, if I need to do something like a salute, I will not do it in a place where the wire is not very strong. So the

sense of touch is extremely important. And then people will ask: OK the sense of touch, but what about taste? The tongue has nothing to do with being on the high wire. Well, actually that is wrong, because when I am on the wire I swallow the air – it's a strange feeling. I taste the air and I can tell if it has a lot of humidity. Once I felt there was going to be a storm, just by tasting the air. Then, I actually went a little bit faster to arrive, it was a long crossing, and the minute I arrived it started pouring down. Actually the greatest wire-walkers should put all their senses to the service of the walk. It's the same thing as if you were a dancer, or a painter. As I said before, starting from your craftsman abilities, you have to put all your senses at the service of the project. Jean Genet, in one of his books, talks about a rope dancer – not a high wire-walker but dancer – who would caress the wire before performing on it, like a human being, a friend, a lover. So, I think the relationship between the craftsman artist and the medium is something that should be taken very seriously and poetically.

EMANUELE ENRIA You have often stated that when walking the wire, the body follows the mind. The mind must know that it will make it. How, then, do mind and body interact when you are balancing on the wire?

PHILIPPE PETIT In my life I have chosen to be a high wire-walker, but I do many other things – I write, I do juggling and magic, but, anyway, my main profession is to walk on the wire. I love that action because you can only walk on the wire if there is a perfect balance not only of you *on* the wire, but balance between you *and* the wire. And, to achieve that balance, it takes years and years of training, it is a matter of knowing how not to impose yourself on the cable. To me, the cable is a living entity:

so, if I impose myself I am going to create vibrations that are not very beautiful. But, if I listen to the wire, if I become friends with it, I am going to listen to the wire breathing. If I manage to go into the rhythm, the breathing of the wire, then my walk is going to look effortless, like a yummy bird. I have spent my life studying this relationship between me and the wire. It's basically half physical and half poetic, half mental. This relationship is like a love story, in a way. First, I don't allow anyone to walk on my wire, or even touch it – it's precious to me. And then I take a long time to prepare the installation of the wire, a long time to make sure that everything is perfect. The wire has to be perfectly straight. So, I became a madman of details, to make everything perfect!

EMANUELE ENRIA In the speeches before your performances you often praise the god of creativity. What do you mean by creativity?

PHILIPPE PETIT Besides being a high wire-walker, I am a writer. This year I have written my tenth book, and the title of that book is *Creativity, the Perfect Crime*. This book is not about creativity in general, it's about *my* creativity. Basically my creativity has a lot to do with intuition, passion, tenacity, being in love, focused. And then many other notions, like surprising yourself, having fun while working. Creativity is a bit of a common denominator of all that I do. Because what is the sense, for anyone, if you repeat yourself all the time… You should never be comfortable in creativity, you should always metaphorically put yourself on the high wire. So, to me, walking the wire and being creative are two sides of the same coin.

From the very beginning, when I started to learn to walk the wire, I was already pretty different from circus people. Because circus people don't walk; they do all kinds of tricks, crazy exercises (and I learned all that, I was able to do it). But, since I was little more than a kid, I wanted to go in the direction of purity. I wanted to impress upon the audience the beauty of simplicity, the impression that it is jolly easy to walk on the high wire. So, I started to learn how to walk. Everybody walks, sure, but there should be a school of walking because people don't know how to do it. Well, on the wire it's very important, because it will make your walk very beautiful. So, first, I found out that when I am walking I need to have my weight on the back leg so that the leg I put on the wire is free and light. And when my leg is light, I can start gliding my foot: my toes first, then my heel. I have tiny little shoes, extremely thin (they are made of buffalo leather), that allow me to feel the wire and enhance my feet's sense of touch. I don't look down, of course, I look horizontal, like a dancer, sometimes I even look up.

First of all I place my toes on the wire and feel if everything is in the right place. Then, I start gliding my feet. When the back of my foot, the heel, is touching the wire, at that moment, I move the weight to the first foot; the back leg is light, and it automatically repeats exactly the same process. In the circus they often use resin or some kind of glue, so that the foot finds a better adherence to the wire. For me, it is the opposite. A few times I even tried to put oil on my slippers to glide better. Now, this will take millions and millions of hours, practicing, which I have done all my life. And then all these beautiful steps are in my system and my nervous system recognizes them as its own. So, walking is a mystery that most people do not understand; that's why it's so interesting to learn new ways of walking on the high wire.

EMANUELE ENRIA From the first step onwards, when do you start to feel that you are approaching the point of arrival?

PHILIPPE PETIT People keep asking me: how do you feel up there? Or, very often, how did you feel in the World Trade Center walk? On the pictures of me on the wire you can see a big smile. It's very important to feel this way. You should not feel fear. When people ask me whether or not I am scared while I'm on the wire, I invariably answer that I am not. I am certain I will get to the other side. And then, of course, you have the feeling of the first step, which is when you are becoming half man and half bird. But, then, you have the feeling of crossing the wire – the point of no return: the middle. Which is very treacherous in your head and in that moment you are very fragile, because you are in the middle of a big space, you see the end but have not arrived yet. Careful, don't ever surrender to feelings like: I've practically arrived, or I have three feet to go but I can feel the victory already. This can be extremely dangerous. Well, you know, many wire-walkers fell on the last two feet because they create a feeling of victory when actually they are still on the wire, still suspended. There is no victory yet. The victory is to touch the building, to arrive, to land.

EMANUELE ENRIA You describe yourself as an explorer, curious about everything that surrounds you. But also as a writer in the sky. The twentieth century has been the century of urbanization, of people moving into cities. And this factor has deeply influenced people's way of perceiving space. How would you describe the perception of space from the line suspended in mid-air on which the wire-walker finds himself?

PHILIPPE PETIT For me, to walk on the wire is like ex-

ploring. You should never stay in the same place, or stay successful or comfortable – you should go on an adventure. So, on the high wire, I very often feel I am exploring. And then, when you explore, comes the notion of territory. You pass through different territories when you are walking in the air. On the high wire I have this experience: beginning, first step, and then the middle (where sometimes the cable is inclined), and then the sky. That's why I say: I write in the sky. So, the notion of territory is very important to me as a walker and as an artist. I kind of direct my space before doing something: for example, if I am juggling in a room, I'm used to facing a wall so that the tracing of my three balls, which is a flat surface, will be parallel to the wall I am facing. And it will help me juggle well. I give you another example on the wire: once or twice in my life, I could not put the wire perpendicular to the towers because these were oriented in different directions. So, my wire was not perpendicular and I hated it. It's very bad for balance, and it's also bad artistically. It doesn't look good.

As far as architecture is concerned here, we should never forget that the wire-walker puts a wire between two monuments – be they architectures of the past or contemporary buildings, or beautiful natural marbles, a beautiful precipice… To me, it's very important where I put my wire: what kind of monuments, and in which precise point of those monuments. You see, sometimes I spend a lot of time deciding where exactly…To me it is really very important, I love architectural monuments and love to play with them. I say play, not impose my wire! It's exactly the same in nature, with its precipices and canyons; nature gives an extraordinary choice to make the work incredible. And inspiring.

EMANUELE ENRIA The perception of limits must be very clear on the wire. On the one hand you can become as light as air, on the other there is the risk of falling. How do you live within these two opposite perceptions?

PHILIPPE PETIT I never pronounce this word: falling. If I have to, I will talk about leaving the wire. Leaving the wire, flying away. But flying away is not falling, it is like what a bird does. Sometimes I dream that I walk on the wire and I fly away. But it is not falling. I am a wire-walker, so I walk. When you perform, there should be no fear. Now, I know perfectly how to deal with it because in life I am fearful of things on the ground. For example, legless animals. Snakes. Or many-legged animals, as spiders. I am scared. Now, why am I scared? Well, I am scared because I don't know those animals. For me, fear is an absence of knowledge. So, if I start studying these animals, then I am going to start loving them. No more fear.

When wire-walkers are afraid of getting off the wire, it's because they don't know the wire enough. My life is attached to that wire: I must make sure the wire is very safe. And I know everything about it. So, this is the absence of fear, when you know everything. All of this is to prove to you that my life is not in danger. I am not taking any risk. And I have no reason to have fear.

Clearly, during the walk there is no constant bal-ance, because balance does not exist. Balance is something you loose and regain continuously, like a juggler who will balance a stick on his head. Great jugglers move very little, bad jugglers move a lot. Nevertheless, to balance something there must be a movement. It's a fight in motion, and when the wire-walker is very talented and elegant, and a poet, then this fight is totally invisible. It is a fight he feels inside. From the outside, all that you see is perfect control and calm, but in the walker's head there's a constant balancing and rebalancing to stay on the rope.

EMANUELE ENRIA What is equilibrium for a tight-rope walker?

PHILIPPE PETIT Sometimes I love to joke, during interviews, and say that equilibrium doesn't exist. I don't really think that. I think I have achieved in my life long moments of equilibrium, but equilibrium, for me, is never gymnastics. It's never to have your body in the right position. It's a balance between the balance of your body and the balance of your mind. I go back to what I know best, the art of walking on a wire: the balance on the wire is a mix of technique and poetry. Body and soul. So, that's what it is for me: balance is a perfect communion between the body and the soul.

PHILIPPE PETIT was born in France. He discovered the art of magic and sleight of hand when he was still a child and took his first steps on a tight-rope at the age of sixteen. A self-taught performer, for the past thirty years he has been living in New York, where he works as artist-in-residence in the Cathedral of Saint John the Divine. His book *To Reach the Clouds*, which describes his adventurous walk on a wire suspended between the Twin Towers, inspired the film *Man on Wire*, directed by James Marsh.